Pharmaceutical Discovery, Development and Manufacturing Forum 2014

Core Programming Area at the 2014 AIChE Annual Meeting

Atlanta, Georgia, USA
16-21 November 2014

ISBN: 978-1-5108-1269-7

Printed from e-media with permission by:

Curran Associates, Inc.
57 Morehouse Lane
Red Hook, NY 12571

Some format issues inherent in the e-media version may also appear in this print version.

Printed by Curran Associates, Inc. (2015)

For permission requests, please contact AIChE
at the address below.

AIChE
120 Wall Street, FL 23
New York, NY 10005-4020

Phone: (800) 242-4363
Fax: (203) 775-5177

www.aiche.org

Additional copies of this publication are available from:

Curran Associates, Inc.
57 Morehouse Lane
Red Hook, NY 12571 USA
Phone: 845-758-0400
Fax: 845-758-2634
Email: curran@proceedings.com
Web: www.proceedings.com

TABLE OF CONTENTS

380580 Development of a Multistep Continuous Process and Manufacturing for a Key Pharmaceutical Intermediate of Dapagliflozin

Abstract:

While conventional batch possessing dominates most API-related pharmaceutical manufacturing operations, continuous processing technology has been attracting increased attention due to its inherent benefits such as enhanced mass and heat transfer, precisely-controlled reaction times, and the use of small reactors. In the synthesis of Dapagliflozin, the preparation of a key pharmaceutical intermediate, BMS-587319, by common batch processing requires cryogenic conditions to ensure acceptable stability and yield of certain reactive species. To overcome potential challenges for scale-up on cryogenics requirement and process efficiency, a continuous process was later developed and demonstrated successfully on scale to convert the low-temperature synthesis to an energy-efficient, non-cryogenic operation. In addition to these manufacturing advantages, the continuous flow production consistently delivered comparable product quality and yield to the batch process over multiple campaigns. In this presentation, laboratory development, scale-up and production of BMS-587319 by continuous processing will be discussed.

373054 Process Intensification Using a Continuous Packed Bed Reactor for Enzymatic Hydrolysis

Abstract:

Due to their ability to afford highly selective chiral transformations under mild conditions, reactions utilizing enzymatic catalysis have been an important part of chemical and pharmaceutical manufacturing for many years. Enzyme immobilization offers the possibility of executing enzymatic reactions in flow by using continuous packed bed reactors (PBR). This mode of operation offers several potential advantages over standard batch operation, e.g. ability of catalyst recycle and improved throughput/cycle time due to process intensification. In this work, process development efforts to prepare a pharmaceutical intermediate using an immobilized enzyme PBR will be presented. It is demonstrated that a significant enhancement in reaction rate can be achieved using a PBR, allowing full conversion in $\tau < 5$ min (vs. 24 hours in batch). Reaction kinetics and reactor performance under a variety of conditions (temperature, concentration, superficial velocity, packing density, etc.) are evaluated and analyzed. A simple PBR kinetic model is constructed; model performance and comparison to experimental laboratory scale-up data will be discussed as well.

381900 Organic Solvent Nanofiltration in Continuous Catalytic Reactions

Abstract:

Introduction

Organic solvent nanofiltration (OSN) is an emerging technology for performing membrane separation/purification processes in organic solvents and has been successfully applied for organometallic catalyst recovery. One major advantage of OSN separation is that it does not require any phase transition or biphasic operation. Thus, the development of a continuous catalytic process combined with continuous catalyst recovery by OSN can potentially offer major economic and process efficiency advantages over the conventional batch-based system and/or biphasic operation processes.

The main hurdle towards implementing OSN in catalytic processes has been the compatibility of the existing OSN membranes (particularly the polymeric ones) with the reaction conditions. Typically the transition metal catalyzed reactions are performed at high temperatures (100 °C and above) in aggressive solvents (e.g. DMF) and at high concentrations of base/acid – quite challenging conditions for the polymeric membranes! In this work we report a continuous process for a Heck coupling reaction combined with *in situ* separation of the organometallic catalyst using polymeric membrane [1,2]. In contrast to the previous works, the reaction and the separation are performed simultaneously in the same vessel, both performed at 80 °C (high temperature) and in dimethylformamide (DMF) containing >0.9 mol.L^{-1} triethylamine (high base content). This work will also demonstrate a continuous ring closing metathesis (RCM) reaction where Ru catalyst is retained by a polymeric OSN membrane. Both processes were operated in two modes - continuous stirred tank reactor mode (CSTR) and plug flow reactor mode (PFR).

Results and Discussion

Two reactor configurations are investigated to perform the Heck coupling reaction: a continuous single stirred tank reactor/membrane separator (m-CSTR); and a plug flow reactor (PFR) followed by m-CSTR (PFR-m-CSTR). Both configurations showed stable performance for more than 250 hours at conversion rates higher than 90 %. The combined PFR-m-CSTR configuration was found to be the most promising, achieving conversions above 98 % and high catalyst turn-over numbers (TONs) of ~20,000. In addition, low contamination of the product stream (~27 mg Pd per kg of product) makes this process configuration attractive for the pharmaceutical industry.

Some potential adverse effects in the combined continuous reactor-membrane separation units will be also discussed. Apart from membrane stability under the reaction conditions, another important issue which is often neglected in the membrane studies is the effect of the membrane material on the reaction kinetics. This could possibly be a crucial factor for a continuous process and several membrane materials were investigated for their inhibitory effect on the Pd catalyst. The results showed that cross-linked polyimide reduces the reaction rate, while polybenzimidazole seems to completely inhibit the reaction (most likely via catalyst inhibition) [2]. The study also demonstrates the excellent performance of polyether ether ketone membranes as separation materials in high temperature catalytic processes.

Finally a continuous RCM reaction successfully performed in a flow-through nanofiltration membrane reactor/separator unit in CSTR and PFR modes will be demonstrated. The PFR reactor ran continuously for 32 hours showing a stable conversion of ~100% throughout the run. The permeate stream contained ~0.34 mgL^{-1} Ru suggesting ~0.2% catalyst losses. The plug-flow configuration seemed the most promising and further investigation should be performed in order to evaluate its through potential.

References

[1] L. Peeva, J. Burgal, S. Vartak, A. G. Livingston, J. Catal. 306 (2013)190-201.

[2] L. Peeva, J. Arbour, A.Livingston, Org. Process Res. Dev. 17 (2013) 967-975.

387625 Development of Continuous Antisolvent Nucleation Processes to Achieve Better Control over Crystalline Particle Attributes

Abstract:

Crystallisation from solution is an important separation and purification process commonly used in manufacturing of a broad range of solid products, including fine chemicals, such as dyes, pigments and explosives, as well as active pharmaceutical ingredients and excipients. Consistent crystalline particle attributes, such as solid form, particle shape and size distribution are desirable for optimising downstream processing steps as well as for uniform dissolution time and good bioavailability for drug delivery. Currently, crystallisation of fine chemical and pharmaceuticals is typically performed in a batch mode which can often lead to problems in achieving consistent product specifications, such as crystalline form and particle size distribution. Moving to continuous crystallisation technologies has the potential for significant increases in efficiency, flexibility and product quality, while aiming to obtain required particle attributes directly without a need for subsequent size reduction. However, development of continuous crystallisation processes requires reliable control over crystal nucleation as well as suitable management of possible encrustation and fouling issues.

We have designed and investigated a continuous antisolvent nucleation unit (a nucleator) in order to generate seeding suspensions for subsequent crystallisation steps. The control of crystal nucleation kinetics was achieved by adjusting the mixing efficiency, solvent-antisolvent ratio, supersaturation and residence time in the nucleator. Crystal suspensions produced in the nucleator were continuously introduced to a cascade of continuous stirred crystallisers or a continuous oscillatory baffled crystalliser, where seed crystals were further grown by cooling crystallisation. Careful control of the crystal seed formation separated from subsequent crystal growth allowed for development of more flexible continuous crystallisation operations and improved ability to obtain desirable final particulate products.

365761 Avoiding Microreactor Clogging By Inorganic Solid Formation in Continuous Chemical Synthesis of Pharmaceutical Compounds

Abstract:

Continuous manufacturing of fine chemicals and pharmaceuticals using microfluidic platforms are currently researched due to their advantageous heat and mass transfer rates compared to traditional batch processes, leading to shorter reaction times and higher yields. Palladium catalyzed C-N coupling reaction in microreactors is desirable for pharmaceutical applications but the reaction generates stoichiometric amounts of inorganic salts, which then precipitate and clog the microreactor. Biphasic reactions that can dissolve the solids can introduce mass transfer limitations or poison the catalyst. In this work, we perform a model Palladium catalyzed C-N coupling reaction in a monophasic system to show enhanced control of solid formation. We avoid microreactor clogging by tailoring the morphology of inorganic crystals, achieved through a combination of nucleation control and other mechanical means. We demonstrate the formation of crystal aggregates that can flow through the reactor without causing clogging or undesirable pressure fluctuations, and study the impact of crystal aggregate size on clogging behavior. We optimize the reaction conditions to ensure adequate conversion levels. Design strategies for ensuring reaction optimization while avoiding microreactor clogging will be presented.

Controlled accumulation of inorganic salts in microreactors for palladium-catalyzed amination

(Yizheng Chen, Ryan L. Hartman*)

Department of Chemical and Biological Engineering, The University of Alabama, Tuscaloosa, Alabama 35487, United States, e-mail: rhartman@eng.ua.edu

Solid particulates as the starting material, intermediates, products or by-products are common in modern synthetic chemistry. Solid particulates can deposit near reactors wall or suspend in the bulk solution which will lead to significant degradation in heat and mass transfer and eventually deviation of desired reaction route. Also, solid particles grow and aggregate randomly which can bring a serious pressure loss or even failure in the flow scenario. Therefore, manipulation of solid particulates in the continuous fine chemical and pharmaceuticals becomes ubiquitously necessary.

Constriction, bridging and random detachment are three hydrodynamic failure mechanisms due to solids formation and aggregation which can lead to plugging in microreactors under laminar flow conditions. Understanding what condition is the dominating hydrodynamic failure mechanism and why it is dominating and how to control micro-accumulation of inorganic salts can bring some insights for manipulation of solids for the upstream continuous fine chemicals.

To address this issue, palladium-catalyzed amination of aniline and 1-methoxy-4-chlorobenzene has been chosen as the model reaction to study the inorganic salt micro-accumulations in microreactors. The palladium-catalyzed amination can find broad utility in fine chemicals, pharmaceuticals and commodities. In this study, 1,4–Dioxane is used as the solvent, NaOtBu as the base and Xphos precatalyst as the catalyst source. The inorganic salts can be sodium chloride and/or palladium black which are insoluble in organic solvent. The micro process is monitored by in-line confocal Raman spectrometer and pressure loss is detected by pressure transducer and recorded simultaneously by data acquisition system, and the offline estimation of reaction yield is analyzed by gas chromatography. The residence time distribution in micro-flow condition is calculated according to the in-line UV spectroscopy. These results can provide the elementary basis for micro-accumulation phenomenon of inorganic salts and the continuous flow assurance of palladium-catalyzed amination in microreactors.

361474 An Integrated Systematic Framework to Assist the Development of Pharmaceutical Processes

Abstract:

An integrated systematic framework to assist the development of pharmaceutical processes.

Emmanouil Papadakis, Gürkan Sin, Krist V.Gernaey, Rafiqul Gani*

CAPEC-PROCESS - Department of Chemical and Biochemical Engineering, Technical University of Denmark, Søltofts Plads, Building 229, DK-2800 Kgs. Lyngby, Denmark, tel. +45 45252882, e-mail: rag@kt.dtu.dk

The pharmaceutical industries are dominated by batch wise processes, which have served well the pharmaceutical industries as well as the regulatory bodies. Batch vessels provide the flexibility of handling different unit operations in a multipurpose plant. However, batch processes may not be very efficient for product quality assurance and may have drawbacks. Over the last decade the regulatory bodies required the pharmaceutical companies to demonstrate more process understanding. Continuous pharmaceutical manufacturing (CPM) has become an attractive option, which naturally eliminates some of the drawbacks of the batch processing while maintaining the flexibility [1].

Pharmaceutical industries are therefore looking for opportunities to evaluate the feasibility to apply continuous manufacturing. Methods and tools, which have been applied to other industries (such as chemical and petrochemical) may not be directly applicable for pharmaceutical processes. Therefore Process Systems Engineering (PSE) tools and methods tailor-made for pharmaceutical applications can have an important role in the transition from batch to continuous manufacturing in the pharmaceutical industries [2].

An integrated computer-aided framework for the development of pharmaceutical processes has been developed. The objective of the framework is to assist the pharmaceutical companies to facilitate the move from batch to continuous manufacturing and the development of batch or continuous pharmaceutical processes. The integrated framework consists of four parts: (Part A) identification of the reaction mechanism, (Part B) reaction analysis, (Part C) separation analysis and (Part D) evaluation of the developed process. Identification of the reaction mechanism is the first part and it consists of two steps. Step A.1 is the selection of the active pharmaceutical ingredient (API) or the intermediate to be produced and step A.2 the identification of the reaction pathway. The reaction analysis (part B) consists of six steps and it aims to identify all the relevant information for the reactor design. Step B.1 is the general analysis of the reaction, in this step the basic information for the reaction (such as reactants, solvent, kind of reaction, catalyst, properties and by-product) are collected (if available). In step B.2 the kinetic analysis is performed and the reaction limitations are identified. In the next step (Step B.3) the reaction variables are evaluated with respect the reaction performance. Based on the information which have been collected (or predicted through reliable models), a decision (Step B.4) whether continuous operation is beneficial or not for the system is taken. In the step B.5 decision on the reactor design is taken considering the collected information and finally, if separation of the reaction stream is needed (step B.6), the separation analysis (part C) is performed. The separation analysis consists of seven steps, in the beginning the objective of the separation is defined. In the Step C.1 the mixture to be separated is analyzed and the separation tasks are defined. In the step C.2 the alternatives of the separation are generated and in step C.3 they are evaluated in order to minimize the number of alternatives. Finally, in the steps C.4-C.7 the separation process is designed (unit operation design parameters and operation mode). The final part of the framework (part D) is the evaluation of the developed process and consists of two steps, simulation/evaluation of the process (Step D.1) and optimization, control and monitoring of the developed process (Step D.2).

Development of the framework requires the development and the use of model-data based computer aided methods and tools such as a knowledge database, a model library, modeling tools, solvent selection tools, process analytical technology (PAT) and quality by design (QbD) approaches. The knowledge library should provide all the relevant information for pharmaceutical product/process development such as reaction mechanisms, reactants (substrates, catalysts, and solvents), reaction conditions, and process efficiency. The model library should contain information to describe many unit operations involved in pharmaceutical processes, reaction kinetic models and properties prediction models. The modeling tools must assist in the development of new models and model-based simulations. Solvent selection tools must assist for the selection of an appropriate solvent when needed and to optimize the use of solvents in pharmaceutical development. PAT and QbD approaches are needed to improve and ensure high the quality of the product.

The application of the proposed integrated framework is highlighted through the BHC (Boots and Hoechst-Celanese) patented synthesis pathway for ibuprofen. The improved BHC synthesis of ibuprofen consists of three reaction steps [3] and it is used to verify the proposed integrated framework. The objective here is to demonstrate that the framework plus the methods and tools can be applied for new products (and their processing paths) as well as validate or retrofit existing ones.

References:

[1] K. Plumb, 2005. Chem. Eng. Res. Des., 83, 730.

[2] K.V. Gernaey, A. E. Cervera- Pardell, J. M. Woodley, 2012. Comput. Chem. Eng., 42, 30

[3] Elango, V. Murphy, M., Smith, B.L., Davenport, K.G., Mott, G.N., Zey, E.G., Moss, G.L. (Hoechst Celanese Corp.), 1990. European Patent 4.981.995

364734 Modeling of Disintegration and Dissolution of Formulated API

Abstract:

The absorption rate and extent of absorption of an active pharmaceutical ingredient (API) in patients often depends strongly on the mechanisms of the API release from the formulation into absorbable form. These mechanisms include coating dissolution, tablet mechanical disintegration into granules or aggregates, granule disintegration into component powder particles, and API dissolution from available surfaces. These phenomena can occur simultaneously and can affect each other. Additionally, each of these phenomena is affected by a number of chemical and physical parameters stemming from the molecular properties of the API and excipients, the nature of the dissolution medium, and the processing parameters for tablet formation such as compaction force and blending shear. In this work, a mathematical model is derived to relate API tablet dissolution profile to API and medium properties, tablet preparation parameters, and measurable tablet properties. Case studies are presented illustrating the application and utility of this model for formulations.

A practical computational tool to predict formulation and process variables during the development of spray-dried amorphous solid dispersions

Íris Duarte, Hovione Farmacência SA & University of Lisbon, Faculty of Pharmacy, Portugal

José Luís Santos, Hovione Farmacência SA, Loures, Portugal

João F. Pinto, University of Lisbon, Faculty of Pharmacy, Lisbon, Portugal

Márcio Temtem, Hovione Farmacência SA, Loures, Portugal

Abstract

The aim of this work is to demonstrate the implementation of a computational tool to guide the early identification of potential physically stable amorphous dispersion formulations. The practical utility of this tool will be demonstrated as part of a new screening protocol that has been developed to accelerate the development of pharmaceutical amorphous solid dispersions produced by spray-drying. This methodology enables the narrowing of the scope of formulations, identifying more rapidly suitable systems to be scaled- up.

Introduction

Today, between 70 to 90% of the new chemical entities (NCEs) in the pharmaceutical pipeline belong to BCS Class II or IV, thus presenting at least solubility constraints [1]. Promising therapeutic molecules with limited aqueous solubility are a major concern, especially when considering the development of oral-dosage forms. To overcome this issue enabling platforms such as, particle-size reduction, use of surfactants, salt formation, complexation with cyclodextrines, production of self-emulsifying drug delivery systems or amorphous solid dispersions have been used.

The development of new amorphous solid dispersions (ASDs) has increased exponentially since its appearance in the late 90's [2]. ASDs have become popular among the pharmaceutical industry and their efficacy is supported by the increased number of marketed products (*e.g.* Incivek® (2011), Zelboraf® (2011), Kalydeco® (2012)). The requirements for the successful development of an ASD are related with performance and chemical/physical stability aspects. In what regards the performance requirements, an amorphous dispersion formulation should present an improved dissolution profile compared with the crystalline reference and should be capable of sustaining drug supersaturation in solution for a longer time. Both parameters will contribute to an increased amount of drug available for absorption. In what accounts physical stability, this concept is related with the inherent perceived risk of drug recrystallization from the amorphous state, followed by consequent loss of the solubility advantage.

The development of ASDs with the desirable long-term stability and performance is a challenging process, due to the wide number of formulation and process variables that influence both physical and chemical properties of the product (*e.g.* several existing polymeric stabilizers, surfactants, different drug-polymer ratios, solvents, preparation methods, temperatures, etc). For a long time, the selection of the best formulations was simply based on trial and error experiments, combined with the experience of researchers themselves, which demonstrated to be costly, time-consuming and API demanding. At a time, in which the pharmaceutical industry demands for fast turnaround times, lower costs and to reduce the risks associated with the development of new drugs, it became critical the development of screening programs for narrowing the scope of formulations and to rapidly identify suitable systems for subsequent clinical evaluation.

In the scope of this work the applicability of a computational tool to guide the early identification of potential physically stable systems, based on molecular miscibility predictions between the drug and the polymer, was evaluated. The final goal is to include this tool in a screening program that has been developed for the production of stable ASDs.

Materials and Methods

Materials

Drug A (Tm~200ºC, poorly water soluble) was evaluated with differently selected polymers, ranging from PVP K29/32, PVPVA 64, Eudragit® L100, HPMCAS and HPMC [drug load up to 65% (w/w)], according to the screening protocol developed. By opposition, Itraconazole (ITZ)-polymers systems were also evaluated, but the results only served to validate the computational tool. ITZ was combined with structurally different polymers, namely PVPVA 64, Eudragit® EPO and HPMCAS up to 85% (w/w) drug load.

Methods

A) Screening Protocol Overview

Figure 1 represents a workflow demonstrating the implementation of the screening protocol under development.

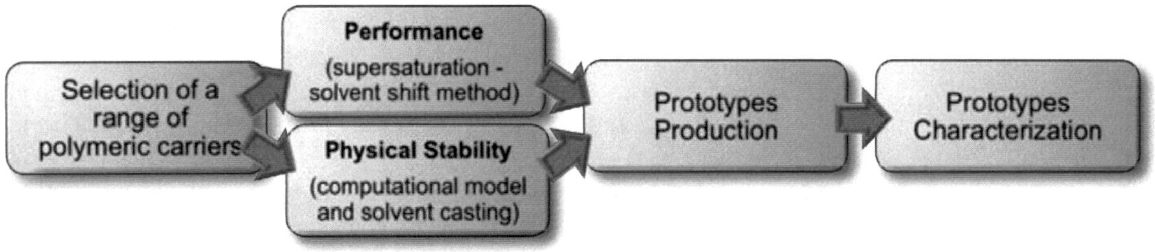

Figure 1: Workflow for the early development of a new spray dried amorphous dispersion.

The implementation of the screening program for a given API first consists in the analysis of drug physicochemical properties, target drug release profile and target dose. According to these characteristics a suitable number of polymeric carriers were selected, based on chemical structure diversity, their physicochemical properties and potential to promote drug-polymer interactions (*e.g.* H-bonding). In order to evaluate the precipitation inhibition effect of the selected polymers on the recrystallization of the API from solution, supersaturation screening studies were performed in a small-scale dissolution apparatus using a solvent-shift method, under fasted biorelevant conditions and at a pH value where the API presents lower solubility [3]. Concomitantly, drug-polymer miscibility estimates were obtained using a computational tool that will be explained in detail in the next section. Still related to the latter, a small-scale solvent casting protocol was implemented. Solvent casting experiments proved to be useful to get early information on experimental stability trends, using small quantities of API [4]. Moreover, it can also be useful to fine-tune formulation variables such as the drug loading, total solids fraction, solvents, among others.

At this point of the development work, polymers selections should be narrowed based on their ability to inhibit drug precipitation in solution, but also based on the drug-polymer miscibility capacity and physical stability. However, in this work and in order to evaluate the applicability range of the screening methods herein presented, different polymers presenting different performances and compatibilities with Drug A were selected to proceed with the solvent casting experiments, lab-scale production of spray-drying prototypes and respective analytical characterization, namely physical stability (DSC, XRPD) and powder dissolution (performance). The analytical results obtained for the spray-dried dispersions were compared with those obtained during the screening phase.

A.1) In silico analysis of drug-polymer miscibility

The computational tool herein presented is based on diffuse interface theories and presents advantages over commonly applied strategies to guide rationale polymer selection and drug load. Examples of current approaches are the analysis of the Hildebrand and Hansen Solubility Parameters (SPs) to infer about drug-polymer compatibility [5] or, more recently, the implementation of lattice-based solution theories, such as the Flory-Huggins (F-H) theory to study the thermodynamics of drug-polymer mixing [6]. The most significant differences of this model over other approaches is (1) the potential to evaluate a ternary system made of drug, polymer and solvent, by comparison with the traditional two-component system and (2) the consideration of time-dependent phenomena, such as components mass diffusion and solvent evaporation (Figure 2).

As can be seen in Figure 2, the model used in this work takes into account different thermodynamic, kinetic and process variables. Those were estimated according to the methods described in Table 1.

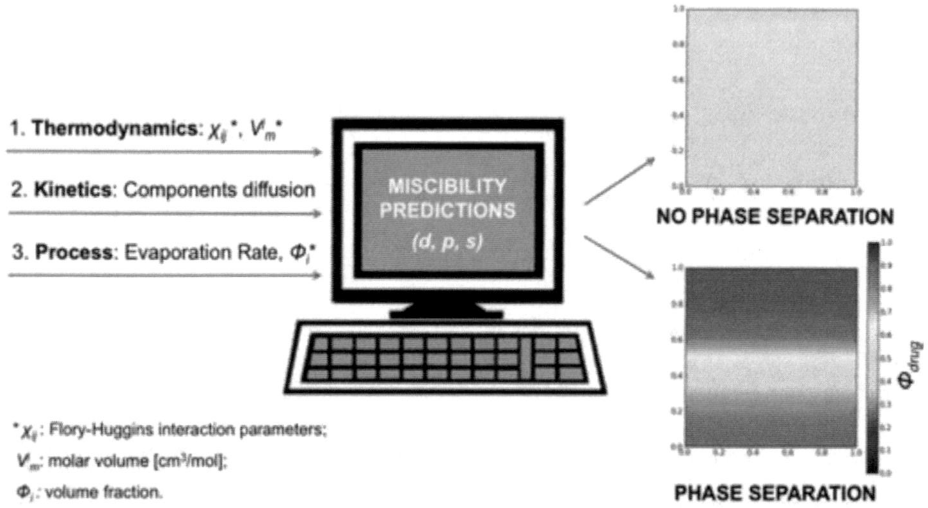

Figure 2: Input parameters and two limiting situations given by the model (outputs).

Table 1: Input parameters and respective methods of calculation.

Input Variable	Type of calculation
Flory-Huggins interaction parameters (χ_{ij})	Inverse gas chromatography (SEA, from SMS) [7]
Molar volume (cm^3/mol)	Molecular weight (g/mol) divided by the true density (g/cm^3)
Solid's diffusivity	Wilke-Chang equation
Evaporation Rate	Correlation for the evaporation of a single droplet in still air

For more details on the derivation and implementation of the computational tool readers are referred to the work of Duarte *et al.* [4].

Results and Discussion

The methodology herein reported is divided in two main parts: performance and stability considerations. First supersaturation profiles were used to assess the potential of the polymers to inhibit Drug A precipitation in fasted biorelevant conditions and at a pH value where the API presents lower solubility. The following polymer ranking by ascending order of precipitation inhibition was obtained: PVP K29/32 < PVP/VA 64 < HPMC < HPMCAS < Eudragit L100. Second, miscibility computational simulations were conducted complementarily to predict the influence of thermodynamic and kinetic material properties and process conditions on the phase behavior limits of Drug A-polymer systems. The following polymer ranking by ascending order of miscibility capacity was obtained: Eudragit L100 < PVP/VA 64 < HPMCAS < HPMC. Both results were compared, and four polymers - PVP/VA 64, Eudragit L100, HPMCAS and HPMC - were selected to proceed with the solvent casting experiments, lab-scale production of spray-drying prototypes and respective analytical characterization.

Validation of the screening methodologies implemented

Figure 3 compiles the predicted and experimental miscibility results given by the computational model, solvent casting and spray drying experiments. In the same figure, the results previously obtained for the ITZ-based systems are also presented [4]. Both results served to validate the screening methodology.

In general, cast films and spray-dried dispersions generated were amorphous, and true solid solutions with drug concentrations up to 65 and 85% w/w of Drug A and ITZ were obtained, respectively. Comparing the experimental and the computational data, the results showed an acceptable correlation with the ability to maintain the rankings initially defined. Indeed, when the simulations indicated the formation of homogenous amorphous systems for higher drug loads the experimental references indicated high miscibility capacity; when the simulations indicated phase separation for lower drug loads the real references presented their maximum of experimental miscibility at lower drug loads.

However, among the systems evaluated two false-negatives were observed (Drug A:PVPVA 64 and Drug A:Eudragit® L100). It should not be neglected the fact that the theoretical basis of the computational model may not fully capture the complexity of drug-polymer particle formation [4].

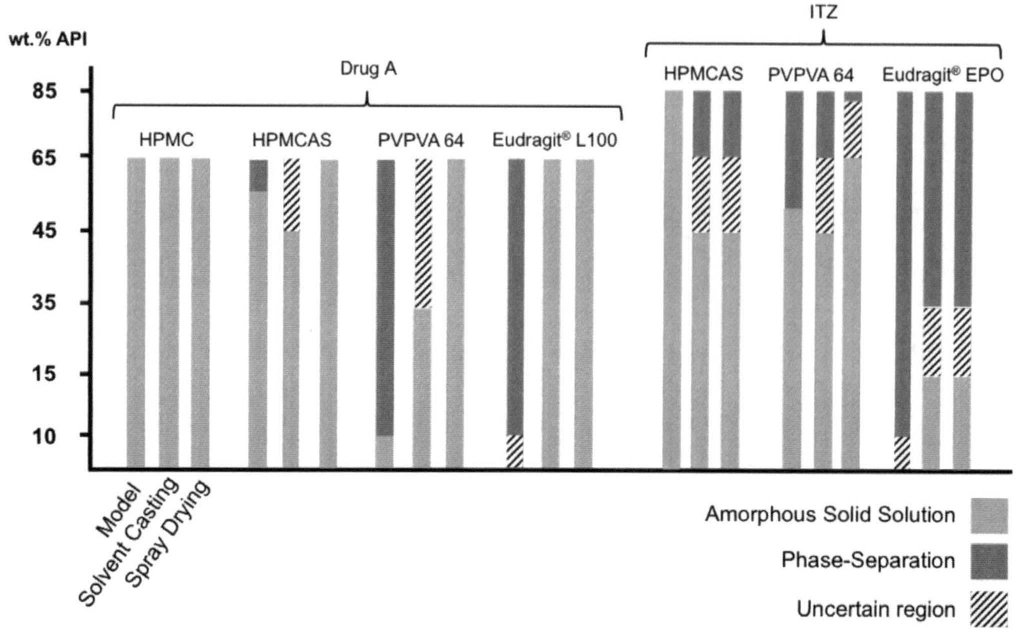

Figure 3: Summary of the results obtained: miscibility and phase behavior predictions (Model) and experimental miscibility assessment by mDSC (solvent casting+spray drying).

The potential occurrence of false-negative results corroborates the importance of implementing a small-scale solvent casting methodology; otherwise one may be neglecting the real solubilization capacity offered by the polymer. Despite these shortcomings, the results show that one can still use the information given by the computational model not only to rank the most stable amorphous formulations with the highest API/polymer ratio, but also to reduce the number of experiments, reducing time and resources in the following phases of product development.

In what regards the evaluation of the performance of the Drug A-based spray dried systems amorphous systems, powder dissolution profiles were obtained for all samples with 65% (w/w) drug load under constant pH (the same pH as in the early supersaturation experiments). The high API load was selected based on the fact that no signs of amorphous-

amorphous phase separation or drug recrystallization were observed. Figure 4 compares the results obtained during the screening phase and the powder dissolution characterization of the spray-dried dispersions. The results are presented in terms of the accumulated drug concentration (AUC) obtained during 240 minutes of testing.

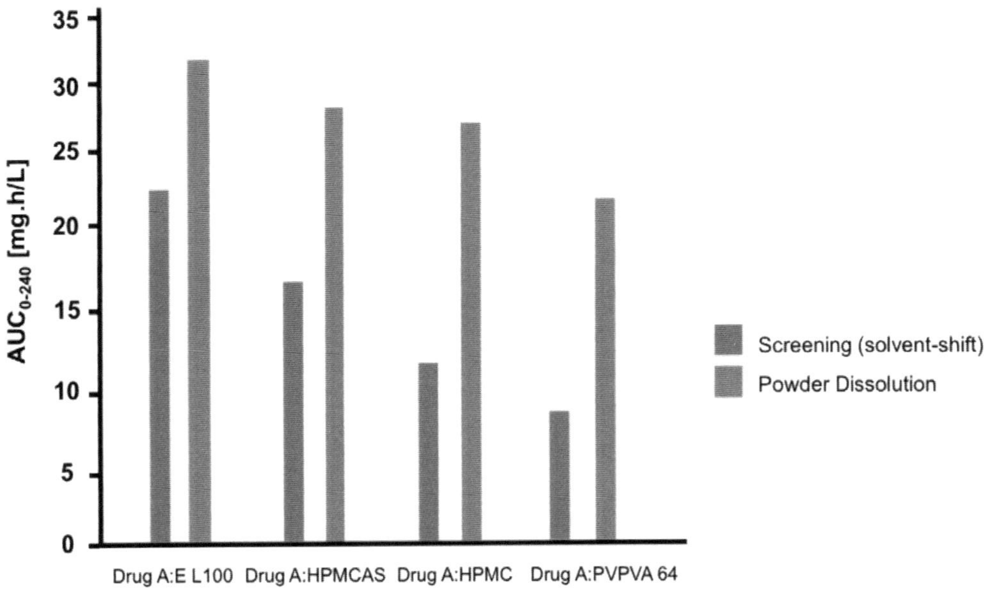

Figure 4: Accumulated drug concentration (AUC) for Drug A systems (up to 240min) in the screening phase and powder dissolution experiments. The AUC was computed using the trapezoidal rule.

As can be seen, both results from the solvent-shift *versus* powder dissolution experiments are aligned. The polymer ranking by ascending order of potential to inhibit Drug A precipitation in solution (PVP/VA 64 < HPMC < HPMCAS < Eudragit L100) agreed with the overall performance and powder dissolution profiles of ASDs of Drug A and the respective polymers.

Conclusions

The implementation of screening programs to support and to expedite the rationale development of new amorphous formulations is of utmost importance. In the present work a screening program was described, which considers performance and physical stability aspects.

The results obtained after implementation of the screening methodologies tested (*i.e.*, solvent-shift method, computational model, solvent casting) provided valuable insights on the performance in solution and miscibility assessment of Drug A amorphous dispersion formulations. Results should be analyzed concurrently to guarantee proper selection of the most promising systems along the different phases of formulation development. The computational model implemented not only allowed the ranking of the most promising amorphous formulations in terms of physical stability, but also to narrow down the range of drug-polymer ratios to be tested in the following phases.

Acknowledgements

The authors would like to thank Anett Kondor, Majid Naderi and Daniel J. Burnett from Surface Measurement Systems Ltd for analyzing the raw materials used in this work using the iGC surface energy analyzer. Íris Duarte would like to thank the financial support from Hovione Farmaciência SA and from Fundação para a Ciência e Tecnologia through the doctoral grant BDE/51422/2011.

References

[1] Löbenberg R. and Amidon GL., *Eur J Pharm Biopharm* (2000), 50(1), pp. 3-12;

[2] Sekiguchi K. and Obi N., *Chem Pharm Bull* (1961), 9, pp. 866-872;

[3] Vandecruys R. *et al.*, *Int J Pharm* (2007), 342, pp. 168-175;

[4] Duarte I. *et al.*, *Pharm Res* (2014), online;

[5] Greenhalgh DJ. *et al.*, *J Pharm Sci* (1999), 88 (11), pp. 1182-1190;

[6] Flory PJ., *Cornell University Press:Ithaca* (1953);

[7] Thielmann F. and Naderi M., Surface Measurements Systems, Application Notes.

375130 Method for Prediction of Pharmaceutical Solubility for Solvent Selection

Abstract:

A method is described for modeling the solubility of organic, non-electrolyte solutes in common solvents and mixtures. This has applications for solvent selection for reaction, crystallization and extraction operations as well as in formulation design. A small number of solubility measurements are required to fit model parameters; thereafter, solubility predictions can be made in a wide range of single, binary and ternary solvent systems (e.g. for identification of conditions for a solubility maximum in mixtures).

These calculations assist scientists and engineers in selecting solvent systems and ensuring that promising candidates that may not be 'on the shelf' are given due consideration.

The method comprises adding an extra group representing the solute molecule to the UNIFAC (ref 1) tables, including its UNIFAC R and Q parameters, then fitting the interaction parameters between the solute and the functional groups of the solvents used when measuring solubility. An alternative tuning of standard UNIFAC has been described elsewhere (ref 2). Heat of fusion and melting temperature are inputs (preferably) and can also be fitted. Predictions can then be made of solubility in any solvent system where these functional groups are present.

The method overcomes some traditional weaknesses of UNIFAC while maintaining the standard table of parameters intact. These include:

- Application for solutes that are pharmaceutically relevant and cannot be fragmented into UNIFAC groups
- Much improved accuracy compared to normal UNIFAC, which precludes regression of parameters.

Some recent applications of the method will be reviewed to indicate its performance relative to other techniques for solubility prediction (refs 3, 4 and 5).

Utilizing the functional group contribution approach (rather than conceptual segments) maintains advantages that may be leveraged in future development of the method to predict solubility of structurally similar impurities as well as primary solutes.

The current implementation considers 163 solvents, enabling potential prediction in thousands of binary solvent mixtures and several million ternary solvent mixtures. Users can limit the scope to solvents that are desirable for e.g. regulatory, cost or environmental considerations.

References:

1. Fredenslund, A., Jones, R. L., Prausnitz, J. M., Group-Contribution Estimation of Activity Coefficients in Nonideal Liquid Mixtures. *AIChE J.*, 1975. **21**(6): p. 1086-1099
2. Modarresi, H., Conte, E., Abildskov, J., Gani, R., & Crafts, P. A. (2008). Model-Based Calculation of Solid Solubility for Solvent Selection A Review. Industrial & Engineering Chemistry Research, 47(15), 5234–5242. Retrieved from http://pubs.acs.org/doi/abs/10.1021/ie0716363
3. E. Sirota and T. Lin, Modelling Small Molecule Synergistic Solubility Behavior, Presented at DynoChem User Meeting, Chicago, May 2011: available online at https://dcresources.scale-up.com/Default.aspx?id=388
4. N. Chennamsetty, O Lyngberg & J. Qiu, Determining Solubility of APIs and Intermediates From Automated Parallel Experiments and Modeling, AIChE Annual Meeting 2011, Paper 56c: http://www3.aiche.org/proceedings/Abstract.aspx?PaperID=236416
5. New solubility prediction utility for solvent selection in DynoChem, Scale-up Systems blog, March 2010: http://blog.scale-up.com/2010/03/new-solubility-prediction-utility-for.html

376951 Impact of Continuous Drug Product Processing on Formulation and Process Development Workflow

Abstract:

Continuous drug product processing can significantly reduce the cycle time from powder dispensing to product formation. However, the output of process development is not product but rather knowledge. The manufacture of product is a relatively small percentage of the knowledge generation cycle time. Non-production related activities include: planning experiments, defining safety measures, collecting samples, analyzing samples, integrating multiple data streams, and analyzing data. This must then be combined with experience and science and compared with experimental objectives. Typically new insights are gained and the next cycle of experiments is planned and executed.

It was observed that the production cycle time sets the tempo for the other aspects of the work. When experiments are measured in weeks (as in batch), the preceding and following steps are also measured in weeks. When it is days or hours (as in continuous), the preceding and following steps are also measured in days or hours. Continuous processing also greatly improves the reliability of the system at the critical interface between sample generation and sample analysis. Improved reliability in delivering the samples to the lab at the promised time improves efficiency in the lab. Taken to its extreme, this becomes real time analytical. This creates the ability to make decisions during the run rather than after the run. This causes decision makers to spend more time in and around the processing suite / lab. Additionally, because of the short time to perform an experiment the impact of a failed experiment is reduced. This means higher risk / higher return experiments can be performed. These experiments can lead to unique insights. Finally, rapid or real-time data enables adaptive design of experiments that can shrink or grow as the runs proceed. If managed well that can be a powerful technique. However, poorly managed adaptations can destroy a well designed experiment.

377737 CFD-Predicted and PIV-Measured Velocity Profiles in a USP Dissolution Testing Apparatus 2 Equipped with an Arch-Shaped Fiber Optic Probe and Their Impact on Tablet Dissolution

Abstract:

Dissolution testing is a critical tool in drug formulation and product quality assessment. This test is a routinely used in pharmaceutical development to evaluate the *in vitro* performance of solid dosage forms, determine the dissolution behavior of formulations and provide an optimization of drug release dosage forms, and as a quality control tool to insure that solid dosage forms have consistent dissolution properties. The USP Dissolution Testing Apparatus 2 is the device most commonly used for this purpose. In a typical test, a tablet is added to the vessel and samples are manually withdrawn from the vessels over time and analyzed for their drug content. An approach to overcome the limitations of manual sampling requires using sampling probes, such as fiber optic probes, permanently inserted in the dissolution medium and continually monitoring drug concentration as the solid dosage form dissolves. Despite their advantages, permanently inserted fiber optic probes can alter the normal fluid flow within the vessel and produce different dissolution testing results.

In this work, the hydrodynamic changes introduced by an arch-shaped fiber optic probe in a USP Dissolution Testing Apparatus 2 filled with 900 mL of dissolution medium were studied by (1) experimentally determining the velocity profiles in the vessel, with and without the probe, using Particle Image Velocimetry (PIV) and quantifying changes in the flow velocities on selected horizontal iso-surfaces; and (2) predicting the velocity distribution using Computational Fluid Dynamics (CFD). These results so obtained were used to analyze the dissolution data from of actual dissolution tests conducted with and without the probe using Prednisone tablets fixed at nine different locations at the bottom of the vessel.

In general, substantial agreement was found between the PIV data and the CFD predictions. The velocities in the upper region of the vessel where the probe is located were clearly affected by the presence of the probe. More importantly, the some velocity components near the vessel bottom where the tablet is usually located, were found to be appreciably affected by the presence of the probe. Specifically, both the simulation results and PIV data show that the axial velocities near the bottom are larger in the probe system than in the standard system. In addition, in the region just under the probe, the differences in axial velocities between the two systems are the largest in both of the simulation and PIV measurement, but they become progressively smaller at locations further away from the probe.

The results of this work indicate that that the arch shaped fiber optic probe introduces as small but measurable baffling effect on the hydrodynamics in the dissolution vessel. This effect results in changes in the flow velocities which parallel the results of dissolution tests, where the dissolution profiles were typically found to be higher in the presence of the probe, but not sufficiently high to fail the tests, according to the FDA criteria (f_1 and f_2 values).

In summary, the hydrodynamic effects generated by the arch-shaped fiber optic probe are small but clearly measurable. The changes in velocity profiles in the dissolution vessel result in in detectable differences in the dissolution profiles, although not high enough to cause test failures per se. However, these differences could contribute to amplify the difference in dissolution profiles in those cases in which tablet has an intrinsically higher release rate.

379767 Reducing Powder Lubrication in an Encapsulation Process for Improved Dissolution Performance

Abstract:

Lubrication of pharmaceutical formulations aids manufacturing by reducing friction and adhesion between the powder and the processing equipment as well as helping to increase the bulk density of the powder for improved flowability. However, if the extent of lubrication becomes too great, adverse effects – including reduced tensile strength of compacts and slowed disintegration/dissolution of dosage forms – can be observed. In this work, a case study is presented where an approach to resolve slow capsule dissolution is described. This approach entails (1) an engineering analysis to identify the root cause and propose alternative equipment design solutions and (2) a process modeling effort using the discrete element method (DEM) to assess the performance of these alternatives *in silico*. Results and outcomes to be described.

Residence Time Distribution Measurement via Video Analysis for Design of Experiment Characterization of Melt Extrusion

P.R. Wahl[1], G. Hörl[1], D. Kaiser[1], S. Sacher[1], J.G. Khinast[1,2]

[1] Research Center Pharmaceutical Engineering GmbH, Graz, Austria

[2] Institute for Process and Particle Engineering, Graz University of Technology, Graz, Austria

The residence time distribution (RTD) is of importance for extrusion, because it strongly influences the product quality, e.g. degradation of the active pharmaceutical ingredient (API) or content uniformity. The first can be reduced by shorter residence times, while the latter improves with a broadening of the RTD, caused by mixing. Therefore in-line video analysis of a melt extrusion process was developed, which has been mainly used off-line up to now. It represents a robust and inexpensive solution, with equipment costs far below that of Raman or NIR, and without the need of calibration model development.

A video of the extrudate was taken with a camera positioned directly behind the die plate (see Figure 1, left). The analysis of the video consisted of four steps: (1) Segmentation of the image to mask the strands, (2) calculation of average color values for the segmented pixels of each frame, (3) fitting of the obtained color curve with an analytical solution [1] and (4) calculating several RTD measures (e.g. minimal and mean RT) for easier comparison of the results (Figure 1, center).

The calculated RTD measures were used as responses of a DoE, consisting of the process parameters: screw design, screw speed and throughput. As an example, the impact of screw speed and throughput on RT_{FWHM} (full-width at half maximum) is shown in Figure 1, right.

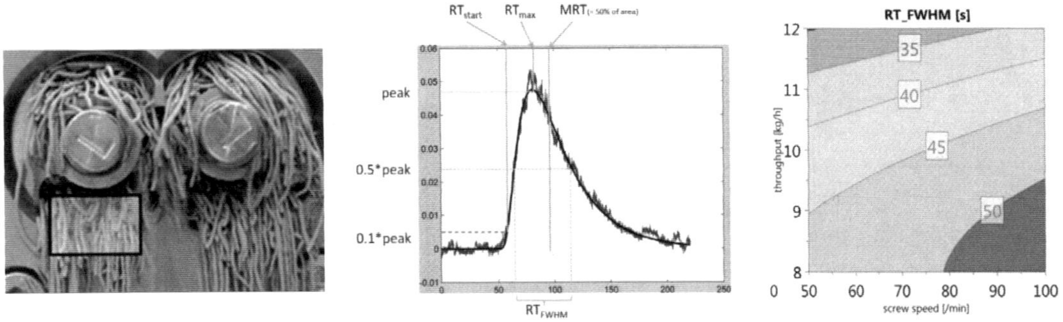

Figure 1: Left: Extrudate and die plate of the extruder. Analysis was performed within the rectangle area. Middle: The measured RTD curve was fit by an analytical solution [1], from which several important RT measures were derived. Right: DoE response for the RT measure full-width at half-maximum (FWHM).

[1] David B. Todd (1975), Residence time distribution in twin-screw extruders, *Polym. Eng. Sci.*, 15:437-443

374567 Tracking the Fate of Agglomerated API in the Drug Product Process with NIR Chemical Imaging

Abstract:

With the accelerated pace of drug product development due to fast changes in the pharmaceutical industry, the quality of the final product and the commercial process relies on the capability of detecting risks early in development and quickly diagnosing issues in routine commercial manufacture. In this particular study, the ability to characterize the behavior of agglomerated API throughout drug product development was enabled by efficient analytical tools to quickly identify risks. Agglomerated API was selected for use over large single particles because the agglomerate manufacturing process offered several advantages such as shorter cycle times and higher yield. However, the use of agglomerated API in the drug product process poses risks to blend uniformity and content uniformity of the final dosage form given that agglomerate properties such as size and strength may vary batch to batch.

One such analytical tool that was employed for understanding risk and found to be critical to monitor agglomerate behavior throughout the process was Near Infrared (NIR) chemical imaging. Near Infrared (NIR) chemical imaging facilitates visualization of the API and excipient distribution based on quantitative estimates of the concentration of the ingredients throughout the sample. The method requires minimal resources and is nondestructive to the sample. More importantly, the turnaround time is short, a much needed element assuring quick but informed decision making.

Embedding NIR chemical imaging throughout development enabled a qualitative understanding of both formulation and process robustness in response to changes in agglomerate size and strength. Using NIR chemical imaging, we can extract the agglomerate distribution and size in all intermediates throughout the drug product process. The visualization of agglomerate size is an advantage of this method in understanding the survivability of agglomerates that cannot be achieved with online NIR monitoring or HPLC based methods for blend uniformity. The NIR images enabled the team to identify the key unit operations and process parameters that produced the size change and further enhanced distribution of the API. This size reduction also provides insight into the agglomerate strength. Given agglomerates of very high strength, little to no size reduction would be observed across the process. However, we demonstrate that the shear forces to which the agglomerates are subjected to during roller compaction are great enough to break down the agglomerates.

Regardless of differences in initial agglomerate size or strength, the final granulation was reproducibly shown to have acceptable uniformity. Thus, we demonstrate that the process is robust against a range of agglomerate properties and there is no risk to content uniformity of the final dosage form.

Behaviour of HPMC and microcrystalline cellulose during a roller compaction process

G.E. O'Mahony[1,2], M.E. Crowley[1,2], R. Dontireddy[1], M.A.P. McAuliffe[3], A.M. Crean[1,2]
[1]School of Pharmacy, University College Cork, Cork, Ireland.
[2]Synthesis and Solid State Pharmaceutical Centre
[3]Centre for Advanced Photonics and Process Analysis, Applied Physics & Instrumentation Department, Cork Institute of Technology, Cork, Ireland.

INTRODUCTION

A critical feature of quality by design (QbD) is the understanding of product variability and how this is influenced by variability in the material and process parameters. The overall objective of this study was to investigate the relationship between variability in the intrinsic/behavioural properties of excipients, hydroxypropyl methyl cellulose (HPMC) and microcrystalline cellulose (MCC)), the resultant properties of ribbons and granules, produced by roller compaction and secondly the properties of the resultant tablets. Ribbons and granules were produced by roller compaction using (1) powder blends composed of different grades of HPMC and (2) MCC stored at different % relative humidities.

A second key objective was to monitor critical quality attributes of ribbon (density) and granulate (particle size distribution) using novel PAT techniques, Eyecon® and Multi-eye®.

METHODS

Blends composed of anhydrous lactose (Supertab 21AN), magnesium stearate and one of three different grades of HPMC (Benecel K100, E15 and E5) were compacted using a CCS 220 roller compactor. MCC was stored at three different % relative humidities; over a lithium chloride salt solution (11% RH), under ambient conditions (41% RH) and in a humidity controlled oven (75% RH).

Envelope and true density measurements were performed using Geopyc 1360 and Accupyc II 1340 respectively. At-line NIR measurements were performed using a Multi-eye® system and in-line particle size measurements were obtained using the Eyecon® system. Off-line analysis of ribbons was performed using a PerkinElmer Spotlight 400 FT-IR in diffused reflectance mode.

RESULTS

Envelope density/porosity of ribbon samples, and the amount of fines produced were strongly influenced by different grades of HPMC and MCC moisture content. In particular a much greater % fines were produced for Benecel K100 than for E5 or E15 grades. Benecel E5 and E15 produced ribbons of lower density compared to Benecel K100 at lower roller pressures. MCC stored at higher % RH produced ribbons with lower porosity. Correlations between at-line NIR measurements (MultiEye®) and off-line NIR measurements with ribbon envelope density (GeoPyc) were obtained. The Eyecon® system detected differences in particle size related to HMPC grade, MMC moisture content and processing parameters.

CONCLUSIONS

Variability in excipient properties was shown to have a significant impact on the critical attributes of roller compacted ribbon (envelope density, porosity) and granulate (particle size). Multieye® can be used for at-line analysis of ribbon envelope density/porosity with potential to be integrated in-line with the roller compaction equipment.

ACKNOWLEDGMENTS

This research is supported by the Synthesis and Solid State Pharmaceutical Centre and SFI under Grant Number 12/RC/2275. We would also like to acknowledge Innopharma Labs for the use of the Multieye® and Prof. Gavin Walker, University of Limerick for the use of the Eyecon®. We would also like to acknowledge Fitzpatrick for their support with the use of the roller compactor. Finally we would like to thank Ashland and FMC Biopolymer for the supply of HPMC and MCC samples respectively.

382598 Impact of Material and Source Variability on the Performance of Roller-Compacted Formulation and Critical Drug Product Attributes

Abstract:

In a pharmaceutical formulation, some excipients are used for their particular functionalities, such as lubricant, superdisintegrant, surfactant, coating material etc. Although present only in small quantities in a formulation, changes in these excipients can sometimes have significant impact on the processiblity and critical quality attributes of the product. The effect can also vary with API properties such as solubility and acidity. In this study, we systematically evaluated this effect using a typical roller compaction formulation with 20% drug loading and two model compounds of different solubility and acidity (acetaminophen and aspirin). Based on risk assessment and prior knowledge on the criticality of the excipients in the formulation, the following are factors selected to be studied in this work:

1. Three different vendors of Magnesium Stearate (Mallinckrodt, Brenntag Specialties Inc, Quimdis S A) and different lubrication durations
2. Three different types of superdisintegrants (Sodium Starch Glycolate, Croscarmellose Sodium, Crospovidone)
3. Presence and absence of the surfactant Sodium Lauryl Sulfate (SLS)
4. Two different types of Opadry coatings (Hypromellose based and Polyvinyl alcohol based)

The responses evaluated include:

1. Thickness and density of ribbons generated by the roller compactor
2. Powder properties (flow, particle size distribution and density) of milled granules
3. Compaction forces used during tablet compression
4. Dissolution and disintegration of core and coated tablets
5. Dissolution behavior of coated tablets upon stressing in accelerated conditions

Among the factors evaluated, the presence of SLS is found to have a significant impact of increasing the compactability of the blend during roller compaction hence decreasing the compactability during tablet compression. SLS also slows down the disintegration and dissolution of both compounds. Differences were observed for the effectiveness of the three superdisintegrants, which is also found to be dependent on the model API. Both types of coatings are found to slow down the dissolution of aspirin formulations, however, this effect is not as significant for acetaminophen formulations. Slow down of dissolution upon stressing in accelerated conditions was observed for aspirin tablets coated with PVA based Opadry. Details of the study and the results along with some mechanistic discussion will be covered in the presentation.

386556 The Synergistic Nature of Chemistry & Chemical Engineering Technology Teams Working Together in the Development of Post-Launch "2nd Generation" API Processes

Abstract:

Pharmaceutical manufacturers are under ever-increasing pressure to continually improve process performance and robustness for products in the post-launch phase due to both regulatory and economic pressures. In many cases novel process chemistries can be developed to manufacture the post-launch API molecules whilst simultaneously unlocking significant environmental, safety ,capacity and cost benefits. Pfizer has a program in place to introduce such new "2nd Generation" API process routes to products in its post-launch phase product portfolio.

This talk highlights the synergistic nature of chemistry and chemical engineering teams working together to develop and commercialise these new API process routes. The roles of batch and continuous processing technologies, biocatalysis technologies, PAT and mathematical modelling from laboratory development to plant commercialisation will be presented. Some examples will be given of where engineering technologies have allowed otherwise inaccessible chemistries to be developed and commercialise. .

384514 Platform Chemical Engineering Approaches to Support Manufacturing Optimisation

Abstract:

Pharmaceutical manufacturers are under ever-increasing pressure to continually improve process performance and robustness for products post-launch, due to both regulatory and economic pressures. The application of fundamental engineering principles to the process at scale has the potential to realise significant improvements, often with the regulatory filing. The key to capitalising on the potential for process improvement is to deepen the understanding of the process behaviour at scale. In this talk, a number of case studies demonstrating the deployment of PAT, modelling and process control for post-launch optimization will be presented. The case studies will focus on the engineering of particle attributes through control of the reaction and crystallization steps with the objective of improving downstream isolation and product performance.

373667 Leveraging Development Knowledge Post-Validation to Control Form and Improve Operational Consistency for a Pharmaceutical Intermediate with Multiple Polymorphs

Abstract:

The importance of controlling crystal polymorph is well-known for active pharmaceutical ingredients (APIs) as the polymorph can influence solubility, crystal morphology, residual solvent content, bioavailability, and formulation. For process intermediates, the need to control the polymorph is determined by the impact of the form on downstream processing and the risk to intermediate and API quality.

This talk provides a case study in crystallization development for a pharmaceutical intermediate with four distinct polymorphs, each with distinct crystal morphologies. These different polymorphs were observed in laboratory and pilot scale batches during development, but had no impact on the intermediate or API quality. This observation, along with the desire to provide the widest possible operating ranges for manufacturing, led to polymorph control being de-emphasized during development.

The manufacturing process consistently demonstrated production of high quality intermediate with observed variations in the polymorph. However, at the manufacturing scale, the variations in polymorph led to variable cycle times for the isolation and drying operations. Further studies were performed post-validation to confirm the impact of crystallization solvent composition on the resulting polymorph. Adjustments to the target operating conditions within the filed acceptable parameter ranges resulted in crystallization of the preferred polymorph to provide operational consistency in the post-crystallization unit operations.

385240 Commercialisation of API Processes in Manufacturing at Eli Lilly

Abstract:

The author will discuss learnings from commercialisation of processes into the Manufacturing setting in Eli Lilly.

The Kinsale manufacturing site in Ireland is the designated Small Molecule API commercialisation site. Commercialisation of both Batch and Continuous processes take place at this site. The author will discuss Eli Lillys strategy for such activity and the learnings from that activity over the last several years

361951 Integration and Application of Multiple Platform Technologies into a Drug Substance Validation Program

Abstract:

In this papers the authors will discuss the integration and application of multiple platform technologies into a drug substance validation program within Abbvie. This is based on their experience of this within Abbvies manufacturing and commercialisation site in Sligo, Ireland.

388737 Novel in Situ Methods for Floc Size Determination in Cell Culture Seperations

Abstract:

Recently flocculation has emerged as a viable option for the improvement of cell culture separations. The controlled flocculation of cells and cell debris can increase filtration and centrifugation speed, enhance throughput and improve clarification performance. Flocculation is particularly important for high density cell cultures where large amounts of biomass, especially in the sub-micron range, can prove challenging to effectively separate.

The particle size of flocculated cell debris is critical to consider when developing and implementing a flocculation step since it is often directly related to separation performance and will define requirements for the separation technology (pore size of depth filters for example). Offline particle size analysis tools have failed to gain broad acceptance for this task since flocs are inhernetly delicate and can break apart easily during the sampling, preparation and analysis phases of an offline analytical method. Also, the time delay between sampling and receipt of results can mean that expensive flocculants are added to a process for longer than needed

In this paper examples of how *in situ* measurement of floc size and count have being used to improve cell culture flocculation and separations will be presented. *In situ* monitoring using ParticleTrack technology will be shown to eliminate the most common errors associated with sampling and provide a more reliable method for the determination of flocculation onset, kinetics and endpoint. Examples relating flocculent dosage and type with floc size and stability will be presented and correlations between floc size and separation performance will be shown.

392128 Progress on a Mixed-Mode Hydrogel Membrane – a New Tool for Mab Purification

Abstract:

In today's high-titer biopharmaceutical mAb processes, there is a need for innovative separation approaches for capture or polish of biomolecules. Challenges exist, for example when the conductivity of the feed stream is high or when the aggregates level needs to be reduced immediately. Natrix is developing a mixed-mode chromatography membrane, which combines the properties of strong cation exchange and hydrophobic interaction. This mixed-mode chromatography membrane offers fast separation in a disposable format without sacrificing process performance. It opens up new opportunities for solving challenging purification problems as an alternative to conventional chromatography columns. This paper discusses the characteristics of the mixed-mode membrane together with performance across the design space to-date. The preliminary data shows high mAb binding capacity (up to 85 mg/mL), good clearance of HCP and aggregates at high conductivity (up to 15 mS/cm) and all at 6 seconds residence time. In addition, the low salt elution capacity (6 mS/cm) facilitates direct load onto a quaternary amine adsorber for mAb polishing.

388551 Rapid Characterization of HCP Removal on Adsorptive Membranes Using Protein Surrogate Biomarkers

Abstract:

Anion exchange membrane adsorbers are used for clearance of impurities such as viruses, host cell protein (HCP), and DNA in flow-through polishing steps for manufacture of monoclonal antibody therapeutics. HCP can be very complex in nature. CHO host cell protein composition shows that the majority of HCP in general, and in Protein A eluates in particular, are acidic proteins with pI ~ 4-6 and molecular weight between 10-75 kDa. In this study, surrogate biomarkers or model proteins that are easily available, easy to quantify and have properties comparable to typical host cell proteins, are utilized for rapid characterization of anionic membranes. Choice of these model proteins allowed for a large number of experiments to be carried out at moderate costs, thus enabling study of a broad range of conditions for determining binding constants and capacity for HCP adsorption for Q membranes. The results of these studies will be presented. Use of surrogate biomarkers for rapid and reproducible scale-up of membrane adsorbers will also be discussed

385271 Optimization of Tangential Flow Ultrafiltration Processing Step with High Concentration Self-Buffering Proteins

Abstract:

High protein concentration biopharmaceuticals are becoming more common due to the increased viability of these formulations for subcutaneous administration. Formulation of a biopharmaceutical product is most often performed using tangential flow ultrafiltration (TFUF) to concentrate the product to the desired level and, then exchange buffer components into a more stable formulation. As protein concentrations increase, self-buffering becomes a contributing factor in achieving the targeted formulation conditions. This case study will examine methods to formulate a self-buffering protein at commercial scale. The developed process demonstrates both robustness to manage expected parameter variation, as well as flexibility to allow potential future changes that may be deliberately implemented upstream of the step.

378644 Purification of Adenovirus Serotype 5 By Two-Column, Open-Loop, Size-Exclusion, Simulated Countercurrent Chromatography

Abstract:

The purification of complex biopharmaceuticals, such as recombinant proteins, antibodies or vaccines, is gaining increasing importance to fulfill the requirements of a reliable downstream purification process with high purity and yield, and, most importantly, cost efficiency. Viruses, in particular, play an important role in the vaccine and gene therapy fields.

The downstream biopurification train of viral vectors has been extensively developed in the past years by combining different chromatography steps, namely ion-exchange (IEX)) and size-exclusion chromatography (SEC), and, less frequently, affinity chromatography, intermingled with concentration and ultra/diafiltration steps.

To be more specific, the classical approach for adenovirus purification consists of three major steps—clarification, concentration/purification, and polishing—applied sequentially: (i) clarification of the harvested bioreaction bulk to remove cell and cell debris; (ii) ultra/diafiltration; (iii) anion exchange; (iv) ultra/diafiltration to concentrate the product; and (v) SEC as a final polishing step. The SEC step is usually carried out last, mainly because of its low productivity and low product titer. Usually in this final step the amount processed is reduced by 50–100 fold.

The current paradigm in the downstream processing of viral vectors is to operate each chromatographic step using classical single-column batch chromatography. However, the cost of goods and overall processing time can be reduced by switching to multicolumn, (semi-)continuous chromatography, which gives higher throughput, has lower buffer consumption, provides for higher capacity of utilization of the stationary phase (or smaller column volume), and hence gives increased productivity. Unfortunately, unlike other industrial fields, biotechnology is late to embrace continuous processing or multicolumn preparative chromatography.

The present work reports on the design and experimental validation of a simple quasi-continuous, open-loop, two-column countercurrent chromatographic process for size-exclusion purification of adenovirus serotype 5 (Ad5). As mentioned above, SEC has been used in the past for both polishing and intermediate purification of adenoviruses. However, it is often claimed that the main drawbacks of SEC, namely low productivity and high product dilution, make it a costly purification step. Here, it is shown that these drawbacks can be eliminated to a large extent by switching from single-column batch operation to two-column SMB-type operation.

Because of the anticipated boost in performance obtained by switching to multi-column (quasi-)continuous operation, the Ad5 purification train employed for assessing the effectiveness of the enhanced SEC step was streamlined: the SMB-SEC step was applied after the first ultra/diafiltration step and replaced the last three steps of the standard purification train. The main reason for working with the streamlined purification train was to challenge the SMB-SEC step with a less purified bulk in order to prove that, once the classical limitations of SEC are alleviated, the standard purification train can be changed to meet specific needs in terms of cost reduction or purity requirements. Moreover, running the process with a less clean bulk gave more confidence on the performance assessment of the SMB-SEC step.

In a two-column SEC process the cycle is divided into two equal-length time intervals, which we shall henceforth refer to as switching intervals. To design the cycle it suffices to define the sequence of flow-line configurations for one of the switching intervals; the sequence for the other twin switching interval is similar, but the positions of the two columns are exchanged. In practice, it is easier to keep the columns fixed in space and to move the positions of the inlet and outlet lines by means of an automated arrangement of two- or three-way valves. Over a full cycle, each column undergoes the same sequence of steps but phased out in time by one switching interval.

In the developed process, each half-cycle is divided into the following sequence of steps:

(1) Elution of the upstream column while the effluent of the downstream column is diverted to the waste line.

(2) Elution of the upstream column but its effluent is diverted to the waste line instead of being directed to the downstream column; the latter column is fed with the clarified bioreaction bulk while is effluent is collected as purified product;

(3) Operation of the system as in step 1, but the effluent of the downstream column is collected as product.

(4) Elution of the upstream column but its effluent is again diverted to the waste line instead of being directed to the downstream column; the latter column is kept frozen, that is, the flow through it is temporarily halted.

At the end of every half-cycle the positions of the two columns are exchanged to implement the simulated moving-bed contact between the solid and the fluid.

Clearance of impurities, namely DNA and host cell protein (HCP), were experimentally assessed. The obtained performance compares very favorably against single-column batch chromatography for the same volume of size-exclusion resin. In particular, our two-column SEC process achieves about the same DNA and HCP clearances as the standard batch process. However, and most importantly, the virus yield is increased from 57% for the batch process to 86% for our two-column because of internal recycling of the mixed fractions of contaminated Ad5 even though the process is operated strictly in an open-loop configuration. And last, but not least, the productivity is boosted by 6-fold with the two-column process. This is a clear demonstration that with this increased productivity, size exclusion can do much more than just final polishing and buffer exchange in the downstream train of a biopurification.

In conclusion, the main drawbacks of size-exclusion chromatography, namely low productivity and low product titer, were easily overcome by an innovative two-column configuration that keeps the mixed fractions inside the system at all times.

377231 Centrifugal Separation of Disk Centrifuge in Downstream Processing

Abstract:

Disk centrifuges[1] have been widely used for downstream harvesting of yeast, mammalian and microbial cells. Heretofore, the operation of the centrifuges is largely based on ad-hoc trial-and-error experiences.

A model of separating cells using a disk centrifuge is presented. The model is validated against experiments on a pilot-scale centrifuge. An interesting example is used to illustrate how the model can be used to predict performance of inclusion-body separation.

Reference

[1] Wallace Woon-Fong Leung, Centrifugal Separations in Biotechnology, Academic Press, UK, 2007.

386204 Managing Residual Risk Via the Control Strategy: A Case Study

Abstract:

There appears to be consensus that a science and risk based development approach such as QbD helps deliver robust, productive processes that assure product quality attributes are consistently met. While QbD has become a more mature concept within the pharmaceutical industry, some questions remain regarding implementation. This presentation will address how ultimately it is within the context of the control strategy that answers are best provided to questions of when continued investment in risk management activities no longer support a proportional benefit to the patient, and how can residual risks (such as scale sensitivity, or the potential process vulnerability to parameters not studied in great detail) be effectively managed. Furthermore, the presentation will highlight how the output of QbD activities, when coupled with considerations for the commercial production facility, can identify multiple options for how to define a control strategy that assures quality attributes are consistently achieved. The important considerations for the commercial production facility can include items such as:

- managing rate limiting steps and vessel utilization towards optimized process time cycle
- balancing the cost of expensive reagents against product yield gains
- evaluating existing equipment capabilities
- requirements for automation systems
- approaches for in-process testing

The case study will focus on one specific quality attribute, namely the impurity profile, as impacted by a GMP coupling reaction. This will include an overview of the iterative nature of the risk assessment process, where assumptions about the risks change as new information regarding processes is gained. The development activities and results driven by the risk assessment process will also be reviewed in order to highlight how information from various steps is compiled to work towards a final, comprehensive control strategy that considers each synthetic step. Finally, it will be reviewed how the multifactor process understanding along with the three key elements of control (raw materials specifications, in process testing, and control by process design) can be formulated into a control strategy that assures final quality attributes are met via a robust, productive process that fits within the anticipated commercial production facility.

391167 Estimation of Product Robustness: Prediction of Manufacturing Variability

Abstract:

This presentation will describe an approach to leverage process models and historical data of potential sources of variability (materials, process parameters, and analytical measurements) to estimate the process performance under defined manufacturing control strategies.

Using Monte Carlo tools, the process performance is estimated as a distribution of the possible outcomes of the measured process quality attributes. The outcomes are generated from a Bayesian network that integrates all the sources of variability associated with the process.

Applied to process models generated during process development, this framework will be useful in providing a statistically sound, quantitative level of risk associated with process performance under different control and monitoring strategies as well as informing technology transfer/validation and process robustness decisions.

373821 Digital Determination of Pars for the Synthesis of a Pharmaceutical Intermediate

Abstract:

This work describes the development of a first principles mathematical model for the digital determination of proven acceptable ranges (PARs) prior to the validation campaign of a pharmaceutical intermediate. The proposed model is capable of predicting the effect of catalyst, solvent composition, and temperature on the reaction kinetics (i.e., it predicts yield, reaction time, and reaction selectivity). Semi-batch addition of the limiting reactant was the preferred approach to minimize the instantaneous excess of the other starting material (genotoxic impurity) needed to maintain the generation of polymeric impurities below a desired limit. In preparation for an upcoming validation campaign, the model will not only be used for digital PAR determination but also will be used to guide the identification of the robust process center point, to describe the overall design space, and to perform the failure modes and effects analysis (FMEA).

375132 Model-Based Process Control in API Manufacturing: A QbD-Vantage!

Abstract:

Quality-by-design (QbD) is a fundamental tenet of any process development group that is engaged in designing robust and efficient processes for commercial API manufacture. The quest for QbD in API synthesis culminates into extensive knowledge generation on various unit operations such as reactions, purifications, and isolations, in the form of fundamental mechanistic understanding, multi-parameter (multivariate) interactions and models. Models bring in predictive ability that is valuable in streamlining R&D work and to develop scientifically sound design space for process parameters. A model could also be applied as an element of process control that satisfies quality requirements or as a process monitoring/trending tool in manufacturing. Presented is a case-study developed for evaluation of model-based process control in API manufacturing.

Ensuring consumption of a genotoxic starting material in API manufacture through the use of kinetics and mixing studies.

The priority of any API process development team is ultimately patient safety. When a known genotoxic starting material must be used in a synthesis, it is imperative to fully understand the behaviour of the mass balance over the design space. The approach used at GSK in a recent late phase asset combines kinetic modelling in DynoChem and mixing studies to ensure maximum consumption of the genotoxic reagent. Worst case spiking studies of the predicted residual were also conducted to confirm the ultimate fate of the genotoxin downstream.

The genotoxic starting material forms a second liquid phase upon addition to the reaction mixture, early scoping work showed reaction time was dependant on stirrer speed. In order to generate accurate time course data for the kinetic model, the reaction was run under dilute single phase conditions, this removed the mass transfer effects and improved sampling. To provide confidence on scale-up that the biphasic reaction would reach completion within the specified stir time, mixing assessments were conducted across a range of vessel geometries. The mixing requirements of this biphasic reaction led to an interesting debate around what critical process parameters would be registered for this stage to allow plant flexibility without compromising patient safety.

380059 Efficacious Polyanhydride Nanovaccines Against White Spot Syndrome Virus in Shrimp

Abstract:

The $10.6 billion shrimp industry is growing at approximately 10% each year. However, viral diseases, such as white spot syndrome virus (WSSV), continue to threaten the socioeconomic aspects of this industry. WSSV infection of shrimp spreads rapidly and can cause 100% mortality to farmed shrimp in 2-10 days. Currently, there is no treatment available for WSSV, and farmers rely on a large number of chemical disinfectants (e.g., iodophor, formalin) to reduce the extent of disease. While vaccination of shrimp with WSSV subunit proteins or dsRNA has been shown to be successful in preventing infection, exposure to water often leads to instability and degradation of such vaccines.

The polyanhydride nanoparticle-based platform is suitable for the encapsulation and release of a diverse range of bioactive compounds such as vaccine antigens, peptides, antibiotics, and nucleic acids. Nanoparticles composed of sebacic acid (SA), 1,6-bis-(*p*-carboxyphenoxy) hexane (CPH), and 1,8-bis-(*p*-carboxyphenoxy)-3,6-dioxaoctane (CPTEG), and their copolymers exhibit surface erosion which limits the bulk exposure to water, and therefore, enhances the stability of encapsulated payloads. The ability to tailor different copolymer chemistries also allows for tunable release kinetics. Finally, previous work has demonstrated that polyanhydride nanoparticle-based vaccines (or nanovaccines) are safe and efficacious.

In this work, polyanhydride nanovaccine formulations encapsulating a WSSV antiviral were designed for reverse gavage administration to shrimp. The polyanhydride nanovaccines demonstrated increased retention within shrimp in comparison to antiviral delivered alone. It was also observed that the nanovaccines were localized to the gills of shrimp, a point of entry for WSSV virions, confirming the pathogen-mimicking nature of the nanovaccines. In addition, examination of tissue sections via histopathology showed no adverse reactions due to nanoparticle immunization. Finally, the efficacy of the polyanhydride nanovaccines was demonstrated *in vivo* in response to viral challenge. Taken together, the studies described herein attest to the promise of the polyanhydride nanovaccine platform as a suitable vehicle for oral delivery of shrimp vaccines.

386398 Understanding What Is Still Patentable in the Life Science Industry

Abstract:

The Supreme Court's recent decisions invalidating patents under section 101 of the Patent Act as patent-ineligible subject matter have affected patent portfolios across all life sciences industries, from biotechnology patents and the development of personalized medicine. These recent decisions have made it essential to reassess your company's patent strategy to ensure protection of your inventions and enable future partnering and marketing success. In *Mayo*, the Court held unpatentable claims to optimizing treatment of an immune-related disorder by determining the level of drug metabolite in a patient's blood based on knowledge of effective levels. According to the Court, the claims recited an unpatentable natural law—the relationship between metabolite levels and the drug effectiveness—applied via well-understood and conventional steps, thus impermissibly preempting all use of that law. Next, in *Myriad*, the Court struck down claims to isolated DNA molecules as directed to unpatentable products of nature, distinguishing between patent-ineligible gene sequences found in nature and patent-eligible genetic sequences, such as complementary DNAs (cDNAs), created by scientists in the lab. In March of this year, the United States Patent and Trademark Office (USPTO) issued its long awaited *Guidance for Determining Subject Matter Eligibility of Claims Reciting or Involving Laws of nature, Natural Phenomena, & Natural Products*. These decisions and subsequent guidelines significantly affect a company's patent portfolio strategy. During this session, we will address how natural product and diagnostic method claims will be examined by the USPTO.

Negatively charged carbon nanohorn supported cationic liposome nanoparticles: a novel delivery vehicle for anti-nicotine vaccine

Hong Zheng,[1*] Yun Hu,[1*] Wei Huang,[1] Sabina de Villiers,[2,3] Paul Pentel,[2] Jianfei Zhang,[4] Harry Dorn,[4] Marion Ehrich,[5] Chenming Zhang[1†]

1. Department of Biological Systems Engineering, Virginia Tech, Blacksburg, VA 24061

2. Minneapolis Medical Research Foundation, Minneapolis, MN 55404

3. Department of Physiology and Pharmacology, Karolinska Institute, Stockholm, Sweden

4. Department of Chemistry, Virginia Tech, Blacksburg, VA 24061

5. Department of Biomedical Sciences & Pathobiology, Virginia Tech, Blacksburg, VA 24061

[†]Correspondence to:

Chenming (Mike) Zhang

210 Seitz Hall

Dept. of Biological Systems Engineering

Virginia Tech

Blacksburg, VA 24061

(540) 231-7601

(540) 231-3199 (Fax)

Chzhang2@vt.edu

*These authors contributed equally to this work.

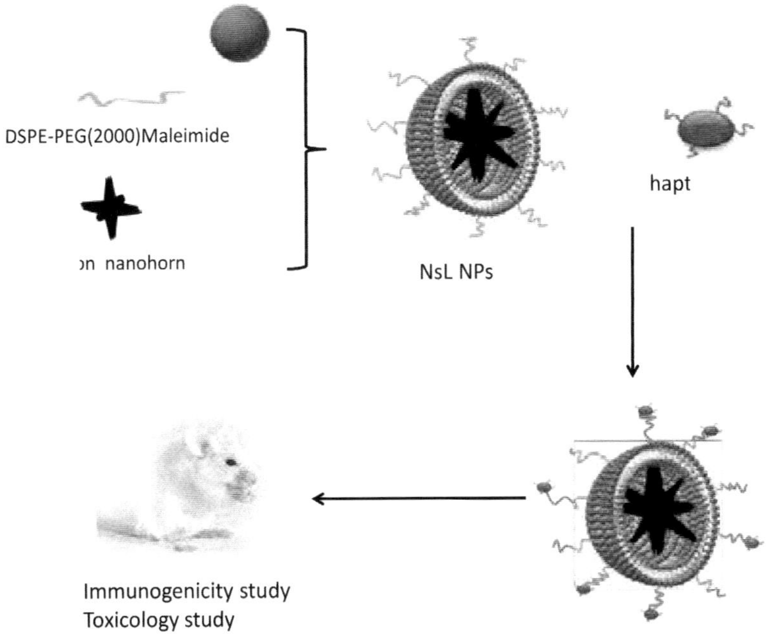

Abstract

Tobacco addiction is the second-leading cause of death in the world. Due to the nature of nicotine (a small molecule), finding ways to combat nicotine's deleterious effects has been a constant challenge to the society and the medical field. In the present work, a novel anti-nicotine vaccine based on nanohorn supported liposome nanoparticles (NsL NPs) was developed. The nano-vaccine was constructed by using negatively charged carbon nanohorns as a scaffold for the assembly of cationic liposomes, which allow the conjugation of hapten conjugated carrier proteins. The assembled bio-nanoparticles are stable. Mice were immunized subcutaneously with the nano-vaccine, which induced high titer and high affinity of nicotine specific antibodies in mice. Furthermore, no evidence of clinical signs or systemic toxicity followed multiple administrations of NsL-based anti-nicotine vaccine. These results suggest that NsL-based anti-nicotine vaccine is a promising candidate in treating nicotine dependence and could have potential to significantly contribute to smoking cessation.

Keywords

Nicotine; nicotine vaccine; anti-nicotine; bionanoparticle; carbon nanohorn; nano-vaccine.

1. Introduction

Tobacco addiction is the second-leading cause of death in the world[1,2]. How to help smokers to quit smoking has been a constant challenge to medical scientists. To date, three medications are FDA-approved for smoking cessation: nicotine replacement therapy, sustained-release bupropion, and varenicline[3,4]. Despite the relative efficacy of these first-line medications, long-term abstinence rates remain disappointingly low[5]. Clearly, more effective treatment options are needed. Immunopharmacotherapy, or vaccination, has shown tremendous promise in fighting against nicotine addiction. Nicotine vaccines have the following advantages: 1) vaccines can have a prolonged effect on the immune system (6-12 months), and could thus reduce the relapse rates, 2) daily administration is not required; only occasional booster shots are needed to maintain an adequate antibody titer, 3) vaccines have minimal to no side effects, and 4) vaccines are potentially compatible with other smoking cessation methods. However, all current nicotine vaccines, particularly the ones that have undergone human clinical trials, demonstrated limited efficacy[6,7,8]. The recently released clinical trial results of NicVax® by Nabi Biopharmaceuticals were particularly disappointing[9]. On the other hand, a recent report described a successful trial in mice treated with adeno-associated virus (AAV) vector expressing a gene encoding a full-length anti-nicotine antibody[10]. However, it is still unclear how this success can be translated to humans, and the long-term safety of the system is still a significant obstacle. Because the success of drug vaccines appears to be directly correlated with the antibody titer in vaccinated subjects[6,7,8,9,10], there is a need for improved vaccines that can elicit a strong immune response.

Over the past decade, particulate carriers have emerged as an attractive means for enhancing the delivery efficacy and potency of vaccines and associated immunomodulatory molecules. Specifically, polymer-based micro and nanoparticles are being extensively studied for this purpose. Polymeric micro and nanoparticles can carry cargo, whether antigens, adjuvants or both, either encapsulated within the particle or on the surface of the particle. They provide an efficient means for delivering antigen to antigen presenting cells (APCs), either in a passive manner through non-specific phagocytosis or in a more active manner through receptor targeting and endocytosis. The latter can be achieved by modifying the particle surface with antibodies against surface receptors of specific APC subsets e.g., DEC205 attachment for targeting to epidermal Langerhans cells[11].

In addition to the particulate nanoparticles, proteins and peptides have demonstrated potential for being effective vaccine carriers. Protein derived particulate systems are biocompatible and degraded into natural byproducts, and they can be easily modified due to the large repertoire of amino acids with a variety of physical-chemical properties. Peptide/protein-based particles made of amino acids[12], denatured collagen (gelatin)[13], and albumin[14] have been shown to be effective in eliciting immune responses. Moreover, liposomes, particulate structures composed largely of natural or synthetic phospholipids, can encapsulate both antigenic and immunomodulatory agents[15,16,17,18], allowing targeted delivery to specific immune cells such as dendritic cells *in vivo*. Depending on their lipid composition, liposomes also may exhibit potent adjuvant-like properties[19,20]. However, one of the main limitations of liposomes is that they are intrinsically unstable and often have a tendency to flocculate[21].

In this study, novel negatively charged (anionic) carbon single-walled nanohorns (SWNHs) were used as a scaffold, providing support to the soft body of cationic liposomes. This support prevented the liposomes from precipitation or flocculation. The SWNHs are nontoxic[22,23] nanoparticle aggregates of single graphene tubules and are spherical dahlia-like-shaped with a narrow diameter range reported as 80-100 nm[24]. For this study, nanohorns with a diameter of around 100 nm were encapsulated into liposomes through a simple freeze-thaw method as described in our previous work[25]. The nanohorn supported liposome nanoparticles (NsL NPs) were used as a vaccine delivery vehicle. The hapten derivative, O-succinyl-3'-hydroxmethyl-(±) nicotine (Nic), was used as the immunogen[26]. The immunogenicity and toxicity of the nano-vaccines were studied in mice.

2. Materials and methods

2.1. Materials

All chemicals were obtained from Sigma-Aldrich (St. Louis, MO, USA) unless otherwise noted. 1,2-dioleoyl-3-trimethylammonium-propane (DOTAP) and 1,2-distearoyl-sn-glycero-3-phosphoethanolamine-N-[maleimide(polyethylene glycol)-2000] (ammonium salt) (DSPE-PEG(2000)-Maleimide) were purchased from Avanti Polar Lipids (Alabaster, AL, USA). EDC (1-ethyl-3-[3-dimethylaminopropyl] carbodiimide) was obtained from Pierce (Rockford, IL, USA). O-succinyl-3'-hydroxmethyl-(±) nicotine (nicotine hapten) was purchased from Toronto Research Chemicals (Toronto, ON, Canada).

2.2. Preparation of nanohorn supported liposome nanoparticles (NsL)

NsL NPs were prepared according to the method described by Huang et al. Briefly, cationic liposomes were made by extruding 1.96 mg hydrated DOTAP and 0.32 mg DSPE-PEG(2000)-maleimide lipid film through polycarbonate membranes with pore sizes of 100 nm. The hydration buffer contained 0.9% NaCl, 5% dextrose, and 10% sucrose in Tris-HCl buffer with pH at 7.4. Single-walled nanohorns (SWNH) were synthesized by Nd:YAG laser vaporization of graphite rods in an argon atmosphere at 1100 °C[27,28]. Negatively charged (anionic) nanohorns with carboxyl groups were prepared by high-speed vibration milling[29]. Anionic nanohorns were briefly sonicated to break up aggregates and incubated with freshly made cationic liposomes at a volume ratio of 1:4. Three freeze-and-thaw cycles were applied using -80 °C and a warm water bath (30 °C). Samples were finally centrifuged under 10,000 × *g* for 10 min, and nanohorn supported liposomes were collected from the top layer of the solution.

2.3. Synthesis and Quantification of Nic-BSA conjugates

The synthesis of Nic-BSA conjugates was based on the procedure described elsewhere[30]. Briefly, a solution of 1-ethyl-3-(3-dimethylaminopropyl) carbodiimide (EDC, 10 molar equivalents compared to nicotine hapten) in distilled H_2O was added to 10 mg nicotine hapten water solution (100 mg/mL) and kept at 0 °C for 10 min. The solution was then mixed with 10 mg bovine serum albumin (BSA) solution (40 mg/mL) and an appropriate amount of distilled water to prepare a total volume of 1 mL. The solution was stirred at 0 °C for 10 min, and stored at room temperature overnight. During this step, pH was checked and adjusted to 6.76 with 0.01 M sodium hydroxide if needed. Nic-BSA conjugates were purified by size exclusion chromatography with a Sephadex G-25 column using an AKTA FPLC system (GE Healthcare, Piscataway, NJ). Nic-BSA was then concentrated to 1 mg/mL using Microcon centrifugal filter units (EMD Millipore Corporation, Billerica, MA) with a 50,000 molecular weight cut-off (MWCO). The number of $-NH_2$ groups used during conjugation of hapten to BSA molecules was determined from the difference between the optical density of the control and the conjugate by a method utilizing 2,4,6-trinitrobenzene-1-sulfonic acid (TNBS)[31].

2.4. Thiolation of Nic-BSA conjugates

Thiol groups were introduced to Nic-BSA by incubating Nic-BSA obtained in the previous step with 1 mg/mL of 2-iminothiolane (Traut's reagent) for one hour in darkness under continuous stirring[32]. The thiolated Nic-BSA was purified using FPLC as described for Nic-BSA purification and concentrated to 2 mg/mL in phosphate buffered saline (PBS, pH 7.4) by Microcon centrifugal filter units (50,000 MWCO). The amount of thiol groups on BSA was quantified by a colorimetric sulfhydryl assay using Ellman's reagent (5,5′-dithiobis(2-nitrobenzoic acid)[33]. Briefly, Ellman's reagent (80 μL of a 4 mg/mL solution in PBS) was added to 0.6 mL aliquots of a control and the samples and allowed to incubate at room temperature for 20 min. Sulfhydryl levels were determined from the absorbance at 412 nm (ε = 14398) using the equation,

$$SH=1.1A_{412}/\ 14398C_{Nic\text{-}BSA}$$

where A_{412} is the absorbance at 412 nm, $C_{Nic\text{-}BSA}$ is the protein concentration, and SH is the number of thiol equivalents.

2.5. Conjugation of thiolated Nic-BSA to NsL

The thiolated Nic-BSA was mixed with the maleimide-PEG–containing nanohorn supported liposomes in 0.15 M NaCl, 0.2 mM EDTA (pH 7.4) at a molar ratio of 1:8 and allowed to react overnight. Unbound protein was separated from the liposomes by dialysis using dialysis membrane (MWCO 1000 kD) from Spectrum Laboratories in 0.15 M NaCl pH 7.4 solution. Nic-BSA associated with liposomes was assayed using a modified protocol described by Ansell[34]. Briefly, 20 μL of prepared liposomes (before and after purification) were diluted with 1 mL 0.15 M NaCl pH 7.4 solution (working dispersion). To assess the total and conjugated protein concentration, 500 μL of working dispersion were mixed with 100 μL of 5% (v/v) Triton X-100, and this mixture was maintained at 65 °C for 5 min to disrupt all lipid vesicles. Both total and conjugated Nic-BSA concentrations were measured using a micro BCA protein assay kit from Pierce (Rockford, IL). The association efficiency (AE) and the loading capacity (LC) of

Nic-BSA to nanohorn supported liposomes were calculated according to the following equations: $AE(\%) = \text{Nic-BSA}_{conjugated}/\text{Nic-BSA}_{Total} \times 100\%$; $LC(\%) = \text{Nic-BSA}_{conjugated}/\text{NPs} \times 100\%$.

2.6. Characterization of NPs

2.6.1. Dynamic light scattering (DLS) and zeta-potential

Particle sizes and distribution of the nanohorn, liposomes, NsL and Nic-BSA-NsL were measured by photon correlation spectroscopy (PCS) using a Zetasizer Nano (Malvern Instruments, UK). Prior to the DLS measurement, each sample was diluted with 0.15 M NaCl pH 7.4 solution until a measurable concentration of particles was achieved. Similarly, the zeta potential of the samples was measured using the same equipment with Laser Doppler Velocimetry[35]. Both measurements were repeated three times for each sample.

2.6.2. Transmission electron microscopy (TEM)

Sample grids (carbon coated copper) were put into one drop of liposomes for 30 s, then in a distilled water drop for washing (10 s) and finally in a phosphotungstic acid drop for staining (10 s)[36]. The excessive stain on the grids was removed using filter paper. The prepared grids were analyzed using a Morgagni™ Transmission Electron Microscope (FEI Corporate, Hillsboro, OR).

2.6.3. Stability of NsL NPs

Stability was monitored using imaging of fluorescent dyes encapsulated following a published method[37]. Briefly, lipid films containing 1.96 mg 1,2-dioleoyl-3-trimethylammonium-propane (DOTAP) and 0.2 mg fluorescein isothiocyanate (FITC) isomer I (molar ratio, 100:2) were hydrated with 250 μL of hydration buffer (section 2.2). The resulting emulsion was incubated at 65 °C for one hour and extruded 14 times through polycarbonate membranes with pore sizes of 100 nm. For nanohorn supported liposomes, the FITC labeled liposomes underwent

5 cycles of freeze-and-thaw at -80 °C and 35 °C when 62.5 μL of a nanohorn solution (1 mg/mL) were added. The resulting mixture was centrifuged at 16,000 × *g* for 30 min, and the nanohorn supported liposomes were collected from the top layer of the solution. Excessive FITC in both liposomes and nanohorn liposomes was removed by 24 h of dialysis using 100 kD Biotech CE Dialysis Tubing (Spectrum Laboratories, Inc., Rancho Dominguez, CA, USA) in distilled water. Morphology was viewed on a LSM510 confocal laser scanning microscope (Carl Zeiss Inc, Thornwood, NY, USA).

2.7. Vaccination

2.7.1. Immunization procedures

All animal studies were carried out following the National Institutes of Health (NIH) guidelines for animal care and use. Animal protocols were approved by the Institutional Animal Care and Use Committee at Virginia Polytechnic Institute and State University. Female Balb/c mice (6-7 weeks, 16-20 g, 8 per group) were randomized into vaccine and control groups. The immunization schedules and administration doses are shown in Table 1. Vaccine groups (Groups 1-5) were immunized subcutaneously (s.c.) with 100 μL of a saline suspension of Nic-BSA conjugates, or Nic-BSA-liposome conjugates or Nic-BSA-nanohorn-liposome conjugates mixed with or without Alum adjuvant (3 mg, Imject, Pierce Biotechnology Inc., Rockford, IL). Control groups (Groups 6 and 7) received liposomes or nanohorn-supported-liposomes formulated with Alum (3mg) in 100 μL saline. All mice were immunized with three injections, given on day 0 with booster immunizations on days 14 and 28. Antibody titers were monitored following vaccine administration using blood samples (~200 μL) taken on days 0, 13, 27, 33, 40 from the retroorbital plexus of mice under isofluorane anesthesia. The blood samples were kept at 37 °C for 30 min and then centrifuged at 4 °C at 2000 × *g* for 30 min, with the supernatant centrifuged a second time. Aliquots of sera were stored at −80 °C prior to analysis.

2.7.2. Measurement of specific anti-nicotine IgG antibodies using ELISA

Mice sera were analyzed according to the ELISA procedure described by de Villiers et al. with appropriate modification[38]. Briefly, the nicotine hapten was conjugated to keyhole limpet hemocyanin (KLH) according to the same method used for the preparation of Nic-BSA conjugate. MICROLON® 96 well plates (Greiner Bio-One, Longwood, FL) were coated with Nic-KLH conjugate (10 μg/mL in carbonate buffer, 0.05 M, pH 9.6, 100 μL/well) and incubated at 25 °C for 5 h. The plates were washed with PBS-Tween (0.1%) for 4 times and distilled water for 2 times, followed by blocking with 300 μL Pierce® protein-free T20 blocking buffer for 12 h. Sera from immunized mice were serially diluted at different dilution (1:25, 1:125, 1:625, 1:3125, 1:15625, and 1:78125) in PBS, and 100 μL of each dilution was added to the plate which was subsequently incubated at 25 °C for 2 h. The plates were washed again, and incubated with 100 μL Anti-Mouse IgG HRP (1:10000) from Alpha Diagnostic Intl (San Antonio, TX) for 1 h. After washing as before, plates were incubated with tetramethylbenzidine (TMB) peroxidase (100 μL/well) for 10 min in the dark. Sulphuric acid was added (0.5%, 100 μL/well) and absorbance was measured at 450 nm, using a Microplate Reader (Synergy HTS Multi-Mode, BioTek Instruments, Inc., Winooski, VT). The ELISA titer was defined as the dilution of the serum which gives a half-maximal optical density signal (OD 50%) in the ELISA.

2.7.3. Determination of of anti-nicotine specific IgG subclasses

For the IgG subtype determination, duplicate ELISA were done for the four antibody isotypes in serum from mice of each group. Horseradish peroxidase conjugated sheep anti-mouse IgG1, IgG2a, IgG2b and IgG3 (Alpha Diagnostic Intl.) were used as the secondary antibodies.

2.7.4. Th1/Th2 index calculation

To determine whether the addition of nanohorn, liposomes and Alum adjuvant in the immunization protocol induced a Th1 (IgG2a and IgG3) or Th2 (IgG1) polarization, Th1:Th2 index was calculated as ([IgG2a+IgG3]/2)/(IgG1) for each immunization groups. According to such calculation, an index value < 1 stands for a Th2 polarization; an index value > 1 stands for a Th1 polarization[39].

2.7.5. Antibody affinity for nicotine evaluation by equilibrium dialysis

In order to assess antibody affinity (K_d) and maximum nicotine binding capacity (B_{max}), sera from immunized mice treated with NsL-based anti-nicotine vaccine in the presence and absence of Alum on day 40 were diluted with control sera (Balb/C mouse serum, Innovative Research Inc., Novi, MI), due to small sample sizes. These were then dialyzed against six different concentrations of nicotine using a 96-well Equil DIALYZER (MWCO 10 kD, the Nest Group, Inc. Southborough, MA). Unlabeled ((-)-nicotine hydrogen tartrate salt, Sigma-Aldrich, St. Louis, MO) and tritiated nicotine, (L-(-)-[N-methyl-3H]-nicotine, PerkinElmer, Boston, MA) were diluted in control sera to total nicotine concentrations ranging from 1 ng/mL to 1024 ng/mL (expressed as free base). The plate was placed vertically on a shaker and incubated for 72 h at room temperature, until equilibrium had been reached. Samples from both sides of the dialysis membrane were collected and counted using a 1450 MicroBeta TriLux Liquid scintillation counter (EG&G Wallac) with MicroBeta workstation software (version 2.7). Disintegrations per minute (DPM) were transformed to total nicotine concentrations based on the radioactivity of the original solutions. Bound versus free nicotine was plotted to create a saturation binding curve from which K_d, the concentration resulting in half of the binding sites being occupied, and B_{max} (GraphPad Prism 5, GraphPad Software, Inc.) were determined. Values for B_{max} were adjusted for the serum dilution.

2.8. Mouse toxicology studies

2.8.1. Animals and animal husbandry

This study was conducted in accordance with the U.S. Food and Drug Administration *Good Laboratory Practice Regulations for Nonclinical Laboratory Studies* [40,41], with the *Guide for the Care and Use of Laboratory Animals* [42], and under a protocol approved by the Institutional Animal Care and Use Committee at Virginia Polytechnic Institute and State University.

The experimental animals were female Balb/cByJ mice (The Jackson Laboratory, Inc., Bar Harbor, ME, USA), 5 weeks old upon arrival. The Balb/c mouse was chosen to conform to the requirements by the National Toxicology Program (NTP)[43,44]. Animals were individually identified by ear notching during a 7-day quarantine. The study began after the quarantine when the mice were 6 weeks old. Mice were housed individually in solid-bottom, polycarbonate cages with stainless steel lids (Laboratory Products, Rochelle Park, NJ) and Certified Sani-Chip® hardwood cage litter (P. J. Murphy, Montville, NJ). Feed (Purina Certified Rodent Chow (#5002), PMI, St. Louis, MO) and tap water (Blacksburg, VA) were provided *ad libitum* throughout the study. The light cycle (12 h light, 12 h dark) and temperature and relative humidity in the animal rooms were controlled and monitored (Siebe/Barber-Coleman Network 8000® System with SIGNAL® Software [Version 4.1], Siebe Environmental Controls

(SEC)/Barber-Colman Company, Loves Park, IL) throughout the study. Temperature and relative humidity were within their respective target ranges (69-75 °F and 35-65%).

2.8.2. Evaluation of body weight, feed and water consumption, and clinical signs

After the 7-day quarantine, the animals were randomly divided into seven groups containing eight animals per group. During the treatment period of seven weeks (see section on Immunization procedures, 2.7.1), body weight, feed consumption, and water consumption were measured by the method of McNair and Bryson[45] and recorded twice per week. In the studies, clinical signs of toxicity included body condition, lethargy, abnormal posture, and ruffled fur.

2.8.3. Evaluation of organ/body weight ratios

On day 45, after 7 weeks of observation described above, all animals were weighed and then euthanized under CO_2. The hearts, lungs, livers, spleens and kidneys were excised and weighed accurately. Organ/body weight ratios were calculated as organ weight/body weight ×100 (%)[46].

2.8.4. Histopathological evaluation

Histopathological analysis of hearts, lungs, livers, spleens, kidneys and skin tissues from the vaccine sites of each experiment group was performed by the method of Iranloye and Bolarinwal[47]. Briefly, organs (hearts, lungs, livers, spleens, kidneys and skins) were fixed in a neutral 10% buffered formalin. Then tissue blocks were embedded in paraffin and routine sections stained as described by Lillie with hematoxylin and eosin[48]. After hematoxylin-eosin (HE) staining, sections were examined by light microscopy on a Nikon Eclipse E600 scope and images captured using a Nikon DS-Fi1 camera run by NIS Elements software (Nikon, Melville, NY).

2.9. Statistical analysis

Statistical analysis was performed using the Student's t-test for the difference between the experimental groups and the control group. Bonferroni correction was used where multiple comparisons were made. Differences between means were considered significant at $P<0.05$, very significant at $P<0.01$ and highly significant at $P<0.001$.

3. Results and discussion

3.1. Conjugating antigen to nanohorn-supported-liposomes

In this present study, nicotine derivative, O-succinyl-3'-hydroxmethyl-(±) nicotine[26] was used as the target hapten. Antibodies induced by this hapten (coupled to virus-like particles derived from the coat protein of the bacteriophage Qb) have been shown to have a high affinity for nicotine (K_d = 35 nM) and no cross-reactivity with acetylcholine, the major nicotine metabolites cotinine and nicotine-N-oxide, and a variety of other neurotransmitters or drugs. Bovine serum albumin (BSA) acted as a model carrier protein to conjugate with the hapten in this study. BSA (67 kD) is a commonly used model carrier protein in vaccine development. For example, the anti-cancer vaccine, in which BSA was conjugated with 3'-fluoro-TF antigen-MUC1, was able to generate high titers of antibodies that could specifically bind to a tumor-associated glycopeptide antigen analog[49].

Covalent attachment of carboxylic acid hapten to the carrier protein (BSA) via available surface amino groups on BSA (26 surface lysines) was confirmed by reacting the prepared conjugates with TNBS reagent. The number of amino groups present in the carrier protein before and after conjugation was quantified with a UV/Vis spectrophotometer (Shimadzu 1601) set at 335 nm. The gradual decrease of the available surface lysines on the protein after reaction with different molar ratios of hapten confirmed the increase in conjugation density (number of hapten molecules per protein molecule) with the increasing hapten-protein molar ratio.

Conjugation of protein to the liposome surface was based on a chemical reaction between the maleimide of the PEGylated lipid with a derivatized sulfhydryl group at the primary amines of a protein by adding Traut's reagent, as previously described for conjugation of lactoferrin to the liposome surface[50]. The thiolation level was measured using 5,5'-dithiobis(2-nitrobenzoic acid) (Ellman's reagent) as described by Benzinger et al[51]. By optimizing the molar ratio of BSA:Nic:Traut's reagent:DSPE-Mal-PEG2000, higher association efficiency of Nic-BSA to NsL can be obtained (Table 2). Conjugates with 15 Nic haptens per BSA were produced by using a 50-fold excess of Nic-hapten. A 200-fold excess of 2-iminothiolane resulted in 5 thiol groups per Nic-BSA conjugate molecule. Moreover, a ratio of 1:4 or 1:8 of Nic-BSA:DSPE-Mal-PEG2000 was used for conjugation, and a strong antigen-NsL association efficiency of 50.4% or 65.8% and loading efficiency of 21.7% or 25.3% were achieved, respectively.

3.2. Physicochemical properties of NsL NPs

3.2.1. Particle size, zeta potential analysis

The particle sizes and zeta potentials of nanohorns, liposomes, NsL and Nic-BSA-NsL are presented in Table 3. The results show that the mean diameter of nanohorns, liposomes, NsL and Nic-BSA-NsL was 107.8± 3.3, 152.0± 6.0, 164.9± 2.5 and 175.3± 3.2 nm, respectively, suggesting there was no significant difference in size after the encapsulation of a nanohorn in a cationic liposome and the conjugation of antigens. In addition, the DLS analysis results of nanoparticle size agree well with the size obtained by TEM (Fig. 1). The zeta potentials of liposomes were around 40 mV regardless of the presence of nanohorn, suggesting that negatively charged nanohorns embedded into the cationic liposomes did not significantly change the zeta potential of the liposome nanoparticles. It has been suggested that full electrostatic stabilization requires a zeta potential >30 mV (ideally >60 mV), potentials between 5 mV and 15 mV result in limited flocculation, and potentials between −5 mV and +3 mV yield maximum flocculation[52,53]. Therefore, the zeta potentials obtained in the present study were within the range of full electrostatic stabilization, and values near 40 mV indicate the presence of strong forces of repulsion among the nanoparticles and little tendency for aggregation. Furthermore, charged nanoparticles adsorbed to the liposome can provide additional steric stabilization of liposomes. Thus the nanohorn-supported liposomes prepared in this study should, therefore, remain stable for a relatively long period of time.

3.2.2. TEM analysis

Figure 1 shows TEM images of nanohorns, empty liposomes, and nanohorn supported liposome nanoparticles. Evidently, carbon nanohorns are nearly spherical particles, approximately 100 nm in diameter, with a so-called 'dahlia flower' structure (Fig. 1A). The empty DOTAP liposomes and the nanohorn supported liposome nanoparticles (NsL) have nearly spherical shapes with sizes of 100-300 nm (Fig. 1B, 1C). The size of the nanoparticles observed by TEM was consistent with the results of dynamic light scattering (Table 3). As shown in Fig. 1C, a nanohorn supported liposome is a combination of the particles shown in Fig. 1A and Figure 1B.

Liposomes with diameters of less than 500 nm were shown to efficiently enhance the immunogenic performance of liposome vaccines over large liposomes (>500 nm)[54]. This is expected because the optimal particle diameter for uptake by APCs is below 500 nm. Particles of small size are also essential if non-liposome vectors are to induce a high immune response. For example poly(D, L-lactic-co-glycolic acid) (PLGA)particles of 300 nm loaded with DNA induced much higher antibody titers than particles with a diameter of 1 μm[55]. Moreover, for the present study, nicotine haptens were designed to bind to the surface of liposomes, and thus small nanohorn supported liposome nanoparticles will not only fit the size requirement of particles preferred for APC uptake but also provide the higher surface/volume ratio that increases capacity for hapten loading.

3.2.3. Stability of nanohorn supported liposome nanoparticles

The efficiency of liposomes as a delivery vehicle is usually compromised after administration into human body by liposomal instability during storage or circulation[56]. The instability can be caused by oxidation, degradation or fusion of liposomes with each other[57,58]. In vaccine development, the desirable size of liposomes should be less than 300 nm due to uptake preference of the major antigen presenting cells (especially dendritic cells). However, the fusion of liposomes tends to result in particles with larger size. Many strategies can overcome this problem, and among them, including PEGylation of the liposomes. Despite the stabilizing effect on liposomes, PEGylation may lead to reduced uptake of liposomes by target cells due to reduced interaction between liposomes and cells[56,59]. In this study, negatively charge nanohorns were the scaffold introduced into positively charged liposomes. Fig. 2 shows that during liposome preparation, nanohorn supported liposomes exhibited better uniformity and smaller size than blank liposomes. Aggregates in Fig. 2A indicate that even during preparation, liposome tended to fuse with each other to form larger particles. However, liposomes supported by nanohorns displayed smaller and more uniform morphology (Fig. 2B). Significant fusion occurred with liposomes alone after storage for 5 days at 4 °C, as shown by the clusters in Fig. 2C. In addition, the presence of smaller particles in Fig. 2C suggests that some degradation also occurred. Although increased size was detected in nanohorn supported liposomes, many still maintained their original size (Fig. 2D). The comparison between liposome images with and without nanohorns suggests that nanohorns exerted a stabilizing effect on liposomes during both preparation and storage.

3.3. Immunogenicity study

3.3.1. Time course for serum IgG antibody responses to Nic-BSA, Nic-BSA-LIP and Nic-BSA-NsL immunization by subcutaneous injection with or without Alum

The development of anti-nicotine immune response was determined by analyzing serum samples collected during animal study using ELISA. ELISA plates were coated with Nic-KLH conjugates. The results demonstrated that all subcutaneously-administered vaccines, each containing 50 μg Nic-BSA, resulted in gradual increase of serum anti-nicotine IgG antibody titers (Fig 3). This occurred both in the presence and absence of Alum. Maximum titers were obtained two weeks after the second booster of a vaccine. Nicotine-specific IgG antibody (NicAb) titers in the presence of Alum were higher than those in the absence of Alum in groups given Nic-BSA-LIP and Nic-BSA-NsL (Fig. 3). No NicAb were detected in sera collected on day 0, showing that Nic vaccine treatment was necessary for the production of antibodies. After the primary immunization and two additional boosters, the titer of NicAb elicited in response to Nic-BSA-NsL with Alum was significantly greater than titers in other groups whether or not the vaccines were given with or without Alum. In addition, even in the absence of Alum, a significant level of nicotine-specific antibodies could be observed in the Nic-BSA-NsL group.

In this study an NsL-based anti-nicotine vaccine delivery system was developed. When administered subcutaneously, the vaccine elicited a strong immune response, inducing 5 and 7 fold increases in total IgG titers compared to the anti-nicotine vaccine (Nic-BSA) formulated with or without liposomes, respectively. It is worth noting that the total antibody titer for Nic-

BSA-NsL, even with the absence of Alum, is similar to the titer of the nicotine vaccine by Selecta Biosciences, SEL-068 (5×10^5 for Nic-BSA-NsL, 1×10^6 for SEL-068), which, based on the limited available information, has haptens conjugated to the surface of the nanoparticles and a universal antigen peptide encapsulated inside the nanoparticles[60]. It has also been reported that the anti-nicotine vaccine, which is based on virus-like particles (VLPs) of the RNA phage Qb to which nicotine haptens are covalently coupled via succinimate linkers (NicQb), induced potent and long-lived antibody responses in mice[61] as well as humans[62, 63,64]. Compared with the hapten delivered by the virus-like particles, NicQb by Cytos AG[61,65], this nano-vaccine was larger in size (~100 nm compared to 25 nm) and contained lipids that induced a stronger immune response.

3.3.2. Detection of subclass distribution of nicotine-specific antibodies after vaccination

Human IgG consists of four subclasses contributing in different ways to humoral immunity against pathogens. Mice, similarly to humans, have four different classes of IgGs, namely IgG1, IgG2a, IgG2b and IgG3, which functionally correspond to the human IgG1, IgG2, IgG4 and IgG3, respectively. Despite some species differences in the IgG subclasses, the overall structure of the humoral IgG pattern in mice and humans are quite similar. In general, mouse IgG1 (as well as IgG4 in humans) is associated with a Th2 profile and the other subclasses are mainly associated with a Th1 profile[66]. In this study, the pattern of IgG subclasses induced by the different anti-nicotine vaccines was evaluated by using secondary goat anti-mouse antibodies specific for each IgG subclass. Sera from animals immunized with Nic-BSA+Alum showed low titers (3182, 16 and 9, respectively) of both Th2- associated IgG1 subclasses and Th1-associated IgG2a and IgG3 subclasses (Fig. 4). Liposome-based anti-nicotine vaccines (Nic-BSA-LIP) in the presence and absence of Alum adjuvant did not significantly enhance IgG1 and IgG3 titers, but induced a 100-fold enhancement in anti-Nicotine IgG2a titers (1859 and 1492, respectively) compared to Nic-BSA+Alum (Fig. 4). In contrast, NsL-based anti-nicotine vaccines (Nic-BSA-NsL) in the presence and absence of Alum adjuvant induced significantly higher titers of all four IgG subclasses (Fig. 4), with the highest titers observed for Th2-associated subclasses (IgG1>>Ig2a>Ig2b>Ig3). Evaluation of IgG subclasses demonstrated that NsL-based anti-nicotine vaccines are capable of inducing stronger humoral immune responses, indicated by the induction of significant antigen specific IgG1 and IgG2a titers.

3.3.3. Th1:Th2 index calculation

To assess whether NsL, LIP and Alum adjuvant formulation elicited a specific IgG subclass profile or induced an increase in all subclasses, a Th1: Th2 index was calculated for each vaccine (Table 4). Such index takes into consideration of both Th1-associated IgG2a and IgG3 subclasses and Th2-associated IgG1 subclass. The Th1:Th2 index shows that, in sera of animals immunized with Nic-BSA+Alum, the antibody response to nicotine was Th2 polarized and Th1: Th2 index was 0.004. However, both Th2-related IgG1 and Th1-related IgG2a and IgG3 antibody responses were all significantly elevated in the mice vaccinated with the Nic-BSA using NsL or LIP-based vaccine delivery ($p < 0.001$) (Fig. 4 and Table 4). In addition, Th1/Th2 balance was improved after vaccination using NsL or LIP-based vaccine delivery even in the absence of Alum adjuvant (Th1: Th2 index 0.085 and 0.160, respectively) (Table 4).

Vaccine-induced immune responses can be generally divided into two different biased effects: Th1 vs. Th2 types. Many factors can contribute to the type of specific immune response of a vaccine, including the type and relative amounts of antigens and adjuvants. Aluminum salts as adjuvant generally are well known to induce a Th2-type immune response[67,68]. On the contrary, liposomes developed as potential adjuvants generally elicit a Th1-type immune response[69]. In addition, Balb/c and C57BL/6 are widely used mouse strains, showing polarized Th cell responses. Balb/c mice typically develop Th2-skewed immune responses, whereas C57BL/6 mice are prone to Th1-dominated immunity[70]. Our results demonstrated that NsL or LIP modulated immune responses from a predominantly Th2 type response to a mixed Th1/Th2 response. Results also demonstrated that NsL-based anti-nicotine vaccine delivered by subcutaneous administration to Balb/c mice can not only significantly enhance both Th1 and Th2 immune responses but also improve Th1/Th2 balance. However, the immunological mechanism of this effect is not clear and requires further investigation.

3.3.4. Antibody affinity to nicotine

Individual serum sample was diluted with control mice serum and dialyzed against 5 different nicotine concentrations. From this saturation binding experiment, the affinity (K_d), and maximum number of binding sites (B_{max}) were determined. The antibody concentration was calculated from the B_{max}, assuming two binding sites per antibody molecule and a molecular weight of the antibody of 150 kD. Both values were obtained by calculating back to undiluted sera. The equilibrium dialysis method does not require any washing steps that might wash away low affinity antibodies, so total binding is measured. At the time the last blood sample was collected, the K_d for nicotine was $5.84\times10^2\pm1.86\times10^2$ nM and the B_{max} was $2.85\times10^3\pm1.26\times10^3$ nM, corresponding to $2.14\times10^2\pm0.95\times10^2$ μg/mL of nicotine-specific IgG in the group of mice immunized with NsL-based anti-nicotine vaccine without Alum. For the group of mice given NsL-based anti-nicotine vaccine with Alum, K_d and B_{max} were $1.27\times10^3\pm0.58\times10^3$ and $4.14\times10^3\pm1.04\times10^3$ nM, respectively, which were equivalent to $3.11\times10^2\pm0.61\times10^2$ μg/mL of nicotine-specific IgG. Our results demonstrated that NsL-based anti-nicotine vaccines prepared in the absence of Alum generated an anti-nicotine antibody with high affinity and a high concentration in the serum of vaccinated mice.

3.4. Preclinical animal toxicity studies

3.4.1. Body weights, feed, and water consumption

Mice continued to grow during their 7-week treatment with anti-nicotine vaccines. Behaviors such as feeding, drinking and activity were normal in all seven groups and there were no overt signs of toxicity. Mean body weights, weight gains, feed consumption and water consumption were all equivalent ($P>0.05$) (Table 5). Although nicotine administration can induce weight loss and change eating behavior in both humans and rodents[71, 72,73], there was no significant difference between the mice in the vaccinated groups and the mice in the control groups. Our vaccines were formulated with low doses of nicotine hapten-BSA conjugates, which may have contributed to this benefit.

3.4.2. Gross pathology

Toxicities may also be indicated by changes in tissue or organ weight, size, color, and gross appearance. Table 6 shows body weight and organ/body weight ratios (%) of the liver, kidney, spleen, lung and heart of mice in the five vaccinated groups and two control groups. No significant differences were measured between the treated and the control groups ($P > 0.05$).

3.4.3. Histopathological assessment

Histopathological review indicated that there were no significant lesions in the heart, lung, liver, spleen and kidney of mice in the two control groups and the five vaccinated groups whether or not Alum was included in the vaccines (examples provided in Fig. 5A-E). However, occasional background lesions were noted, especially in mice vaccinated with Alum. For example, all of these mice had subcutaneous necrosis with associated granulomatous inflammation (Fig. 5F), some of which included focal epicardial calcification[74]. A pulmonary embolus was also seen in the group of mice vaccinated with LIP with Alum. This finding of subcutaneous lesions is in agreement with the report of Goto et al. suggesting that Alum injection was responsible for the induction of granulomas at the injection sites of guinea pigs[75]. Aluminum adjuvants have been commonly used both in veterinary and human vaccines for many years[76,77], but they occasionally produce subcutaneous nodules[78,79], granulomatous inflammation[69,80, 81,82] and sterile abscesses[72] as local side reactions. Although seen in Alum-treated mice in this study, NsL-based anti-nicotine vaccines without Alum did not induce any histopathologic lesions at injection sites.

4. Conclusions

A novel anti-nicotine vaccine, based on NsL NPs, has been successfully developed. The carbon nanohorns are negatively charged (anionic) and act as a scaffold for the assembly of cationic liposomes. NsL NPs were stable after an extended period of storage. The nano-vaccine induced strong, anti-nicotine antibody responses in mice when administrated subcutaneously, leading to production of high titer and high affinity anti-nicotine antibodies for treatment of nicotine dependence. The total antibody titer for NsL -based anti-nicotine vaccine, even without the presence of Alum, reached ~5×10^5. The K_d for nicotine was $5.84\times10^2\pm1.86\times10^2$ nM and B_{max} was $2.85\times10^3\pm1.26\times10^3$ nM corresponding to $2.14\times10^2\pm0.95\times10^2$ μg/mL of nicotine-specific IgG for this nano-vaccine without Alum. In addition, evaluation of IgG subclasses demonstrated that NsL-based anti-nicotine vaccine was capable of inducing a more balanced Th1 and Th2 response. Furthermore, the safety evaluation of NsL NPs in mice revealed no apparent toxicity.

Acknowledgements

This research was supported by National Institute on Drug Abuse (R21 DA030083, PI: C. Zhang). We are grateful to Dr. B. Jortner for his assistance in histopathological evaluations.

References

[1] L. Esson, S.R. Leeder, The millennium development goals and tobacco control: an opportunity for global partnership. World Health Organization (WHO); Geneva: **(2004)**

[2] World Health Organization Why is tobacco a public health priority? A Tobacco Free Initiative. **(2005)**

[3] H.J., Aubin, L. Karila, M. Reynaud, Pharmacotherapy for smoking cessation: present and future. Curr. Pharm. Des. 17, 1343–1350 **(2011)**

[4] J.T. Hays, J.O. Ebbert, Adverse effects and tolerability of medications for the treatment of tobacco use and dependence. Drugs. 24, 2357–2372 **(2010)**

[5] The Health Consequences of Smoking, Centers for Disease Control and Prevention USA. Surgeon's General Report **(2004)**

[6] A.Y. Moreno, K. Janda, Immunopharmacotherapy: vaccination strategies as a treatment for drug abuse and dependence. Pharmacol. Biochem. Behav. 92, 199-205**(2009)**

[7] J. Cornuz, S. Zwahlen, W.F. Jungi. J. Osterwalder, K. Klingler, G. van Melle, Y. Bangala, I. Guessous, P. Müller, J. Willers, P. Maurer, M.F. Bachmann, T. Cerny, A vaccine against nicotine for smoking cessation: a randomized controlled trial. PLoS One 3, e2547**(2008)**

[8] E.H. Cerny, T. Cerny, Vaccines against nicotine. Hum. Vaccin. 5, 200-205 **(2009)**

[9] News release, Nabi Biopharmaceuticals, Nov. 7, (http://phx.corporate-ir.net/phoenix.zhtml?c=100445&p=irol-news) **(2011)**

[10] M.J. Hicks, J.B. Rosenberg, B.P. De, O.E. Pagovich, C.N. Young, J. Qiu, S.M., Kaminsky N.R. Hackett, S. Worgall, K.D. Janda, R.L. Davisson, R.G. Crysal, AAV-directed persistent expression of a gene encoding anti-nicotine antibody for smoking cessation. Sci. Transl. Med. 4, 140ra87 **(2012)**

[11] Y.J. Kwon, E. James, N. Shastri, J.M. Fréchet, In vivo targeting of dendritic cells for activation of cellular immunity using vaccine carriers based on pH-responsive microparticles. Proc. Natl. Acad. Sci. U.S.A. 102, 18264-18268 **(2005)**

[12] T. Akagi, X. Wang, T. Uto, M. Baba, M. Akashi, Protein direct delivery to dendritic cells using nanoparticles based on amphiphilic poly(amino acid) derivatives. Biomaterials. 28, 3427-3436 **(2007)**

[13] C. Coester, P. Nayyar, J. Samuel, In vitro uptake of gelatin nanoparticles by murine dendritic cells and their intracellular localisation. Eur. J. Pharm. Biopharm. 62, 306-314 **(2006)**

[14] K.G. Yeboah, M.J. D'souza, Evaluation of albumin microspheres as oral delivery system for Mycobacterium tuberculosis vaccines. J. Microencapsul. 26, 166-179 **(2009)**

[15] G. Chikh, M.P. Schutze-Redelmeier, Liposomal delivery of CTL epitopes to dendritic cells. Biosci. Rep. 22, 339-353 **(2002)**

[16] M.J. Copland, M.A., Baird T. Rades, J.L. McKenzie, B. Becker, F., Reck P.C. Tyler, N.M. Davies, Liposomal delivery of antigen to human dendritic cells. Vaccine. 21, 883-890 **(2003)**

[17] E. Shahum, H.M. Thérien, Effect of liposomal antigens on the priming and activation of the immune system by dendritic cells. Int. Immunopharmacol. 2, 591-601**(2002)**

[18] K. Moribe, K. Maruyama, Pharmaceutical design of the liposomal antimicrobial agents for infectious disease. Curr. Pharm. Des. 8, 441-454 **(2002)**

[19] G.D. Sprott, C.J. Dicaire, K. Gurnani, L.A. Deschatelets, L. Krishnan, Liposome adjuvants prepared from the total polar lipids of Haloferax volcanii, Planococcus spp. and Bacillus firmus differ in ability to elicit and sustain immune responses. Vaccine. 22, 2154-2162 **(2002)**

[20] C.R. Alving, V. Koulchin, G.M. Glenn, M. Rao, Liposomes as carriers of peptide antigens: induction of antibodies and cytotoxic T lymphocytes to conjugated and unconjugated peptides. Immunol. Rev. 145, 5-31 **(1995)**

[21] R.G. Laughlin, Equilibrium vesicles: fact or fiction? *Colloids Surf.* A, 128, 27–38 **(1997)**

[22] Miyawaki, J., Yudasaka, M., Azami, T., Kubo, Y., and Iijima, S, Toxicity of single-walled carbon nanohorns. ACS Nano. 2, 213–26 **(2008)**

[23] R. M. Lynch, B. H. Voy, D. F. Glass, S. M. Mahurin, B. Zhao, H. Hu, A. M. Saxton, , R. L. Donnell, and M. D. Cheng, Assessing the pulmonary toxicity of single-walled carbon nanohorns. Nanotoxicology. 1, 157–66 **(2007)**

[24] S. Iijima, M. Yudasaka, R. Yamada, S. Bandow, K. Suenaga, F. Kokai, K. Takahashi, Nano-aggregates of single-walled graphitic carbon nanohorns. Chem. Phys. Lett. 309, 165-170 **(1999)**

[25] W. Huang, J. Zhang, H.C. Dorn, D. Geohegan, and C. Zhang, Assembly of single-walled carbon nanohorn supported liposome particles. Bioconjug. Chem. 22, 1012-1016 **(2011)**

[26] P. Maurer, G.T. Jennings, J. Willers, F. Rohner, Y. Lindman, K. Roubicek, W.A. Renner, P. Müller, M.F. Bachmann, A therapeutic vaccine for nicotine dependence: preclinical efficacy, and phase I safety and immunogenicity. Eur. J. Immunol. 35, 2031-2040 **(2005)**

[27] D. Kasuya, M. Yudasaka, K. Takahashi, F. Kokai, and S Iijima, Selective production of single-wall carbon nanohorn aggregates and their formation mechanism. J. Phys. Chem. B. 106, 4947–4951 **(2002)**

[28] J. Zhang, J. Ge, M. D. Shultz, E. Chung, G. Singh, C. Shu, P. P. Fatouros, S.C. Henderson, F.D. Corwin, D.B. Geohegan, A.A. Puretzky, C.M. Rouleau, K. More, C. Rylander, M.N. Rylander, H.W. Gibson, and H.C. Dorn, In vitro and in vivo studies of single-walled carbon nanohorns with encapsulated metallofullerenes and exohedrally functionalized quantum dots. Nano Lett. 10, 2843–2848 **(2010)**

[29] C. Shu, J. Zhang, J. Ge, J.H. Sim, B.G. Burke, K.A. Williams, N.M. Rylander, T. Campbell, A. Puretzky, C. Rouleau, D.B. Geohean, K. More, A.R. Esker, H.W. Gibson, and H.C. Dorn, A facile high-speed vibration milling method to water-disperse single-walled carbon nanohorns. Chem. Mater. 22, 347– 351 **(2010)**

[30] Y. Hu, H Zheng, W Huang, C.M. Zhang, A novel and efficient nicotine vaccine using nano-lipoplex as a delivery vehicle. Hum. Vaccin. Immunother. 10, 1-9 **(2014)**

[31] AFSA. Habeeb, Determination of free amino groups in proteins by trinitrobenzene-sulfonic acid. Anal. Biochem. 14, 328–336 **(1966)**

[32] R. Jue, J.M. Lambert, L.R. Pierce, R.R. Traut, Addition of sulfhydryl groups to Escherichia coli ribosomes by protein modification with 2-iminothiolane (methyl 4-mercaptobutyrimidate). Biochemistry. 17, 5399-5406 **(1978)**

[33] S.M. Ansell, P.G. Tardi, and S.S. Buchkowsky, 3-(2-pyridyldithio)propionic acid hydrazide as a cross-linker in the formation of liposome-antibody conjugates. Bioconjug. Chem. 7, 490-496 **(1996)**

[34] S.M. Ansell, T.O. Harasym, P.G. Tardi, S.S. Buchkowsky, M.B. Bally, and P.R. Cullis,Antibody conjugation methods for active targeting of liposomes. Methods Mol. Med. 25, 51-68 **(2000)**

[35] M. Kaszuba, J. Corbett, F.M. Watson, and A. Jones, High-concentration zeta potential measurements using light-scattering techniques. Philos. Transact. A Math. Phys. Eng. Sci. 368, 4439-51 **(2010)**

[36] T. Kodama, N. Tomita, S. Horie, N. Sax, H. Iwasaki, R. Suzuki, K. Maruyama, S. Mori, F. Manabu, Morphological study of acoustic liposomes using transmission electron microscopy. J. Electron. Microsc. (Tokyo) 259, 187-96 **(2010)**
[37] L. Zhang, S. Granick, How to stabilize phospholipid liposomes (using nanoparticles). Nano Lett. 6, 694-698 **(2006)**
[38] S.H. de Villiers, N. Lindblom, G. Kalayanov, S., Gordon, I. Baraznenok, A. Malmerfelt, M.M. Marcus, A.M. Johansson, and T.H. Svensson, Nicotine hapten structure, antibody selectivity and effect relationships: results from a nicotine vaccine screening procedure. Vaccine. 28, 2161-2168 **(2010)**
[39] M.L. Visciano, M. Tagliamonte, M.L. Tornesello, F.M. Buonaguro, L. Buonaguro, Effects of adjuvants on IgG subclasses elicited by virus-like particles. J. Transl. Med. 10, 4 **(2012)**
[40] U.S. FDA. U.S. Food and Drug Administration, Good Laboratory Practice Regulations for Nonclinical Laboratory Studies. CFR (April 1), **(1988)** p. 229–243.
[41] U.S. FDA. U.S. Food and Drug Administration. Good Laboratory Practice Regulations for Nonclinical Laboratory Studies. CFR (March 21), **(1994)**. 56FR12300.
[42] NRC Guide for the care and use of laboratory animals, National Academy Press, Washington, DC. **(1996)**
[43] T.F. Collins, R.L. Sprando, M.E. Shackelford, D.K. Hansen, J.J. Welsh, Food and Drug Administration proposed testing guidelines for developmental toxicity studies. Regul. Toxicol. Pharmacol. 30, 39-44 **(1999)**
[44] WHO guidelines on nonclinical evaluation of vaccines, WHO Technical Report Series. 17-21, November **(2003)**. p. 9-12
[45] E. McNair, R. Bryson, Effects of nicotine on weight change and food consumption in rats. Pharmacol. Biochem. Behav. 18, 341-344 **(1983)**
[46] L.J. Pecora, B. Highman, Organ weights and histology of chronically thiamine-deficient rats and their pair-fed controls. J. Nutr. 51, 219-30 **(1953)**
[47] B.O. Iranloye, A.F. Bolarinwa, Effect of nicotine administration on weight and histology of some vital visceral organs in female albino rats. Niger. J. Physiol. Sci. 24, 7-12 **(2009)**
[48] R.D. Lillie, Histopathologic Technic. The Blakiston Co., Philadelphia, Pa. **(1948)**
[49] A. Hoffmann-Roder,, and M. Johannes. Synthesis of a MUC1-glycopeptide-BSA conjugate vaccine bearing the 3'-deoxy-3'-fluoro-Thomsen-Friedenreich antigen, Chem. Commun. (Camb.) 47, 9903-9905 **(2011)**
[50] F.Y. Huang, W.J. Chen, W.Y. Lee, S.T. Lo, T.W. Lee, J.M. Lo, In vitro and in vivo evaluation of lactoferrin-conjugated liposomes as a novel carrier to improve the brain delivery. Int. J. Mol. Sci. 14, 2862-2874 **(2013)**
[51] P. Benzinger, G. Martiniy-Baron, P. Reusch, G. Siemeister, J.T. Kley, D. Marmé, C., Unger and U. Massing, Targeting of endothelial KDR receptors with 3G2 immunoliposomes in vitro. Biochim. Biophys. Acta 1466, 71−78 **(2000)**
[52] C. Schwarz, W. Mehnert, J.S. Lucks, R.H. Muller, Solid lipid nanoparticles (SLN) for controlled drug delivery. I. Production, characterization and sterilization. J. Control. Release 30, 83-96 **(1994)**
[53] C. Alves, N., de Melo, L. Fraceto, D. de Araújo, M. Napimoga, Effects of 15d-PGJ_2-loaded poly(D,L-lactide-co-glycolide) nanocapsules on inflammation. Br. J. Pharmacol. 162, 623-632 **(2011)**
[54] M. G. Carstens, M. G. Camps, M. Henriksen-Lacey, K. Franken, T. H. Ottenhoff, Y. Perrie, J.

A. Bouwstra,, F. Ossendorp, and W. Jiskoot, Effect of vesicle size on tissue localization and immunogenicity of liposomal DNA vaccines. Vaccine. 29, 4761-4770 **(2011)**

[55] M. Singh, M. Briones, G. Ott, and D. O'Hagan, Cationic microparticles: A potent delivery system for DNA vaccines. Proc. Natl. Acad. Sci. U.S.A. 97, 811-816 **(2000)**

[56] D. Pornpattananangkul, S. Olson, S. Aryal, M. Sartor, C.M. Huang, K. Vecchio, and L. Zhang. Stimuli-responsive liposome fusion mediated by gold nanoparticles. ACS Nano. 4, 1935-1942 **(2010)**

[57] B. Pietzyk, and K. Henschke. Degradation of phosphatidylcholine in liposomes containing carboplatin in dependence on composition and storage conditions. Int. J. Pharm. 196, 215-218 **(2000)**

[58] B. Stark, G. Pabst, and R. Prassl, Long-term stability of sterically stabilized liposomes by freezing and freeze-drying: Effects of cryoprotectants on structure. Eur. J. Pharm. Sci. 41, 546 -555**(2010)**

[59] P. Milla, F. Dosio, and L. Cattel, PEGylation of proteins and liposomes: a powerful and flexible strategy to improve the drug delivery. Curr. Drug Metab. 13, 105-119 **(2012)**

[60] K. Kishimoto, D. Altreuter, L. Johnston, P. Keller, L. Pittet. SEL-068 a fully synthetic nanoparticle vaccine for smoking cessation and relapse prevention. Society for Research on Nicotine and Tobacco's 18th Annual Meeting, Abstract Book, **(2012)** POS1-1.

[61] A. Jegerlehner, A. Tissot, F. Lechner, P. Sebbel, I. Erdmann, T. Kündig, T. Bächi, T. Storni, G. Jennings, P. Pumpens, W.A. Renner, M.F. Bachmann. A molecular assembly system that renders antigens of choice highly repetitive for induction of protective B cell responses. Vaccine. 20, 3104–3112 **(2002)**

[62] P. Maurer, G.T. Jennings, J. Willers, F. Rohner, Y. Lindman, K. Roubicek, W.A. Renner, P. Müller, M.F. Bachmann A therapeutic vaccine for nicotine dependence: preclinical efficacy, and Phase I safety and immunogenicity. Eur. J. Immunol. 35, 2031–2340 **(2005)**

[63] P.M. Ambühl, A.C. Tissot, A. Fulurija, P. Maurer, J. Nussberger, R. Sabat, V. Nief, C. Schellekens, K. Sladko, K. Roubicek, T. Pfister, M. Rettenbacher, H.D. Volk, F. Wagner, P. Müller, G.T. Jennings, M.F. Bachmann. A vaccine for hypertension based on virus-like particles: preclinical efficacy and phase I safety and immunogenicity. J. Hypertens. 25, 63–72 **(2007)**

[64] T.M. Kündig, G. Senti, G. Schnetzler, C. Wolf, B.M. Prinz Vavricka, A. Fulurija, F. Hennecke , K. Sladko, G.T. Jennings, M.F. Bachmann, Der p 1 peptide on virus-like particles is safe and highly immunogenic in healthy adults. J. Allergy Clin. Immunol. 117, 1470–1476 **(2006)**

[65] J. Cornuz, S. Zwahlen, W.F. Jungi, J. Cornuz, S. Zwahlen, W.F. Jungi, J. Osterwalder, K. Klingler, G. van Melle, Y. Bangala, I Guessous, P. Müller, J. Willers, P. Maurer, M.F. Bachmann, T.A Cerny. A vaccine against nicotine for smoking cessation: a randomized controlled trial. PLoS One 3, e2547 **(2008)**

[66] Banerjee K, Klasse PJ, Sanders RW, Pereyra F, Michael E, Lu M, Walker BD, Moore JP, IgG subclass profiles in infected HIV type 1 controllers and chronic progressors and in uninfected recipients of Env vaccines. AIDS Res. Hum. Retroviruses. 26, 445-458 **(2010)**

[67] Sokolovska A, Hem SL and HogenEsch H. Activation of Dendritic Cells and Induction of CD4(+) T Cell Differentiation by Aluminum-Containing Adjuvants. Vaccine. 25, 4575-85 **(2007)**

[68] O'Hagan DT and Rappuoli R, Novel Approaches to Vaccine Delivery. Pharm. Res. 21, 1519-1530 **(2004)**

[69] Rosenkrands I, Agger EM, Olsen AW, Korsholm KS, Andersen CS, Jensen KT, Andersen P, Cationic liposomes containing mycobacterial lipids: a new powerful Th1 adjuvant system. Infect. Immun. 73, 5817-5826 **(2005)**

[70] M. Becker, S. Reuter, P. Friedrich, F. Doener, A. Michel, T. Bopp, M. Klein, E. Schmitt, H. Schild, M.P. Radsak, B. Echtenacher, C. Taube, M. Stassen, Genetic variation determines mast cell functions in experimental asthma. J. Immunol. 186, 7225-7231 **(2011)**

[71] E. McNair, R. Bryson, Effects of nicotine on weight change and food consumption in rats. Pharmacol. Biochem. Behav. 18, 341-344 **(1983)**

[72] N.E. Grunberg, S.E. Winders, K.A. Popp, Sex differences in nicotine's effects on consummatory behavior and body weight in rats. Psychopharmacology (Berl.). 91, 221-225 **(1987**)

[73] J. Gross, M.L. Stitzer, J. Maldonado, Nicotine replacement: effects of postcessation weight gain. J. Consult. Clin. Psychol. 57, 87-92 **(1989)**

[74] C.E. Nabors, C.R. Ball. Spontaneous calcification in hearts of DBA mice. Anat. Rec. 164, 153-61 **(1969)**

[75] N. Goto, H. Kato, J. Maeyama, M. Shibano, T. Saito, J. Yamaguchi, S. Yoshihara, Local tissue irritating effects and adjuvant activities of calcium phosphate and aluminium hydroxide with different physical properties. Vaccine. 15, 1364-1371 **(1997)**.

[76] N.W. Baylor, W. Egan, P. Richman, Aluminum salts in vaccines--US perspective. Vaccine. 20, S18-23 **(2002)**

[77] E.B. Lindblad. Aluminium adjuvants--in retrospect and prospect. Vaccine. 22, 3658-3668 **(2004)**

[78] N.R. Butler, M.A. Voyce, W.L. Burland, M.L. Hilton, Advantages of aluminium hydroxide adsorbed combined diphtheria, tetanus, and pertussis vaccines for the immunization of infants. Br. Med. J. 1, 663-666 **(1969)**

[79] L. Frost, P. Johansen, S. Pedersen, N. Veien, P.A. Ostergaard, M.H. Nielsen, Persistent subcutaneous nodules in children hyposensitized with aluminium-containing allergen extracts. Allergy. 40, 68-72 **(1985)**

[80] R.G.White, A.H. Coons. J.M.Connolly, Studies on antibody production. III. The alum granuloma. J. Exp. Med. 102, 73-82**(1955)**

[81] M. Erdohazi, R.L. Newman. Aluminium hydroxide granuloma. Br. Med. J. 3, 621-623 **(1971)**

[82] N. Goto, K. Akama, Histopathological studies of reactions in mice injected with aluminum-adsorbed tetanus toxoid. Microbiol. Immunol. 26, 1121-1132 **(1982)**

Table 1. Immunization scheme of mice (Balb/c). Each group had 8 mice. All vaccines (suspended in 100 μL saline) were administered subcutaneously with one primary immunization and two boosters two weeks apart.

Group No.	Vaccine name	Dose of immunogen (μg)	Dose of Alum (mg)
1	Nic-BSA	50	1.2
2	Nic-BSA-LIP	50	-
3	Nic-BSA-LIP	50	1.2
4	Nic-BSA-NsL	50	-
5	Nic-BSA-NsL	50	1.2
6	LIP	-	1.2
7	NsL	-	1.2

BSA: bovine serum albumin; LIP: liposome; NsL: nanohorn supported liposome.

Table 2. Determination of Nic-hapten density, Nic-BSA thiolation grade and AE and LC of Nic-BSA to NsL NPs.

Molar ratio of additive (BSA:Nic:Traut's reagent: DSPE-Mal-PEG2000)	Approximate no. of amino groups used	Approximate no. of thiol groups used	AE (%)	LC (%)
1: 50: 200: 4	15.1 (15)	5.69 (5)	50.4± 1.0	21.7±2.4
1: 50: 200: 8	15.1 (15)	5.69 (5)	65.8± 1.4	25.3±4.6

Results are expressed as means (SD) for n=3. AE: association efficiency; LC: loading capacity.

Table 3. Size distribution and zeta potential of nanohorn, DOTAP liposomes, NsL and Nic-BSA-NsL in 0.15 M pH 7.4 NaCl solution or ultrapure water.

	Size (nm)	Zeta potential (mV)	Polydispersity
nanohorn	107.8 ± 3.3	-29.5 ± 1.29	0.18 ± 0.02
DOTAP liposomes	152.0 ± 6.0	42.8 ± 0.96	0.20 ± 0.03
NsL	164.9 ± 2.5	47.6 ± 1.10	0.23 ± 0.02
Nic-BSA-NsL	175.3 ± 3.2	52.2 ± 1.38	0.19 ± 0.01

Results are expressed as means ± SD (n=3).

Table 4. Th1:Th2 index for nicotine vaccine.

	Nic-BSA+Alum	Nic-BSA-LIP	Nic-BSA-LIP+Alum	Nic-BSA-NsL	Nic-BSA-NsL+Alum
IgG1 (Th2)	$(3.18\pm0.48)\times 10^3$	$(4.25\pm1.32)\times 10^3$	$(5.99\pm1.37)\times 10^3$	$(1.91\pm0.03)\times 10^4$	$(2.94\pm0.07)\times 10^4$
IgG2a (Th1)	16.3 ± 4.5	$(1.49\pm0.29)\times 10^3$	$(1.86\pm0.20)\times 10^3$	$(2.17\pm0.45)\times 10^3$	$(2.39\pm0.34)\times 10^3$
IgG3 (Th1)	9.4 ± 4.2	32.4 ± 7.7	56.5 ± 6.5	$(1.57\pm0.96)\times 10^3$	$(2.85\pm0.92)\times 10^3$
Th1/Th2 index	0.004	0.180	0.160	0.085	0.076

Results are expressed as mean ± SD, n=8.

Table 5. Effect of subcutaneous administration of nicotine vaccine on weight gains, feed consumption and water consumption of Balb/c mice.

	Body Weight (g)			Food Consumption (g/day)	Water Consumption (g/day)
	Initial	Final	Change		
LIP+Alum	18.0±1.6	20.9±1.9	2.9±0.6	2.8±0.2	4.6 ±0.5
NsL+Alum	17.9±1.4	20.7±1.7	2.8±0.6	2.9±0.2	4.3±0.5
Nic-BSA+Alum	17.4±0.7	20.4±1.1	3.0±0.7	2.6±0.1	4.2 ±0.5
Nic-BSA-NsL	18.4±1.1	21.3±1.5	2.9±0.5	2.7±0.3	4.4± 0.6
Nic-BSA-NsL+Alum	19.0±0.9	21.7±2.3	2.7±1.7	2.8±0.3	4.3± 0..7
Nic-BSA-LIP	17.9±1.3	20.4±1.6	2.5±0.3	2.7±0.4	4.6± 0.7
Nic-BSA-LIP+Alum	18.2±1.4	21.0±1.4	2.8±0.9	2.7±0.2	4.4±0.6

Results of the mean body weight gains, feed consumption and water consumption of seven groups of mice are expressed as means ± SD. Data were analyzed by paired t tests and no differences were significant. (**P**>0.05, n=8)

Table 6. Effect of of nicotine vaccines on body weight and organ/body weight ratio of Balb/c mice.

Group	Body weight (g)	Organ/Body Weight Ratios (%)				
		Heart	Lung	Liver	Spleen	Kidney
LIP+Alum	21.21±1.70	0.66±0.08	1.40±0.53	5.11±0.56	0.47±0.08	1.47±0.24
Nic-BSA+Alum	20.36±1.29	0.65±0.12	1.62±0.27	5.02±0.17	0.47±0.07	1.53±0.13
NsL+Alum	20.68±1.73	0.74±0.06	1.50±0.26	5.17±0.55	0.52±0.07	1.54±0.14
Nic-BSA-NsL	21.28±1.52	0.74±0.08	1.39±0.22	5.28±0.53	0.47±0.06	1.53±0.15
Nic-BSA-NsL+Alum	21.68±2.27	0.75±0.12	1.53±0.35	5.25±0.20	0.52±0.04	1.56±0.26
Nic-BSA-LIP	20.43±1.58	0.65±0.13	1.59±0.41	4.92±0.39	0.46±0.05	1.51±0.18
Nic-BSA-LIP+Alum	21.03±1.37	0.74±0.14	1.60±0.27	4.85±0.30	0.50±0.08	1.49±0.15

Results of the mean body weight and organ/body weight ratio of seven groups mice are expressed as means ± SD. Data were analyzed by paired t tests and no differences were significant. (**P**$>$0.05, n=8)

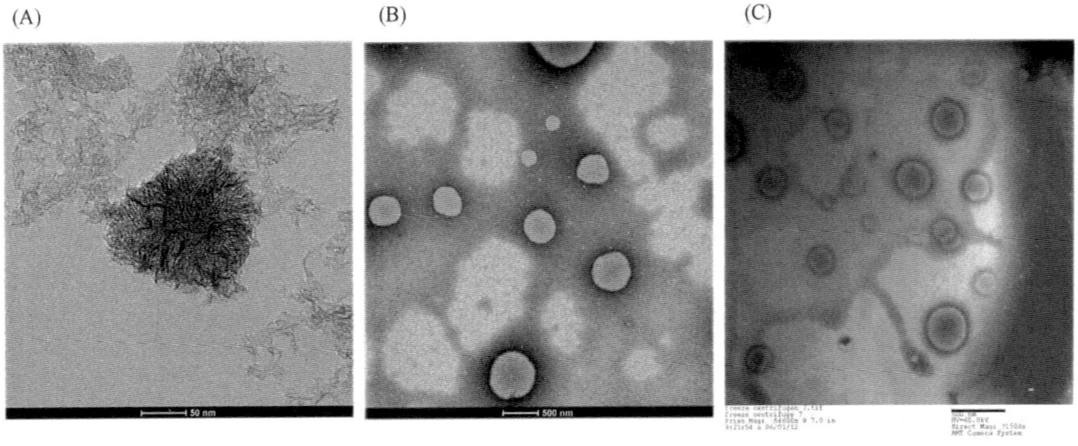

Fig. 1. TEM images of (A) Nanohorn, (B) liposomes prepared with DOTAP, and (C) NsL. Scale bar in (A) =50 nm. Scale bar in (B) and (C) = 500 nm.

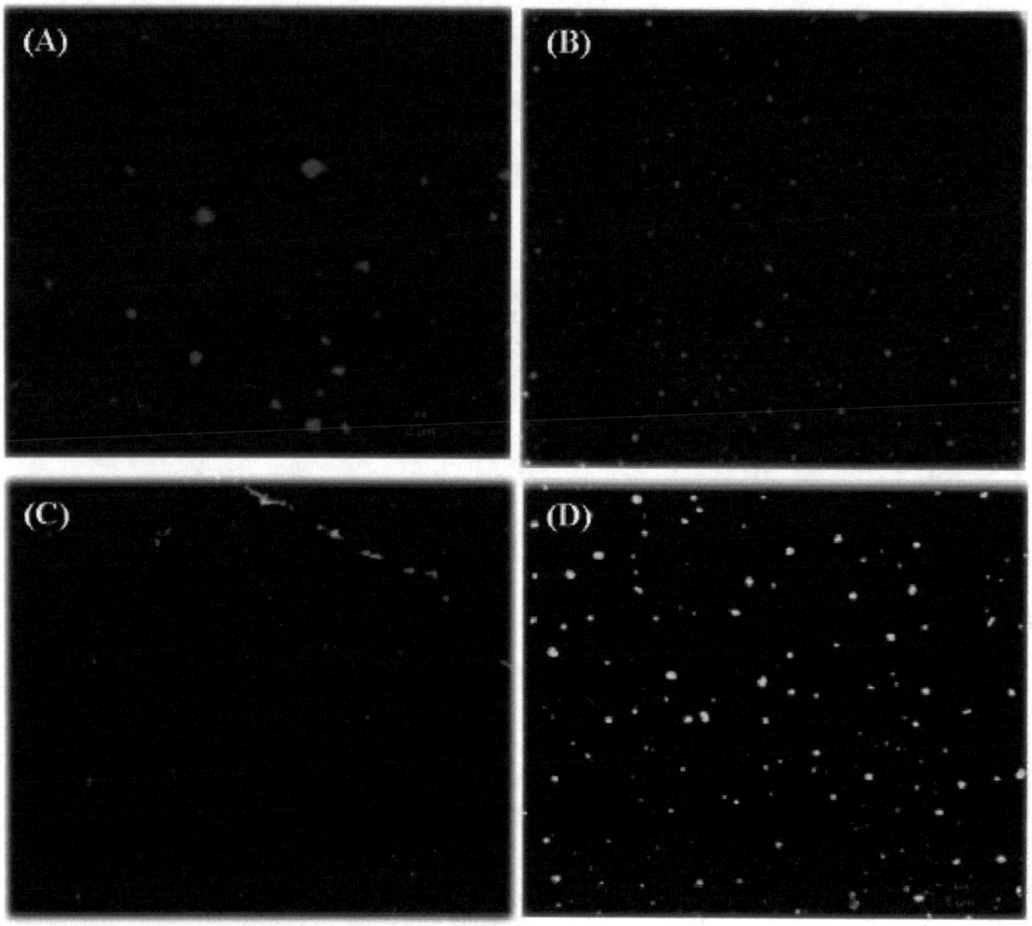

Fig. 2. Confocal microscopic images of liposomes and nanohorn supported liposomes labeled with FITC. (A) newly synthesized liposomes, (B) newly synthesized nanohorn supported liposomes, (C) liposomes after storage for 5 days, (D) nanohorn supported liposomes after storage for 5 days.

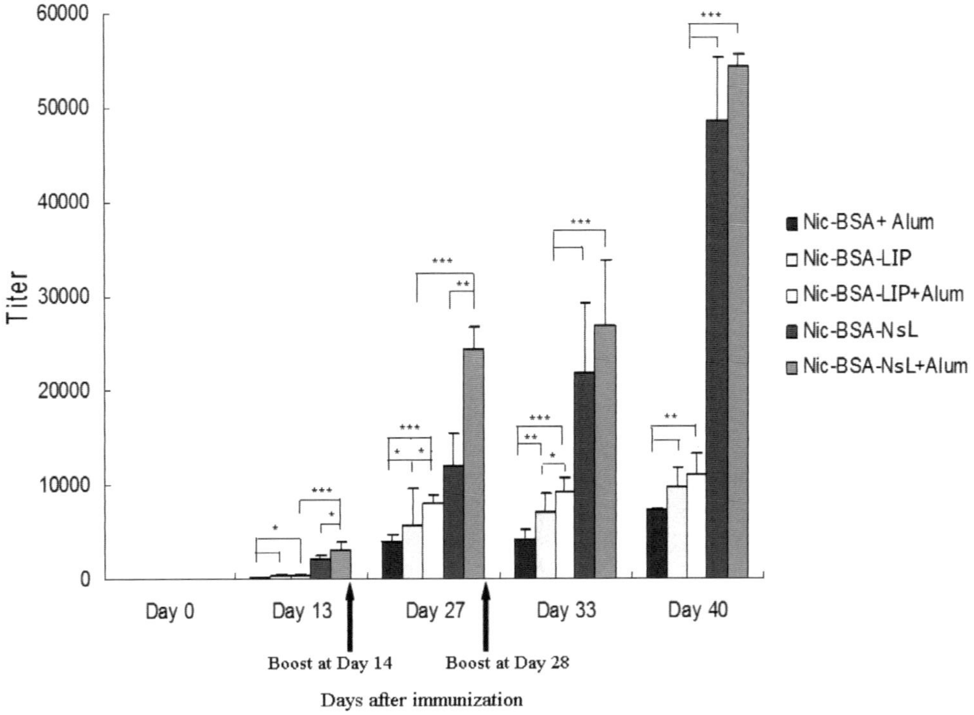

Fig. 3. Time-course of IgG nicotine-specific response to subcutaneous immunization with nicotine vaccines. Mice were given 50 µg of Nic-BSA with Alum and Nic-BSA-LIP or Nic-BSA-NsL with or without Alum adjuvant on days 0, 14 and 28. Two groups of control animals received either LIP with Alum or NsL with Alum, respectively. Serum samples from eight animals per group were taken at two weekly intervals. Control animals showed no detectable nicotine-specific IgG. Mean nicotine-specific serum IgG antibody titers and the corresponding standard deviations are represented by columns and error bars (**P*< 0.05; ***P*<0.01; ****P*<0.001).

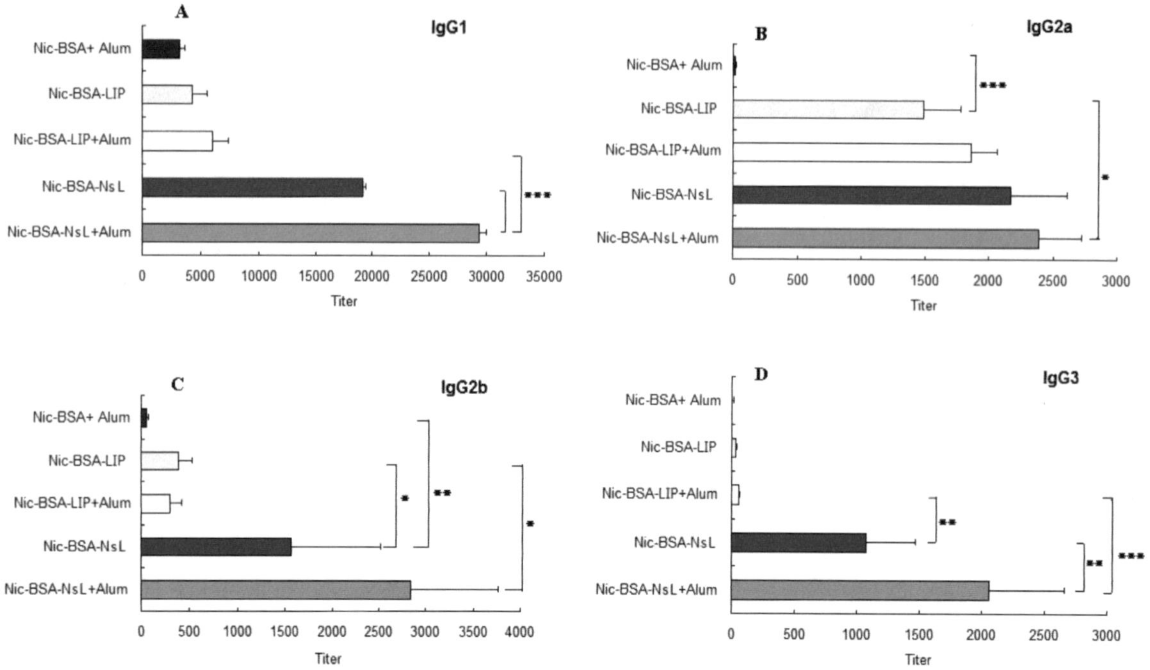

Fig. 4. Responses of IgG subclasses to subcutaneous immunization with nicotine vaccines. Mice were given 50 μg of Nic-BSA with Alum and Nic-BSA-LIP or Nic-BSA-NsL with or without Alum adjuvant on days 0, 14 and 28. Serum samples from eight animals per group were taken on day 40. Mean nicotine-specific serum IgG subclass antibody titers and the corresponding standard deviations are represented by columns and error bars (*$P< 0.05$; **$P<0.01$; ***$P<0.001$).

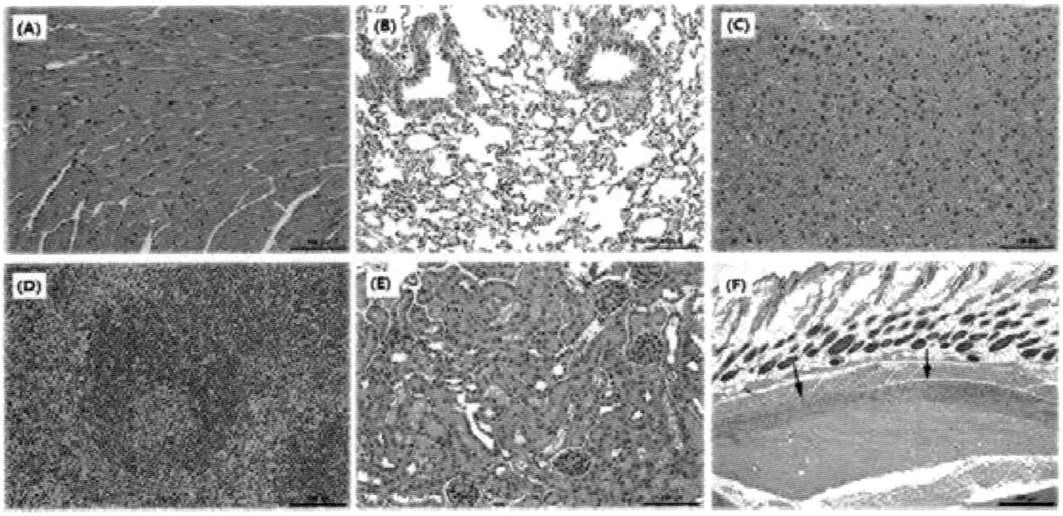

Fig. 5. Representative histopathological images of different mouse tissues after subcutaneous administration of nicotine vaccines. No lesions were noted in (A) heart, (B) lung, (C) liver, (D) spleen, and (E) kidney. Figure 5F is from a mouse given a vaccine with Alum. Arrows indicate subcutaneous lesions. Scale bars in (A) - (E) are 100 μm; for (F) the scale bar is 500 μm.

389420 Protein Pegylation Using Micro-Reactors

Abstract:

PEGylation, which involves the conjugation of polyethylene glycol (or PEG) with a protein, is one of the main chemical modification techniques used for improving properties of biopharmaceutical proteins. Protein PEGylation is typically carried out as liquid-phase batch reactions. The mono-PEGylated protein is usually the preferred form and various strategies have been used to improve both its yield and selectivity. A large number of reactions have to be carried out to optimize such protein-polymer conjugation reactions. Scale-down or miniaturization of reaction has commonly been successfully used in such situations. This work focuses on the design and development of reactor systems suitable for studying PEGylation at a micro-scale where the protein and the PEGylating reagent are reacted in a micro-channel. Miniaturization of protein PEGylation using microfluidics would enable parallel, high-throughput study of factors likely to affect reaction yield and selectivity such as PEGylation chemistry (e.g. aldehyde reaction, NHS reaction), reactor volume, residence time, reagent concentrations, polymer to protein molar ratios, pH and quencher addition.

Industrial-scale protein PEGylation is usually carried out in stirred tank reactors in which turbulent flow predominates. However, the flow of liquid within the micro-channels is laminar. In order to make the micro-reactor system more representative of the actual macro-scale process, various active and passive mixing strategies available in the domain of microfluidics were examined. Amongst these, chaotic mixing by the using grooved-channels was found to be most promising. In the micro-reactors used in this study, reactants are efficiently mixed by the chaotic advection due to groove cycles embedded in the floor of the channel. Small volumes of the fluid are stretched and folded over the cross section of the channel and the mixing length is thereby reduced. The micro-reactor was fabricated by casting a single PDMS layer on a 3D-printed acrylic-based mold. The PDMS layer was attached to a glass slide using air plasma to create the micro-channel. The PEGylating reagent and protein solutions entered the micro-reactor through a Y-shaped terminal and the product stream leaving the reactor was quenched in-situ using glycine solution. The PEGylated protein containing product stream was analyzed for conversion and selectivity using appropriate analytical techniques.

367989 Optimization of Valuable Intermediates By 11-Alpha-Hydroxylation of Steroid DHEA By Solvent-Enhanced Beauveria Bassiana

Abstract:

This research evaluates the oxidative capacity of fungus *Beauveria bassiana*; a versatile whole-cell biocatalyst used in the biotransformation of chemicals. The goal of this research is to standardize, control and optimize the biocatalysis capacity of *B. bassiana* in the biotransformation of DHEA. To extend use of *B. bassiana* to commercial applications, the optimization of reaction conditions and accurate prediction of biotransformation products are necessary. This work enhances the selective hydroxylation capacity of Strain ATCC 7159, resulting in a cost effective and eco-friendly process for the synthesis of valuable 11α-hydroxy steroids. Our work establishes the metabolic pathway of DHEA and defines the optimum reactor arrangement, substrate concentration, reaction temperature and pH. NMR and LC-MS analysis were performed to characterize reaction products. Higher substrate conversion, selectivity and yield of desired product was achieved in a "Resting Cells" arrangement. As well, higher volumes of growing medium compared to reaction buffer, increase the activity of the biocatalyst. The apparent rate of reaction fits a Michaelis-Menten kinetics model with a maximum reaction rate of 4.45mM-day. Interestingly, when a diluted amount of substrate is used, a higher yield of 11α-hydroxy steroids is achieved. Also, reactions at 26°C with pH ranges between 6.0 and 7.0 resulted in the highest conversion (70%) and the higher product yield (45.8%). *B. bassiana* has the capacity to metabolize DHEA and similar steroids in different set-ups and has a promising future as biocatalyst to be used in the production of drug metabolites.

Corrosion protection properties of polybenzoxazine/organoclay nanocomposite coatings on mild steel

Changlu Zhou, Zhong Xin*

State key laboratory of chemistry engineering, School of Chemical Engineering, East China University of Science and Technology, Shanghai, 200237, P. R. China

Recently, polybenzoxazines are considered to be a better candidate over epoxies and traditional phenolic resins as a good organic coating material for corrosion protection due to their unique properties such as low water absorption, low surface free energy, near zero shrinkage, excellent dielectric properties, and so on.. In our previous studies, we prepared a novel silane-functional polybenzoxazine (PB-TMOS) coating with silane groups that could bond to the substrate, and the corrosion protection properties of PB-TMOS coated specimens was considerably enhanced by its high hydrophobicity and water resistance resulted from the dual cross-linking network of polybenzoxazine and Si-O-Si in PB-TMOS matrix[1]. In order to improve the corrosion resistance of PB-TMOS coating, nanometer size fillers had been incorporated into PB-TMOS to prepare hybrid coating. Considerable enhancement on corrosion protection performance was obtained by introducing SiO_2 nanoparticles through improving the coating's barrier properties against gas, water or other corrosion medium[2]. In comparison of SiO_2 nanoparticles, organoclay is considered as a good filler to reduce gas permeability of polymers due to its high ratio of diameter to thickness. In the polymer/organoclay composite system, the organoclay pillaring agents' functional groups can interact with the continuous phase matrix polymer to form different dispersion state in matrix polymer, which must play an important role on the coating's performance. But few researches focus on the relation between corrosion resistance of composite system and dispersion structure of organoclay, which is determined by the pillaring agent utilized.

Therefore, two types of organoclays, Nanocor's Nanomer I.44P and I.30P, at various concentrations were tested as nanofillers in PB-TMOS coatings applied on mild steel (MS). Coated steel panels were characterized in terms of microstructure and corrosion performance in 3.5 wt% NaCl solution. Dimethyl dioctadearyl quaternary ammoniums (in I.44P) promoted an intercalated structure for composites, and the *d*-spacing value of organoclay was increased by compositing with PB-TMOS, which was evident from X-ray diffraction (XRD) measurements (Fig. 1). An improvement on corrosion protection abilities of PB-TMOS was also observed at all studied amounts, evidently due to a favorable aspect ratio and orientation of the organoclay I.44P. Addition of organoclay I.30P with octadearyl quaternary ammonium as pillaring agent yielded coatings where the organoclays were relatively randomly oriented which introduced an exfoliated morphologies. The barrier and corrosion protection properties were clearly compromised for clay I.30P.

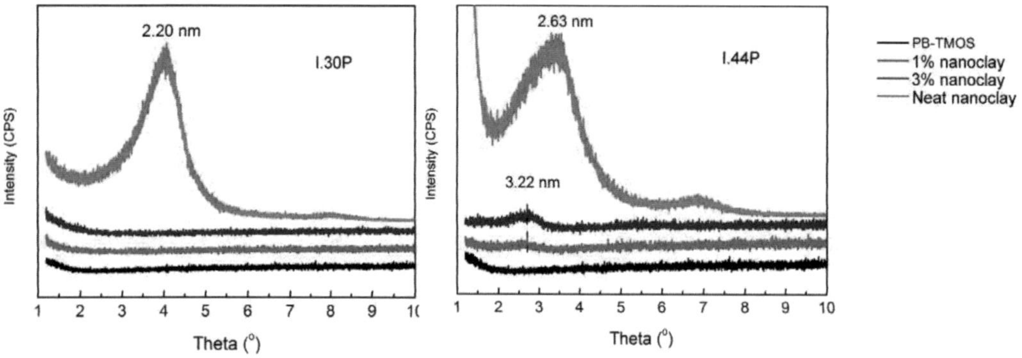

Figure 1 X-ray diffraction patterns of neat organoclay, neat PB-TMOS and PB-TMOS/organoclay nanocomposites

Keywords: Materials; Polymers; Polybenzoxazine; Coating; Nanomposite

Changlu Zhou, E-mail: changluzhou@gmail.com

*Correspondence Zhong Xin, Professor, Tel.: +86 21 64240862; Fax: +86 21 64251772, E-mail: xzh@ecust.edu.cn

1. Zhou, C.; Lu, X.; Xin, Z.; Liu, J., *Corrosion Science* ***2013,*** *70, 145-151.*
2. Zhou, C.; Lu, X.; Xin, Z.; Liu, J., *Corrosion Science* ***2014,*** *80, 269-275.*

372713 Viscosity and Short Time Dynamics of Concentrated Solutions of Proteins Interacting with a Short Range Attractive and Long Range Repulsive Interaction

Abstract:

Concentrated colloidal dispersions interacting with a short range attractive and long range repulsive interaction are currently of significant scientific and technological interest as they can exhibit intermediate range order as well as significant viscosities. Such systems with competing interactions are capable of producing, among a wide array of microstructures, clustered fluid phases in which reversible, thermodynamically stable aggregates of finite size exist in equilibrium with monomers. Clustered fluids are formed by the balance of these competing potential features and are thought to be especially prevalent in concentrated protein solutions. Of particular importance in biotechnology is the effect of this fluid structure on the viscosity and diffusivity of proteins in solution. Recent studies have linked the formation of clusters in solutions of lysozyme protein and monoclonal antibodies (mAbs) to a substantial increase in solution viscosity (Biophys J. 105:720, 2013). The goal of our work is to quantify the effect of intermediate range order and cluster formation on the viscosity of model protein solutions. Here we present experimental results for lysozyme, which is selected because of its availability, stability, and globular structure, the latter enabling quantitative comparison to models. Zero shear viscosity is obtained by microrheological measurements to avoid artifacts of interfacial rheological effects. A strong divergence of zero shear viscosity is observed for volume fractions significantly below that typical for hard sphere dispersions. The fluid microstructure and protein short time-self diffusion are measured across a broad range of conditions by small angle neutron scattering (SANS) and neutron spin echo (NSE), respectively. Previously validated models that include explicit hydrodynamic, Brownian, and interaction contributions to the viscosity fail to account for the large viscosity rise with concentration. However, this excessive viscosity rise can be semi-quantitatively predicted when protein clustering is properly accounted for and effective cluster-cluster interactions properly included. The implications for understanding anomalous viscosities observed in some concentrated monoclonal antibody solutions are discussed.

374345 Ensuring Success of Broad Industrial/Academic Collaboration

Abstract:

This presentation will address general aspects of precompetitive collaboration and then focus on a specific example, the SSPC.

The Synthesis and Solid State Pharmaceutical Centre (SSPC), is a **Global Hub of Pharmaceutical Process Innovation and Advanced Manufacturing**, funded by the SFI and industry, is a unique collaboration between 17 companies, 8 academic institutions and 12 international academic collaborators. There are three strands to the research program, covering chemical synthesis, crystallization and formulation. Following a successful first phase, the second phase of this program commenced recently with €40M in funding.

Collaboration between companies in the same or different industries has evolved over the past forty years. Using the current SSPC collaboration in the pharmaceutical sector as a case study together with the existing literature, the critical success factors and barriers associated with current collaborations are assessed. Areas such as trust, culture, knowledge sharing, proximity, control, metrics, and uncertainty are all seen to play a role in the success of collaborations. Communication and people are found to be the most significant overarching themes for success.

The presentation will indicate how to get involved in this current collaboration and also provide insight into broader funding opportunities, especially those funded centrally by the European Commission.

Exploring Opportunities for Information Exchange and Mutual Development of Enabling Technologies

Srinivas Tummala,[1] *Kevin D. Seibert,*[2] *Margaret Faul,*[3]

[1] Chemical Development, Bristol Myers Squibb, New Brunswick, [2]Small Molecule Design and Development, Eli Lilly & Company, Indianapolis, IN; [3]Chemical Process Research & Development, Amgen Inc., Thousand Oaks, CA 91320

The mission of the IQ Consortia is to advance science-based and scientifically driven standards and regulations for the pharmaceutical industry. Given that mission, an exploratory group has been established to pursue opportunities of member companies to engage in information exchange focused on the advancement of pre-competative technologies. Opportunities exist to advance a variety of technologies, methods, and equipment setups essential to advancing drug process development. A pre-competative collaborative exchange of information additionally opens up opportunities to develop vendors to focus on desired product enhancements for potential holistic solutions.

This discussion will focus on the collection of survey data and a summary of the phases of the project including Phase 1: identification of areas of focus, Phase 2: refinement and prioritization, and Phase 3: establishment of the working groups around the identified technologies.

379915 The Center for Pharmaceutical Development (CPD) – an I/U CRC for Late-Stage Pharmaceutical Development

Abstract:

The pharmaceutical industry currently faces simultaneous challenges in form of both increased cost of drug development and increased price pressure in most markets. The Center for Pharmaceutical Development (CPD), established in February 2010, provides a forum for academic and industrial scientists to develop novel approaches for the improvement of pharmaceutical API **manufacturing**, drug product **formulation**, and **analytical methods**. The CPD seeks to address the needs of our industrial partners through the development of selective and robust manufacturing processes, formulations with enhanced performance and stability, and better analytical methods for characterizing products. The Center's mission is to create more selective and robust biological and chemical catalysts, to develop methods for stabilizing drugs and vaccines, to design of new techniques for nondestructive evaluation of pharmaceutical products.

The presentation will discuss how university-industry collaboration within the CPD fits into the framework of options generated by a recent Council for Chemical Research (CCR) workshop.[1] As the CPD is a representative of the Industry/University Cooperative Research Center program by the National Science Foundation (NSF), the NSF established most rules under which the CPD operates. The presentation will conclude with some of the success stories of the CPD over the last few years; selected publications are listed here.[2-8]

References

[1] Precompetitive Collaboration on Enabling Technologies for the Pharmaceutical Industry, C.J. Welch, J.M. Hawkins, and J. Tom, *Org. Process Res. Dev.* **2014**, *18*, 481-487

[2] "Development of a Novel Amine Dehydrogenase for Synthesis of Chiral Amines", M.J. Abrahamson, E. Vazquez-Figueroa, N.B. Woodall, J.C. Moore, A.S. Bommarius, *Angew. Chem. Intl. Ed.* **2012**, *51*, 3969-3972

[3] "The Evolution of an Amine Dehydrogenase Biocatalyst for the Asymmetric Production of Chiral Amines", M.J. Abrahamson, J.W. Wong, A.S. Bommarius, *Advanced Synthesis & Catal*ysis **2013**, *102*, 377-386

[4] "Molecular Dynamics Simulation of Amorphous Indomethacin–Poly(Vinylpyrrolidone) Glasses*:* Solubility and Hydrogen Bonding Interactions", Tian-Xiang Xiang and Bradley Anderson, *J. Pharm. Sci.* **2013**, *102*, 876–891

[5] "Water Uptake, Distribution, and Mobility in Amorphous Poly(D,L-Lactide) by Molecular Dynamics Simulation", T.-X. Xiang, and B.D. Anderson, *J. Pharm. Sci.,* DOI 10.1002/jps.23855.

[6] "Investigating miscibility and molecular mobility of nifedipine-PVP amorphous solid dispersions using solid-state NMR spectroscopy", X. Yuan, D. Sperger, and E. J. Munson, *Molecular Pharmaceutics,* **2014**, *11*, 329-337

[7] "Salt-Induced Aggregation of a Monoclonal Human Immunoglobulin G1", J. Rubin, L. Linden, W.M. Coco, A.S. Bommarius, S .Behrens, *Journal of Pharmaceutical Sciences,* **2013**, *102*, 377-386

[8] "Gauging Colloidal and Thermal Stability in human IgG1 – Sugar Solutions through Diffusivity Measurements", J. Rubin, A. Sharma, L. Linden, A.S. Bommarius, S.H. Behrens, *J. Phys. Chem. B* **2014**, *118*, 2803-2809

387757 Collaborative Development of an Auto-Sampling Probe for HPLC

Abstract:

Pre-competitive collaborations allow pharmaceutical companies to work with each other and with instrument vendors to develop new and valuable enabling technologies. Here, we describe the evolution of a collaboration between five pharmaceutical companies and an instrument vendor to develop an auto-sampling probe for HPLC. This started as an in-house technology development project at Pfizer with the goal of automatically sampling reactions for profiling by HPLC where the liquid and solid components of heterogeneous reactions are sampled representatively. The project evolved from working with an internal machine shop to collaborating externally with Mettler which provided key engineering expertise. The collaboration further evolved to include valuable input and support from the pharma partners. As the probe was developed from proof of concept to an alpha prototype to a beta prototype and ultimately the commercialization phase at Mettler, each of the pharma partners contributed to testing the system in real life project applications and provided design input and feedback.

392437 Systems-Based Pharmaceutics - An industrial pre-competitive alliance

Abstract:

The Systems-Based Pharmaceutics industrial alliance seeks to develop a model-based holistic platform for the development and optimization of drug manufacture and delivery. This holistic approach aims to address the various drawbacks associated with the silo thinking approach that is prevalent in many pharmaceutical companies.

Given the many technical challenges in such an endeavor, the current phase of implementation does *not* primarily focus on the development of new fundamental scientific knowledge (e.g. on drug manufacturing unit operations, or on drugs pharmacokinetics/pharmacodynamics); but rather on the implementation of the current state of the art as published by academic and professional scientists.

SBP is also not prescriptive about the type of mathematical models that it will support. Some of these will be based on first-principles descriptions while others will be statistics-based models or even simple correlations. Furthermore, component models of different types may be integrated within the same system-level model.

Overall, SBP aims to be an integrating environment facilitating the accelerated transfer of new fundamental knowledge from sources of new science (i.e. academic research or R&D efforts) to industrial practice, while supporting the effective sharing of non-proprietary knowledge across the pharmaceutical industry.

Jack Huang, Henry Lin and Jerry Yang

A novel and universal method for high mannose Fc glycan modulation in recombinant therapeutic antibodies

High-mannose (HM) Fc glycoforms are critical quality attributes for therapeutic antibodies. HM Level of Fc glycan affects antibody effector function, such as antibody-dependent cell-mediated cytotoxicity, antibody clearance rate and its therapeutic efficacy. Hence, it is imperative to control HM level for both developmental and commercial antibody therapeutics during manufacturing processes. Methods to control HM, however, have been empirical and it is challenging to target a desired HM level during antibody process development. In this study, we demonstrated a novel method to modulate antibody HM glycoforms using multiple antibodies and cell lines, without adversely impacting cell culture performance including viable cell density, viability, and protein yield. This approach utilizes an alternative carbohydrate as an additional carbon source, and the ratio of this carbohydrate to total hexose in feed medium linearly determines antibody HM glycan level. Scale-up of this strategy from 24 deep-well plates to 2L bioreactors was also demonstrated with comparable results. Possible hypotheses for the HM glycan modulation using this method will also be discussed.

388444 Towards Controlling the Glycoform: Bridging the Gap from the Extracellular Environment to Antibody Glycosylation

Abstract:

Glycoproteins are the largest and most valuable group of biologically-derived medicines and account for 77 of the 643 therapeutics approved by the EMA [1]. These large glycoprotein molecules are characterised by their glycan moieties, which are complex post-translation modifications formed by a series of enzymatic reactions in the ER and the Golgi. The synthesis of glycan structures is driven by the availability of nucleotide sugar donors (NSDs), which are the co-substrates to the glycosylation process. As a result of this the distribution of glycan structures is strongly influenced by process conditions such as temperature, nutrient availability, dissolved oxygen, operation schedule and culture mode. This work focuses on nutrient availability as well as feeding of NSD precursors to the culture medium, which are known to influence intracellular availability of NSDs, which in turn is known to impact the glycan profile of biotherapeutics [2]. In the context of biotherapeutics glycan structures are known to play an important role in pharmacokinetics such as serum half-life, therapeutic efficacy and most importantly drug safety.

In this work a modelling platform has been developed linking the extracellular environment via intracellular metabolites and their availability in the cytoplasm and Golgi to the glycosylation of the heavy chain of a monoclonal antibody [3]. The presented dynamic modelling framework is split into four interlinked parts: 1) a cell growth model based on modified Monod kinetics, 2) a simplified purine and pyrimidine synthesis network, 3) a mechanistic representation of the nucleotide sugar donor synthesis pathway and 4) a model describing the N-linked glycosylation of the antibody Fc region as developed by del Val et al [4]. Particular attention was paid to the mammalian NSD pathway by means of a bottom-up mechanistic *in silico* reconstruction of the network. The model comprises 34 mass balances corresponding to the equivalent number of species that make up the metabolic network and the balances are connected through a total of 60 individual reaction rates. All reaction steps and NSD transporter outlets were modelled as saturation kinetics and described as individual rate expressions based on enzyme mechanisms and kinetics found in the literature. Where necessary model parameters were estimated from experimental data. This included cell growth dynamics, extracellular nutrient availability, dynamic intracellular NSD data, antibody titer and the antibody product glycoform.

The result of this study is a modelling framework, which is able to link the extracellular environment via intracellular metabolites to the glycosylation profile of the IgG product. The linked outputs of all parts of the model are in good agreement with experimental data and the model was able to capture the dynamic impact of nucleotide and sugar additions to the media on the availability of intracellular NSDs. Ultimately this modelling framework would allow to assess the impact of genetic engineering and process operation strategies such as NSD precursor feeding strategies *in silico* and guide experimentation and finally process development. This framework is a first step towards a platform for the *in silico* optimisation of bioprocess conditions with respect to product quality and safety in line with the Quality by Design (QbD) paradigm.

While this model describes the dynamic distribution of the conserved glycan of the IgG heavy chain, the modular nature of the framework allows it to be coupled with any other dynamic model using NSD flux and/or concentrations as input. Furthermore the modular nature will allow for the framework to be easily translated to other operation modes, culture conditions or expression systems.

References

1. Kyriakopoulos, S.; Kontoravdi, C., Analysis of the landscape of biologically-derived pharmaceuticals in europe: Dominant production systems, molecule types on the rise and approval trends. *Eur J Pharm Sci* **2013**, *48*, 428-441.

2. P.M., J.; I., J.d.V.; K.M., P.; C., K., Applying quality by design to glycoprotein therapeutics: Experimental and computational efforts of process control. *Pharmaceutical Bioprocessing* **2013**, *1*, 51-69.

3. Jedrzejewski, P.M.; del Val, I.J.; Constantinou, A.; Dell, A.; Haslam, S.M.; Polizzi, K.M.; Kontoravdi, C., Towards controlling the glycoform: A model framework linking extracellular metabolites to antibody glycosylation. *International journal of molecular sciences* **2014**, *15*, 4492-4522.

4. del Val, I.J.; Nagy, J.M.; Kontoravdi, C., A dynamic mathematical model for monoclonal antibody n-linked glycosylation and nucleotide sugar donor transport within a maturing golgi apparatus. *Biotechnol Progr* **2011**, *27*, 1730-1743.

Optimization of a Novel CHO Cell Bank Production Method

A new platform process flow was recently developed to replace traditional methods for generating CHO master and working cell banks (MCBs and WCBs). The process, referred to as ICeBOUND (Improved Cell Banking Operations Utilizing Novel Disposables), utilizes a series of single-use equipment in place of conventional glass/stainless steel equipment for cell accumulation, harvest and concentration. Implementation of the novel process has several drivers:

- To mitigate risk of using of nearly obsolete equipment (lacking replacement parts/ service support)
- To standardize commercial cell bank generation processes (for MCBs and WCBs)
- To reduce open processing and risk of contamination
- To reduce time and labor associated with cleaning and sterilization operations and validation
- To keep pace with industry trends in single-use technology application

Initial proof of concept studies comparing ICeBOUND derived cell banks to traditional MCBs or WCBs did not meet acceptance criteria. Poor thaw performance and comparability was observed for several cell lines, where viabilities were offset from controls during primary culture. The investigation that followed took a systematic approach to optimize each phase of the process. Several modifications were made to improve growth and viability of cultures during the cell accumulation phase. In the end, it was found that simple, yet important, enhancements could be made to the harvest and cell pooling steps to provide better control over the cell pool environment up to and during ampoule filling. A summary of the troubleshooting strategy, process enhancements and final process confirmation results will be presented.

372670 Developing a High Titer Defined Medium and Feed for E. coli Fermentation Systems

Abstract:

E. coli has been a workhorse for the biotechnology industry for numerous years and holds many advantages. The foremost of the advantages is the cell line's ability to grow on many types of nutrients from complex components to minimal salts. Complex growth components offer assured and robust growth with high titers, but little in the way of development opportunities. Minimal salts are well defined and allow for maximal avenues of investigation, but generally come with less robustness and lower titers. There is need for a balanced approach where all of the nutritional components are known, but all of the benefits of the complex supplements are maintained. By defining the components of complex media, fine tuning processes to modify metabolic rates in *E. coli pr*ocesses can be undertaken to increase titer. Through the analysis of the complex medium used in a fermentation process, and the investigation of its constitutive components, the development of such a medium and feed for E. coli can be achieved. For this study, concentrates were formulated and blended to generate fermentation medium and feeds which achieved commensurate growth and production for both insoluble and soluble recombinant protein production when compared to complex formulations. The study method and results are described in this presentation.

390127 Directional Oscillating Shear in Conventional and Novel Culture Dishes

Abstract:

Wall shear stress (WSS) on anchored cells affects their responses, including cell proliferation, morphology, and gene expression. WSS is two-dimensional. We previously developed a new directional oscillatory shear index (DOSI) to quantify the directionality of oscillating shear. DOSI is unity for uniaxial flow for any nonzero amount of shear reversal, and zero for bi-axial oscillations of equal magnitudes. Results indicated that more, smaller and less elongated cells grew in bi-axial shear with nearly equal oscillatory fluctuations (low DOSI), and fewer, larger cells with greater elongation were promoted in uniaxial oscillatory shear (high DOSI). Statistical analyses of the individual and interacting effects of multiple factors showed that DOSI significantly affected all the responses, indicating that directionality is an important determinant of cellular responses.

A widely-used, convenient platform for studying these responses is flow in a cylindrical dish in orbital motion on a shaker table, such as in multiple well plates. Computational modeling shows that WSS varies substantially with radial location within the dish. DOSI approached zero for biaxial oscillatory shear of equal magnitudes near the center and approached one for uniaxial pulsatile shear near the wall, where large tangential WSS dominated a much smaller radial component. We introduce here novel-geometry dishes designed to reduce the variability of WSS across the dish, which was quantified by comparing DOSI characteristics with that in conventional cylindrical dishes.

A Materials Science Approach to Bilayer Tablet Compaction

Jasmine M. Rowe, Keirnan LaMarche, Brian Delancy, Brian Carroll, Preetanshu Pandey, Faranak Nikfar

Fixed dose combination products (FDC) consist of two or more active pharmaceutical ingredients (API) combined within a single dosage form. FDCs are becoming increasingly popular forms of treatment for several therapeutic areas because it offers convenience, a reduced dosing burden, and cost savings for patients. In many cases, when the APIs are chemically incompatible or multiple release strategies are desired, the dose must be formulated as a layered tablet. However, layered tablets pose unique development and manufacturing risks. A fine balance exists between maintaining sufficient physical separation between the layers to mitigate chemical incompatibilities and yet also maintain sufficient cohesion between the layers so that the tablet remains intact (i.e. avoid layer delamination) throughout the manufacturing process and product shelf life. Challenges during the manufacture of layered tablets can be difficult to resolve due to the limited mechanistic understanding of the compaction process. Thus, layered tablet formulations have been historically developed through trial-and-error, which requires significant resource investments, especially when considering that layered tablet production on a rotary press requires kilogram-quantities of materials.

In this work, a materials-science-based approach to bilayer compaction is utilized. Because failure of layered tablets typically occurs due to delamination at the interface, a fundamental appreciation of the governing mechanisms that impact interface strength is crucial in order to improve the overall mechanical integrity of that layered tablet. The authors leverage knowledge about the plastic/elastic properties of the materials of interest to predict the likelihood for that material combination to yield a feasible bilayer tablet. An empirical heuristic has been developed in which less than 20 g of material can be used to predict the feasibility for a particular material combination to result in successful bilayer tablets. Three case studies will be highlighted to illustrate the utility of this predictive tool to guide efficient formulation development of bilayer tablets. The case studies will focus on the ability of this tool to evaluate the following scenarios:

- Over-lubrication: Impact to bilayer tablet strength when the blend making up one of the layers is over-lubricated
- Change in wet-granulation conditions: Impact to bilayer tablet strength when granulation conditions for one of the layers is changed
- Change in excipient grades, i.e. particle size differences: Robustness of bilayer tablet strength in response to changing excipient grade, especially when different grades possess different particle size distributions

386474 Design of Component Particle Attributes to Optimize the Processibility of Pharmaceutical Blends

Abstract:

Predictive measurements of pharmaceutical powder blends intended for tabletting or encapsulation are crucial for development of robust formulations associated with few manufacturing operational issues. While assessments should ideally be conducted early in product development when the attributes of components controlling performance are still being established, processing concerns are typically manifested only at larger scales when flexibility to modify particle attributes is limited. Blend flowability is an example of an issue for which improved understanding of how component properties translate to bulk behavior in full-scale equipment is desirable.

By analysis of particle morphology, particle size, and coordination number, we present case studies revealing unexpected transitions in flowability and cohesion with changes in material properties and composition. We explore the nature of interactions between the active pharmaceutical ingredient and the excipients in the blend to explain these transitions, and optimal concentrations are identified. Hypotheses are tested by investigating the influences of geometric constraints, contact mechanics, and Van der Waals forces on glidant and lubrication effects, which in turn determine flowability performance. The resulting understanding should translate to improved design of active pharmaceutical ingredient physical properties to limit pharmaceutical manufacturability issues earlier in pharmaceutical product development.

389352 Development of a Novel Method for the Solubilization of Poorly Water-Soluble Drugs

Abstract:

In recent years, large portion of new drug candidates and drugs currently on the market have poorly water-soluble nature. As the low solubility in water reduces bioavailability of the drugs, improvements in the solubility as well as solubilization rate are the most important for the drug formulation. There are many techniques that improve the solubility such as mixing with an aqueous surfactant solution, dissolution with an organic solvent and successive solubilization and/or spray drying, and formation of suspension. However, the use of surfactants as well as organic solvents might suffer the residual chemicals harmful to the human body. In this study, a novel method for the solubilization of poorly water-soluble drugs that are environmentally benign was developed. The method is based on the hydrophobic interaction of poorly water soluble drugs with the hydrophobic core of water soluble polymer micelle which is formed by a lower critical solution temperature (LCST) nature of the polymer.

In this study, Soluplus® (SOL) was used as the water-soluble polymer having a surfactant activity and LCST nature. Model drugs used as the poorly water soluble drugs is indomethacin (IND:acidic drug). One weight percent of aqueous SOL solution was maintained at a predetermined temperature, and the drug powder (as received) was added to the solution. After the predetermined duration of time with stirring at 500 rpm using a magnetic stirring tip, a white suspension was obtained. The concentration of drugs in the suspension was examined by a UV-VIS spectrophotometer, and freeze dried to examine the crystallinity of drugs in the hydrophobic core by a X-ray powder diffraction (XRD) analysis and differential scanning calorimetry (DSC). Infrared spectroscopy (FT-IR) was used to examine the interaction of drugs with the SOL. Long term stability of the suspension was also examined after predetermined time periods up to 1 month.

Suspensions were prepared with different periods of mixing (1 ~ 24 h) and different temperatures (room temperature ~ 70 °C). It was found that turbidity of the solution increases with the increase in stirring time regardless of the preparation temperature. At the same time, higher the temperature, shorter the time for the formation of milky white suspension of IND. Suspensions prepared at 70 °C and 50 °C achieved the IND concentration of 1000 mg/10 g in the dispersion with the 1 hr of stirring. On the other hand, suspension prepared at room temperature could achieve the value by increasing the stirring time. Further, maximum drug loading was found to be higher prepared at the higher temperatures. The mechanism of the solubilization could be explained as follows: even the poorly water soluble IND dissolves into water, and the saturated solubility increases with the increase in temperature. The incorporation of IND into the hydrophobic core of the SOL occurs via the distribution equilibrium of IND solid/water/SOL core. Thus higher the temperature, higher the water phase concentration of IND which leads to the higher distribution equilibrium concentration of IND in the SOL hydrophobic core. All samples thus prepared were stable up to 1 month and over.

The XRD measurement of the suspension prepared at 70 °C and 50 °C gave the halo pattern, which shows the amorphous state of IND/SOL mixtures. Further, no endothermic peak was observed in the DSC measurement. Therefore, it was found that the IND was solubilized in the hydrophobic core of the SOL with the amorphous state.

It could be concluded that this method is effective as a solubilization method of poorly water-soluble drugs in the suspension.

386485 Experimental & Numerical Investigation of Bipolar Charging in a Hopper-Chute Assembly

Abstract:

Bipolar charging, which is a special class of tribocharging is defined as procuring two opposite polarities on different sized particles of the same material (usually, smaller particles getting negatively charged and ,bigger particles getting positively charged). In an attempt to explain this unusual bipolar charging behaviour, various groups have presented hypothesis based on the differences in the adsorbed impurities and surface roughness of different sized particles. The purpose of the study is to investigate the same, using various analytical tools ranging from X-ray photoelectron spectroscopy (XPS), Time of Flight- Secondary Ion Mass Spectroscopy (ToF-SIMS), Energy Dispersive Using X-Ray analysis (EDEX), Interferometer-Profilometer study, Scanning Electron Microscopy (SEM), Thermogravimetric Analysis (TGA) and Dynamic Vapor Sorption (DVS). After thoroughly mapping the surfaces of various materials, Density Functional Theory (DFT) calculations were performed on these materials to calculate the work function values and the effect of adsorbed impurities on the same. Hopper-chute flow experiments were conducted to study the possibility of bipolar charging due to third body interaction.

A major difference in the amount of water sorption (smaller particles sorbing ca. 10 times more water as compared to bigger particles) was observed. No significant differences in the surface or bulk impurities (XPS, ToF-SIMS, EDEX) or in surface roughness (Interferometer-profilometer and SEM) was found. Accordingly DFT calculations were performed on molecular slabs with and without water to study the effect of adsorbed water on the work-function values. Water sorption was found to lower the work function values of the smaller particles due to formation of dense electron cloud on the surface from the oxygen's lone pair of electrons as follows. Work-function of; MCC: 6.15eV, MCC with adsorbed water: 5.94 eV, Lactose: 6.29 eV and Lactose with adsorbed water: 6.06 eV. Water sorption on smaller particles leads to lowering of work function values, which suggests that electron transfer is not the sole or most dominant mechanism behind bipolar charging and there is a significant involvement of other mechanisms such as ion transfer in the process. Hence bipolar charging might be a result of preferential adsorption of negatively charged OH- ions on smaller particles due to higher surface area and greater water sorption on the same.

THE EFFECT OF PROCESS PARAMETERS INCLUDING BLEND SHEARING AND COMPACTION PRESSURE ON PHARMACEUTICAL TABLET PROPERTIES

P. Pawar, H. Joo, Y. Wang, A. Cuitino, F. Muzzio

Rutgers University, New Brunswick, NJ

The purpose of this study was to gain an understanding of the effects of process parameters on tablet properties in the case of blends with lubricants. Magnesium Stearate has been widely used as a lubricant in pharmaceutical formulations for the past 50 years. Different studies have been published to explain the effect of concentration levels of magnesium stearate on tablet and blend properties. However, as much is the concentration of magnesium stearate important, so is the amount of energy that goes into its blending with the excipients.

We adopted a full factorial design of experiment approach to study the effect of blend shearing and tablet compaction force on final tablet properties. The data reveals the effect of strain on the in-die tablet porosity as well as the porosity after relaxation, and how this along with the bonding capacity ultimately affects the tensile strength and finally the tablet dissolution.

A pre-blend consisting of lactose monohydrate, acetaminophen and magnesium stearate was prepared in a V-blender and subdivided into equal fractions. Each of the fractions was subjected to different levels of strain, 0 rev, 160 rev, 640 rev and 2560 rev respectively in a shearing device. Each of these blend fractions was then compressed into tablets using nominal compression forces of 8, 12 and 16, 24 and 30 kN. The responses studied were tablet porosity, tensile strength and dissolution time. Statistical analysis was carried out on each of these responses to understand the significance of these process parameters. Compaction force had a significant effect on tablet porosity, while total strain level was insignificant, contrary to the expectations. On the other hand, as expected, both shear and force had a significant effect on tensile strength, while the effect of strain on the percentage of drug dissolved was insignificant during the first few minutes of the dissolution test but became more significant as the dissolution progressed.

In order to use API (active pharmaceutical ingredient) particles intended to target the tiny airways of the deep lung in dry powder inhalers (DPIs), the range of an aerodynamic diameter of 1 µm to 5 µm is highly preferred. Particles in this size regimen are typically cohesive and possess poor flow properties [1], leading to difficulties concerning volumetrically dosing. To overcome this flowability problem, it is a general practice to formulate a carrier based formulations wherein the API particles are attached to the surface of a larger carrier particle (50 µm – 200 µm).

This study aims to investigate the efficiency of unprocessed and engineered lactose and mannitol as carriers to improve the aerosolization and in vitro deposition of the model API. Surface engineering was performed by decantation (Figure 1).

Figure 1 Scanning electron micrographs of lactose carrier particles a) before and b) after decantation and mannitol carrier particles c) before and d) after decantation

Both carriers were extensively characterized (e.g. particle size, particle shape, flow properties, surface topography) to compare the different particulate properties. Spray dried salbutamol sulfate (Selectchemie, Zurich, Switzerland) was chosen as model API and prepared on a Nano Spray Dryer B-90 (Buechi Labortechnik AG, Flawil, Switzerland). The spray drying conditions were chosen according to our previous work. [2].

Drug-carrier interactive mixing was performed in a tumble blender TC2 (Willy A. Bachofen Maschinenfabrik, Muttenz, Switzerland) using the sandwich method. The mixing time was 60 min at 60 rpm. Subsequent capsule filling was performed with different process setting on a dosator nozzle capsule filling machine (Labby, MG2, Bologna, Italy) with a target fill weight of 20mg. To evaluate the performance of the different mixtures, in vitro lung deposition experiments were carried out with a next generation impactor (NGI, Copley Scientific, Nottingham, United Kingdom) and the fine particle fraction (FPF) was calculated based on the specification of the European pharmacopoeia. The inhalation device used for these experiments was the Aerolizer®/Cyclohaler®, a capsule inhaler.

Due to the possible relationship of various formulation and process parameters on the final product properties, we specifically investigated the influence of carrier morphology and dosator capsule filling process on the performance of DPIs (i.e. the ease of drug detachment and consequently the FPF). Moreover, the effect of carrier types (lactose versus mannitol) as well as the particle manipulation (unprocessed versus engineered) on the capsule filling efficiency and on the DPI performance were evaluated. The obtained data are highly useful to improve the understanding of the relationship between carrier morphology and carrier type, drug detachment and capsule filling efficiency and will help to generate DPI formulations with the desirable performance.

References

[1] G. Pilcer, N. Wauthoz, K. Amighi, Lactose characteristics and the generation of the aerosol., Adv. Drug Deliv. Rev. 64 (2012) 233–56. doi:10.1016/j.addr.2011.05.003.

[2] E.M. Littringer, S. Zellnitz, K. Hammernik, V. Adamer, H. Friedl, N. a. Urbanetz, Spray Drying of Aqueous Salbutamol Sulfate Solutions Using the Nano Spray Dryer B-90—The Impact of Process Parameters on Particle Size, Dry. Technol. 31 (2013) 1346–1353. doi:10.1080/07373937.2013.793701.

[3] P.M. Young, D. Roberts, H. Chiou, W. Rae, H.-K. Chan, D. Traini, Composite carriers improve the aerosolisation efficiency of drugs for respiratory delivery, J. Aerosol Sci. 39 (2008) 82–93. doi:10.1016/j.jaerosci.2007.10.003.

Design and Administration of Dissolving Microneedle Patches In Human Subjects

Jaya Arya, Haripriya Kalluri, Devin McAllister, Winston Pewin, Xin Dong Guo, Mark Prausnitz

School of Chemical and Biomolecular Engineering, Georgia Institute of Technology, Atlanta, GA, USA

jaya.arya@gatech.edu

Dissolving microneedle patches contain micron-sized needles made of water-soluble biodegradable polymers that dissolve in the skin to deliver encapsulated vaccine. They offer the simplicity of patch application and the possibility to mitigate the logistical and safety challenges associated with using conventional hypodermic needles. To ensure reliability in the dose being delivered, it is important to understand the biomechanics of the skin and the parameters that affect insertion and dissolution of microneedles into skin. We are studying these parameters in a human study with placebo dissolving microneedle patches (i.e., containing no vaccine) to prepare for a planned phase 1 clinical trial of influenza vaccination using microneedle patches of similar design.

We fabricated microneedle patches suitable for use in humans using a micromolding process that employs excipients (polymers and sugars) found in FDA-approved products. The study is approved by the Georgia Tech IRB and informed consent is obtained from all participants.

The parameters studied are:

1. **Effect of microneedle geometry on efficiency of insertion into skin:** We expect that microneedle size and needle to needle spacing will be critical parameters affecting biomechanics of insertion into the skin. We therefore designed microneedles of different geometries and inserted them using spring-loaded applicators that apply varying amounts of forces. Initial findings show that a force of 45 N or greater was sufficient for reliable insertion, as determined by staining the skin with a dye to image sites of microneedle puncture, microscopically inspecting microneedle patches after insertion into the skin and assaying for the amount of needle dissolved.

2. **Kinetics of microneedle dissolution in the skin:** We inserted the patch in the forearm by pressing down with the thumb and left the patch on for varying amounts of time from 30 seconds to 15 minutes. Initial studies have shown microneedle dissolution in the skin within 5 – 10 min of insertion. [Fig 2 and 3]

3. **Skin tolerability of dissolving microneedles in terms of local irritation in the skin after patch insertion:** Skin tolerability has been good, with mild erythema observed localized to the site of patch application for up to two days

Ongoing studies are carrying these studies out using a range different microneedle patch geometries and using larger numbers of human subjects.

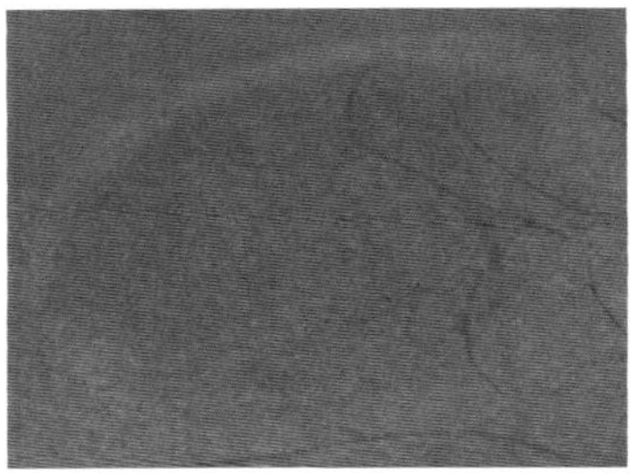

Fig 1. Skin insertion site showing minimal irritation immediately after microneedle patch insertion

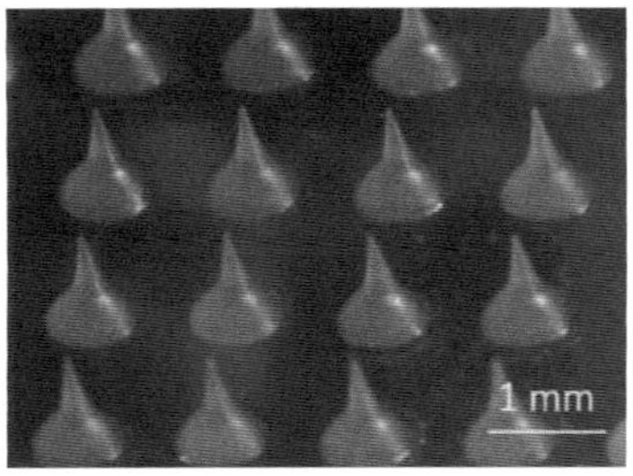

Fig 2. Dissolving microneedle patch before insertion

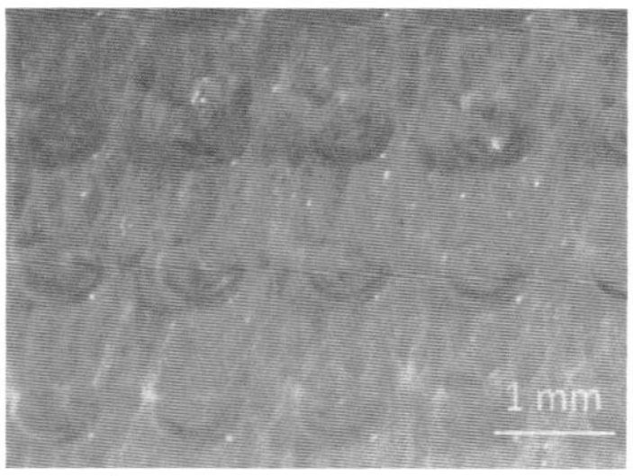

Fig 3. Dissolving microneedle patch after insertion and patch wear time of 10 min

pH-Sensitive P(IA-*co*-NVP) Hydrogels as Oral Delivery Vehicles for High Isoelectric Point-Exhibiting Therapeutic Proteins

Michael C. Koetting[1*], Annie Zhang[1], and Nicholas A. Peppas[1,2,3].

[1]McKetta Department of Chemical Engineering, [2]Department of Biomedical Engineering, and [3]College of Pharmacy
The University of Texas at Austin, Austin, TX 78712, USA
*Presenting author, email: mkoetting@utexas.edu

Introduction: Although protein therapeutics are a cornerstone of modern medicine, they are currently delivered primarily by injection due to the numerous obstacles presented by the harsh, degradative environment of the gastrointestinal tract. Delivery via the oral route is preferable to avoid the pain and inconvenience of injection and to improve patient quality of life. Our group has previously developed pH-responsive hydrogel systems based on poly(methacrylic acid) to overcome the barriers of the GI tract and enable the oral delivery of insulin. However, due to coulombic interactions in the small intestine, diffusive release of proteins which exhibit a high isoelectric point, pI, is comparatively limited within the small intestine. High pI proteins account for approximately half of the human proteome and over 50 current FDA-approved therapeutics, so improvements are imperative for effective delivery of numerous therapeutics. In this work, we present improvements in the material chemistry and protein-loading procedure which yield improved release of proteins exhibiting high pI.

Methods: Copolymer hydrogels comprised of itaconic acid (IA) as the pH-responsive moiety, and N-vinylpyrrolidone (NVP) or poly(ethylene glycol) monomethyl ether monomethacrylate (PEGMMA), were prepared by UV-initiated free radical polymerization. The hydrogels were washed pure in deionized water, dried under vacuum, and crushed and sieved into microparticles 90-150 μm in size. Hydrogel swelling characteristics were determined at equilibrium by placing 12 mm diameter hydrogel disks in PBS buffer at pH 7.4 or 0.01 N HCl at pH 2.0 for 72 h and weighing before and after swelling. Dynamic swelling was measured by swelling in 10 different buffers of successively increasing pH ranging from 3 to 7.6 for 7 minute intervals with weighing after each step. Protein loading and release characteristics were determined by loading high pI proteins, including salmon calcitonin and urokinase, into microparticles via imbibition in a pH 7.4 PBS buffer and subsequent collapse by acidification to encapsulate the protein. Proteins were then released in 0.150 M PBS buffer solutions at pH 7.4 and 37 °C for 4 h. Protein loading and release levels were determined by MicroBCA assay.

Results: Equilibrium swelling studies show that all prepared gels remain collapsed at low pH and swell up to 23 times their dry weight (q=23) at neutral pH, with IA-based gels exceeding a MAA-based gel by 21-68% in swelling ratio. In dynamic swelling studies, IA-based gels achieved near-equilibrium swelling ratios (q=11.2-15.4) in a time-limited system, whereas an MAA-based gel showed minimal swelling (q=1.5). The faster swelling response is expected to aid in diffusive release during the limited time through the small intestine. In loading and release experiments using salmon calcitonin, a 1:2 P(IA-co-NVP) formulation showed up to 2.7 times the release level of a 1:1 P(MAA-g-EG) formulation and 16.8 times the release level of a 2:1 P(MAA-co-NVP) formulation. Finally, by altering the ionic strength of the loading solution used during protein loading, we show that reduced ionic strength yields significant improvements in protein loading and release, as shown in Table 1, achieving up to an 83-fold improvement in release level if a 1.5 mM PBS solution is used during loading rather than a 150 mM standard PBS solution. We also show the ability to delivery urokinase (2.7 μg/mg), although delivery levels are reduced due to urokinase's much larger size.

Table 1: Salmon Calcitonin Release from P(IA-co-NVP) Microparticles Loaded in Various Ionic Strength

Loading Solution	sCT Released (μg/mg hydrogel)	% Release (% of total sCT loaded)
0.01x PBS	48.4	45.8
0.1x PBS	16.1	9.4
1x PBS	0.6	4.8
10x PBS	0.6	1.1

Conclusions: In this work, we show that P(IA-co-NVP) hydrogels demonstrate beneficial material properties for oral protein delivery, exhibiting greater swelling in both equilibrium and time-limited conditions compared to P(MAA-co-NVP) hydrogels, and superior protein loading and release behavior with high pI proteins. Additionally, we show that significant improvement in protein delivery is available through procedural change by utilizing a reduced ionic strength protein loading solution.

Acknowledgements: This work was funded by the National Institutes of Health, grant R01-EB000246-20, and by the Fletcher S. Pratt Foundation.

388013 Self-Assembled Peptide Amphiphile Micelles Improve Group a Streptococcal Vaccination

Abstract:

Streptococcus pyogenes(Group A Streptococcus, GAS) is a Gram-positive bacterium restricted to natural growth in humans where it frequently elicits diseases that range in severity from mild infections of the pharyngeal mucosa and dermis to life-threatening invasive infections of connective and muscle tissues leading to necrotizing fasciitis, myonecrosis, and toxic shock. Additionally, post-infection sequelae diseases like acute rheumatic fever and glomerulonephritis arise following localized infections of the nasopharynx and skin, respectively. Epidemiological studies estimate that each year greater than 500,000 worldwide deaths are attributable to GAS infections, placing it among the top ten leading causes of death from infectious pathogens. In the United States alone, more than $600 million is spent annually treating diseases caused by this organism with no effective preventative method established short of prophylactic antibiotic usage.

Vaccines against GAS remain unavailable despite decade's worth of research and development. Whole-killed and live-attenuated vaccines have been tremendously effective in preventing pathogenic infections, but can be associated with undesirable side effects. Subunit vaccines that deliver just the peptide antigen of interest have been shown capable of stimulating an immune response, but are generally weak immunogens on their own requiring strong adjuvants (non-specific immunostimulants) to be effective. In order to enhance the immunogenicity of peptide vaccines, new delivery systems must be designed. Peptide amphiphiles are unique biomaterials comprised of peptide-lipid conjugates that undergo self-assembly into micelles in water and have been shown capable of delivering biologically active peptides for a variety of applications. Since peptide vaccines are often delivered in low local concentrations, peptide amphiphile micelles provide a novel platform to improve the host immune response to peptide vaccines.

The J8 peptide is a 29 amino acid, conformationally dependent B cell epitope that has been shown to generate an opsonophagocytic, high titer antibody response in mice. J8 was covalently tethered to a di-palmitic acid tail (J8-diC_{16}) and fabricated into peptide amphiphile micelles in water. When delivered to mice subcutaneously, J8-diC_{16} was found to induce J8-specific high antibody titers greater than soluble J8 delivered with commercially available adjuvants. To further enhance the antibody response, mixed micelles comprised of J8-diC_{16} and amphiphilic adjuvants were synthesized. Mixed micelles induced a strong immune response after a single vaccination and higher titers than all other formulations regardless of the vaccine regimen. In addition to the J8 peptide, other GAS-associated peptides are currently being investigated with the micelle platform.

381619 Advanced Structuring of Lipids for Controlled Drug Release; polymorphic crystallizations and curing effects

Abstract:

A recent strategy for improving patient adherence to medication is the development of "direct to mouth" multi-particulate drug delivery systems, which strike a good and balanced compromise between taste masking and controlled drug release, taking both immediate and extended release into account. Such systems play a key role in the medication of patients from the geriatric or paediatric population suffering from dysphagia. Hot-melt coating in a fluid bed coater is a suitable technology for manufacturing of such products, offering numerous advantages for the pharmaceutical industries. For instance no need of solvents leads to significantly shorter processing times by sparing time and cost consuming solvent evaporation and recovery steps. The materials commonly used are lipid-based and hence offer further advantages such as toxicological harmlessness and natural availability at a low price point. The challenge is however, their polymorphic stability.

This abstract presents the advanced structuring of lipids by controlling their polymorphic behavior utilizing hot melt technology to provide products with controlled release profile.

Extended release profile: Highly water-soluble drug model, citric acid, was hot melt coated with tristearin. Different inlet temperatures were successfully used for re-crystallization of lipid to provide unstable α polymorph (process A) and stable β polymorph (process B) respectively. An additional third structure was created by curing the coating A into β-form in a heating oven with blooming occurring in the inner and outer surfaces of microcapsules (process AB). The low temperatures used in process A permitted to recover nearly 100% of coating in the α form. All three coatings exhibited the drug release mechanisms expected: an initial slight burst; a zero-order phase; and a Higuchi diffusion process based on the Fick's law, square root time dependent. However the choice of the polymorphic crystallization had a tremendous impact on the release kinetic slowing the drug release but improving the partition form core to coating.

Immediate release profile: A polymorphic stable formulation containing tripalmitin and additives was successfully developed, which showed an immediate release profile (85% of release within 30 minutes), an adequate taste masking (pH 3.5 in more than 60 s) and no changes in the polymorphism and the dissolution profile during storage for six months at different. A further advantage is the low process temperature of 25°C, which is necessary for the processing of thermolabile substances.

We showed the successful controlling of polymorphic behaviour of lipids, to use them as advanced safe excipients for development of multi-particulate systems with various release profiles.

Optimal Startup of a Continuous Pharmaceutical Process

Sahlodin, Ali M., Barton, Paul I.

Department of Chemical Engineering

Massachusetts Institute of Technology

Batch processing has long been the major paradigm adopted by the pharmaceutical industry. Recently, however, pharmaceutical companies are considering continuous manufacturing alternatives in order to keep pace with the increasing demand for efficient and low-cost production [1]. Continuous manufacturing is known to improve process yield over batch manufacturing [2]. Moreover, it offers potentials for a lower capital cost and better responsiveness to the market by allowing quicker product switches in multipurpose plants [2,3].

The operation time in continuous pharmaceutical processes is expected to be short [1]. As a result, the startup phase may constitute a considerable portion of the process runtime, thereby having a significant impact on the overall economy and quality of the products. This greatly motivates the use of optimization techniques to find an optimal startup strategy. In this work, startup optimization of a continuous pharmaceutical plant is considered. Startup optimization is addressed by dynamic optimization, and is inherently associated with a number of complexities. First, the process model used for optimization is a dynamic model, often represented in the form of differential-algebraic equations (DAEs). The presence of DAEs usually hampers efficient solution of the optimization problem. Second, startup scenarios are typically associated with abrupt changes within the flows and unit operations due to, e.g., equipment switchings, empty vessel startups, flow diversions, etc. These changes often cause continual switches in the process model, and give rise to so-called hybrid dynamic models.

Although there exist established methods for simulating hybrid dynamic models (e.g. [4,5]), this is not quite so for optimizing such models with gradient-based algorithms. In particular, the hybrid behavior can introduce non-differentiability and even discontinuity in the process model. To mitigate these effects in the startup optimization, reformulation of the switching behaviors into continuous changes is considered where possible. Also, the use of so-called nonsmooth optimization algorithms [6] to tackle the remaining non-differentiabilities is investigated. Finally, different formulations are proposed for the startup optimization, and the potential of each for improving the overall performance of the plant is studied.

[1] Benyahia, B., Lakerveld, R., Barton, P.I. “A Plant-Wide Dynamic Model of a Continuous Pharmaceutical Process,” Industrial & Engineering Chemistry Research 51 (47), 15393-15412, 2012.

[2] Mascia, S., Heider, P.L., Zhang, H., Lakerveld, R., Benyahia, B., Barton, P.I., Braatz, R.D., Cooney, C.L., Evans, J.M.B., Jamison, T.F., Jensen, K.F., Myerson, A.S., Trout, B.L. “End-to-

End Continuous Manufacturing of Pharmaceuticals: Integrated Synthesis, Purification, and Final Dosage Formation," Angewandte Chemie International Edition 52 (47), 12359–12363, 2013

[3] Schaber, S.D., Gerogiorgis, D.I., Ramachandran, R., Evans, J.M.B., Barton, P.I., Trout, B.L. "Economic Analysis of Integrated Continuous and Batch Pharmaceutical Manufacturing: A Case Study," Industrial & Engineering Chemistry Research 50 (17), 10083–10092, , 2011.

[4] PARK, T., Barton, P.I. "State Event Location in Differential-Algebraic Models," ACM Transactions on Modeling and Computer Simulation 6 (2), 137–165, 1996.

[5] Galán, S., Feehery, W.F., Barton, P.I. "Parametric Sensitivity Functions for Hybrid Discrete/Continuous Systems," Applied Numerical Mathematics 31, 17–47, 1999.

[6] KIWIEL, K.C. "Methods of Descent for Nondifferentiable Optimization," Lecture Notes in Mathematics 1133, Springer-Verlag, 1985.

Operation Strategy Development for Grignard Reaction in a Continuous Stirred Tank Reactor

Sze-Wing Wong[1], Shujauddin M. Changi, Richard Shields, Willis Bell, and Bernard McGarvey

Small Molecule Design & Development and Process Engineering Center, Eli Lilly and Company, Indianapolis, IN 46285

Abstract: A case study in establishing the operation space of a continuous Grignard reaction in a continuous stirred tank reactor (CSTR) will be presented. The operation space is the multivariate space with boundaries defined by the proven acceptable range (PAR) of every CSTR process parameter, such as flow rates and temperature. The mapping of the operation space is obtained through the understanding of reaction kinetics, magnesium sequestration efficiency, equipment characterization, and the impact of process disturbances on steady state. A fit-for-use reaction kinetic model was developed to parameterize the reaction and mass transfer rates of the Grignard reaction across different scales from 250mL to 500 gallons. The reaction kinetic model was then applied to design the Mg recharge frequency by taking operational variability into account. Furthermore, the model was coupled with equipment heat transfer characteristics to understand the process dynamics of the CSTR in the presence of process disturbances. Based on the kinetic model and energy balance calculations, reactor temperature was determined to be suitable to detect process failures and ensure process safety at manufacturing. Computational fluid dynamics (CFD) models were also applied to aid equipment design to improve Mg sequestration in the CSTR. Based on the optimal equipment design, the unit operation was scaled-down to test the sequestration efficiency. The resulting process understanding enabled the team to define the final operation strategy to ensure a safe and robust process to be scaled up to commercial scale production.

[1] Corresponding author
E-mail address: wongsz@lilly.com

A Computational Analysis of a Combined Cooling Anti-Solvent Continuous Crystallization Cascade: A Paracetamol/Methanol/Water Case Study

Kieran Hutton, Patrick Frawley.
SSPC, MSSI, University of Limerick, Ireland.

Abstract

Presented here are the results of computational investigations into the performance of a two tank continuous crystallization cascade. The system of interest is a combined cooling, anti-solvent crystallization of Paracetamol from methanol and water. The kinetic parameters used in the model were obtained from isolated cooling and anti-solvent batch experiments. The input parameters of interest here are the feed mass flow rate, relative sizes of the two tanks and the distribution of anti-solvent and feed across the tanks. Preliminary findings indicate that the distribution of feed and anti-solvent across the cascade has some impact on the solute recovery but the largest response is seen in the particle size distribution. The maximisation of recovery is generally seen where all feed is directed to the first tank and a portion of the anti-solvent is directed to the first tanks with the remainder directed to the second tank. For each configuration (flow rate and volume) there exists an optimum distribution of streams to maximise recovery and/or to target a particular particle size.

Introduction

With increased attention being given to the development of continuous crystallization systems to replace existing batch crystallizations, the need for a deeper understanding of their behaviour has arisen and as such much research has been conducted into this field [1] – [7]. Experimental determinations and characterisations of continuous processes are notoriously labour, time and material intensive as compared to batch investigations. To combat this, complex computational models of continuous crystallizations are needed to compliment the experimental investigations. Developed here, in the MATLAB computing environment is a numerical model of continuous combined cooling and anti-solvent crystallization of Paracetamol in methanol and water. The model incorporates the effects of particle growth, nucleation (primary and secondary), aggregation and breakage as well as convection through the vessel, assumed to be perfectly mixed. The kinetic parameters used in the model were obtained and validated from batch cooling and anti-solvent experimental investigations [8] [9]. The methods and technologies developed and implemented here are very valuable in that the model can be applied to or adapted for any crystallization once kinetics are known. Due to the highly non-linear behaviour and physical inter-dependence of such processes, it is necessary to be able to determine the underlying driving mechanisms for particle dynamics. For example, is the system growth, aggregation or nucleation dominated? Analysis techniques for simple growth/nucleation systems have existed for some time and are well understood, however techniques to elucidate this complex cascade behaviour is lacking if non-existent. In this work, methods have been developed to analyse the underlying particle dynamics for cascaded anti-solvent/cooling crystallizations. The algorithms developed in this work allow for the determination of the relative contributions of each of the physical phenomena to the underlying dynamics. For example, it may be determined what the dominant source of particle size increase or surface area generation is. This decoupling of the physics affords the capacity to design/ modify continuous systems to maximise or minimise any one or more desirable or undesirable mechanisms of crystallization. Optimisation of crystallization systems' operation is vital for their continued efficient operation. Outlined in this text are methods for determining the operating conditions which will give either a maximum particle size or maximum recovery for a given feed mass flow and tank configuration. It must be noted that this is not limited to the maximum particle size but may be used to target a specific particle size.

Material & Methods

Model Outline

The current research involves investigations into the performance of a two tank, combined anti-solvent/cooling CSTR cascade. The set-up involves two vessels in series with continuous feed and anti-solvent streams to each vessel as well as a recycle line (not investigated here). The temperature of each vessel may be individually controlled. As a result of this, a large number of configurations may be investigated. For example, vessel 1 may be a pure cooling stage followed by a pure anti-solvent stage or perhaps a combined anti-solvent/cooling crystallization may be desirable in both vessels. The determination of the optimum configuration is one of the desired outcomes of this research.

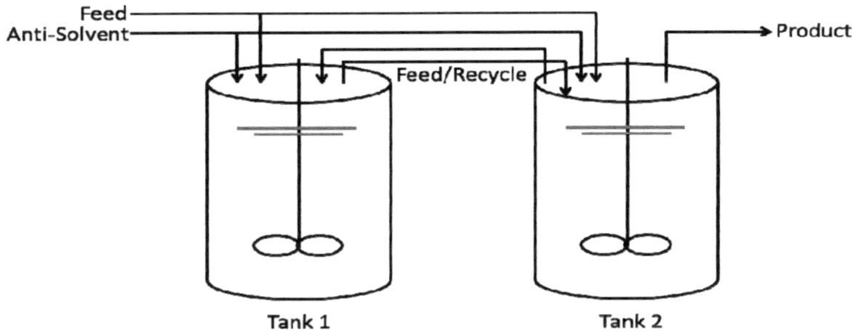

Figure 1. Simple Pictorial Representation of Crystallization Cascade.

The feed and anti-solvent splits are utilised here to distribute the feed or anti-solvent streams between the two vessels in the cascade. A feed split of 1 results in the entire feed stream being pumped into tank 1 with no feed directly to tank 2. A feed split of 0.4 results in 40% of the feed being fed directly to tank 1 with the remaining 60% being fed directly to tank 2. The same logic applies to the splitting of the anti-solvent stream.

The total anti-solvent flow rate is determined based on a user-defined final solvent phase composition. To clarify, if the feed stream comes in with a certain fixed composition then there exists an anti-solvent flow rate that will result in the final outflow from tank 2 having a solvent phase composition equal to that specified by the user.

$$\dot{m}_{AS} = \frac{\dot{m}_F[(1-R)\phi_{AS} - R\phi_s]}{R-1} \tag{1.1}$$

where f_{AS}, f_S and R are the feed anti-solvent mass fraction, feed solvent mass fraction and the optimum (final) solvent phase water mass fraction. The final solvent phase water mass fraction used in these simulations was 0.85. It must be noted that the mass flow rate of anti-solvent required to obtain a certain desired mass fraction tends to infinity as one tends towards R=1. The figure of 0.85 was chosen as it gave a good trade-off between optimum recovery and "sensible" use of anti-solvent.

Solubility & Kinetics

The solubility of Paracetamol in methanol/water mixtures was determined through gravimetric analysis. This data led to the generation of a solubility vs. composition and temperature surface and is shown in Figure 2 below.

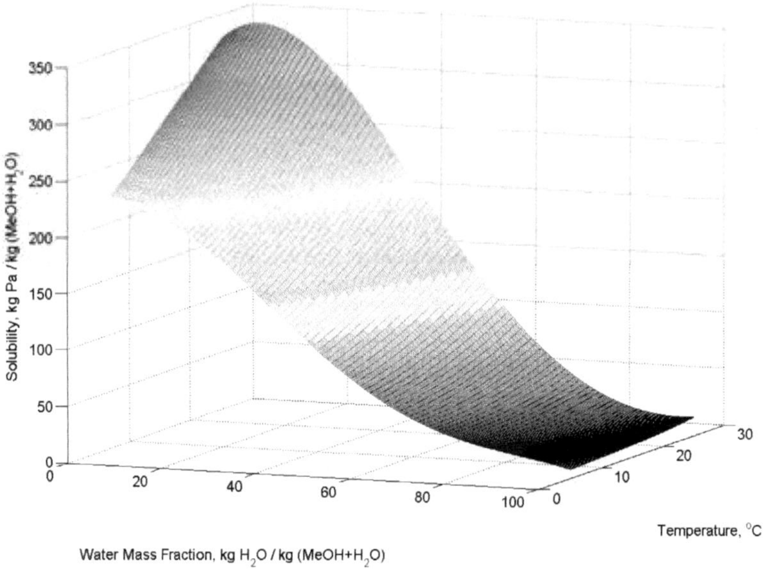

Figure 2. Solubility of Paracetamol as a function of temperature and composition.

The kinetic parameters used in this work were obtained from batch cooling and anti-solvent experiments conducted in isolation. Temperature and composition effects are accounted for in the solubility surface and no temperature term is seen in any of the kinetic expressions governing growth, nucleation etc.

Growth Rate

$$G = k_G \Delta C^g \qquad (1.2)$$

where G, k_g, ΔC and g are the linear growth rate (m/s), growth constant (m/s), supersaturation (kg Pa/kg mixture) and growth order (-) respectively. $k_g = 14.14 \times 10^{-6}$ m/s, g = 1.19.

Nucleation Rate

$$J = k_N \Delta C^{n_1} \mu_2^{n_2} \qquad (1.3)$$

where J, k_N, m_2, n_1 and n_2 are the nucleation rate (#/m^3/s), nucleation constant (#/m^3/s), second moment (m^2/kg mixture), nucleation order, surface area dependence. $k_N = 5.38 \times 10^7$ #/m^3/s, $n_1 = 0.97$, $n_2 = 0.72$.

Aggregation Kernel

$$k_{i,j}^a = 1.29\sqrt{\frac{\varepsilon}{\nu}\left(L_i + L_j\right)^3} \qquad (1.4)$$

where $k^{a}_{i,j}$, ε, n and L are the aggregation kernel (#/m^3/s), turbulent energy dissipation rate (m^2/s^3), kinematic viscosity (m^2/s) and particle diameter (m) respectively.

Aggregation Efficiency

$$\psi_a = G^a \qquad (1.5)$$

where a is the aggregation growth dependence. a = 0.71.

Breakage

As particle breakage was not identified as a contributory factor in the original batch investigations [8] it was not included in this continuous model.

Population Balance: Model Equations

The dynamics of crystallization processes are described by the population balance equation. This equation describes the rate of change of the particle size distribution over time with the effects of particle growth, nucleation, aggregation/agglomeration and breakage accounted for.
The spatially distributed formulation of this equation is given as [10]

$$\frac{\partial n}{dt} = \nabla \cdot v_e n + \frac{\partial}{\partial L}(Gn) + \delta(L - L_{min})J + B - D \qquad (1.6)$$

where n ,v_e, L, G, J, B and D are the number density function, flow velocity (convective term), particle diameter, growth rate, nucleation rate and aggregation/breakage related birth and death functions respectively. This equation is suitable for resolving the time and spatial evolution of the particle size distribution of particles in a flow. Such a formulation is ideal for Computational Fluid Dynamics (CFD) computing environments. For non-CFD applications, the spatial derivative terms are taken to be zero and the so-called 1-dimensional model results. This equation may be given as

$$\frac{\partial n}{dt} = \frac{\partial}{\partial L}(Gn) + \delta(L - L_{min})J + B - D \qquad (1.7)$$

The birth and death terms on the right hand side of equation 1.2 above represent the impact of both aggregation/agglomeration and attrition/breakage on the PSD [11].

High precision numerical solutions to this equation tend to be very computationally expensive. For very fast crystallizations, particularly systems with high nucleation rates resulting in sharp fronts or gradients in the Particle Size Distribution (PSD), solution of this equation may be prohibitively expensive. For example, depending on the dynamics of the system, up to 2000 classes may be required [12]. This means that Equation 1.7 must be solved up to 2000 times per time step for such a system.
Many alternative solution techniques exist and are reported in the literature. These include non-linear discretisation techniques [13] [14] [15], the Moving Pivot Technique [16], Method of Characteristics [17] [18], Method of Moments [10] and the Quadrature Method of Moments [19] [20]. For the simulations presented here, the Quadrature Method of Moments was utilised. Selection of this method was justified by the fact that it is one of the fastest solution methods available. Its main disadvantage, being that the PSD is not conserved, was not seen as a serious drawback as the major statistics of interest are available from the moment sets. If required the PSD may be reconstructed using one of many published reconstruction techniques [21] [22] [23].

The simulation of crystallization systems requires that the solution to the above set of equations be advanced in time. Simple linear or Eulerian time stepping algorithms may be suitable for slow dynamic systems provided a small enough time-step size is utilised. This may lead to excessive computational times and resource use if numerical instability is to be avoided. In order to solve this problem a dynamic time stepping algorithm was developed in which the size of the time-step is adapted every time-step in order to slow down the simulation (reduce time-step size) and capture the system's behaviour during fast-evolving stages of the process. As the system slows down the algorithm increases the size of the time-step resulting in a reduction in the number of iterations required. To combat this issue an adaptive time-stepping algorithm was developed based on the Bogacki-Shampine [24] butcher tableau and was found to provide a good compromise between function evaluations and accuracy.

Decoupled Physics

The main outputs of interest are the average particle diameter and product recovery. These quantities are typical descriptors of many crystallization systems. As well as these standard outputs, the relative impacts of the underlying physics can be uncovered. What is meant by this is, for a given product particle size, how much of that size is due to growth? How much is due to aggregation? What impact does the feed stream (to either tank) have on the average size? These issues may be of importance if, for a given system, solvent entrainment within aggregates is a problem or only weak aggregates are formed which may break-up easily during further processing resulting in changing size distributions. The quantity of interest for each of the underlying phenomena is its contribution to $d[d_{10}]/dt$ at steady state. For example, growth and aggregation will always increase the average diameter, whereas nucleation will always decrease the average diameter. These phenomena occur continuously and are perfectly balanced at steady state.

The rate of change of the average particle diameter due to any one mechanism can be defined as

$$\frac{d}{dt}d_{10} = \frac{d}{dt}\left(\frac{m_1}{m_0}\right) \tag{1.8}$$

$$\frac{d}{dt}\left(\frac{m_1}{m_0}\right) = \frac{1}{m_0}\frac{dm_1}{dt} - \frac{m_1}{m_0^2}\frac{dm_0}{dt} \tag{1.9}$$

where m_0 and m_1 are the zeroth and first moments respectively.

We have $\frac{dm_0}{dt}$ and $\frac{dm_1}{dt}$ for each of nucleation, growth, aggregation and breakage from the population balance. i.e. the contribution of each phenomenon to the rate of change of the first two moments.

For nucleation

$$\frac{dm_k}{dt} = JL_{min}^k \tag{1.10}$$

$$\frac{d}{dt}d_{10} = \frac{JL_{min}}{m_0} - \frac{m_1 J}{m_0^2} \tag{1.11}$$

Where L_{min} is the nucleus size. This takes a value of 1mm as this is the smallest lower bin limit on the FBRM probe and was also used in the acquisition of the nucleation kinetic parameters in [2].

For growth

$$\frac{dm_k}{dt} = kGm_{k-1} \tag{1.12}$$

$$\frac{d}{dt}d_{10} = G \qquad (1.13)$$

The same principles apply to aggregation and breakage where their respective contributions to the rate of change of moments have already been calculated from the QMOM routine.

Results & Discussion

Presented here are the results of some of the numerical investigations conducted into the operation of the two-tank cascade. The typical outputs seen as the system approaches steady state from a fixed initial condition are presented. These cover the evolution of the moments of the particle size distribution in the system as well as concentration of each species. The effects of distributing the feed and anti-solvent streams between the two tanks for a fixed mass flow rate and volume ratio are examined.

Transient Simulations – Start Up & Path to Steady State

Depending on the chosen operating conditions the final steady state outputs of the system can be varied significantly. Once at steady state, the outputs from the system remain constant. However in the initial, or start-up, phase, the system's outputs can vary or oscillate quite dramatically, resulting in the possibility of product being "out of spec" for the next phase of the process.
For the proceeding simulations the operating conditions are given as:

Table 1. Operating Conditions used in Sample Simulations.

Solute Mass Fraction: 0.3
Solvent Mass Fraction: 0.3
AS Mass Fraction: 0.4
Temp. Tank 1 = 15°C, Temp. Tank 2 = 5°C

Both tanks are initially seeded with a seed loading from 0-1% with a mean of 5μm and standard deviation of 5%. This is done to avoid zero-valued moments in the initial phase. Below, are typical outputs from the simulations as the system goes from its initial condition, through the start-up phase to the final steady state operation. For the proceeding example the feed and anti-solvent splits are set to 1.0 which means that all streams are fed directly to tank 1.

Upon start-up, the system starts evolving almost immediately. It was observed that the gradient on the 0th moment is greatest with the lowest (absolute) gradient being that of the 3rd moment (corresponding to the solute concentration). It may also be noted that Tank 1 comes to steady state at approximately 2.5-3 hours. It was observed that Tank 2 reaches steady state after approximately 6 hours; roughly twice the transient time of tank 1.. This may be explained by the fact that Tank 1 is feeding directly to Tank 2 and until Tank 1 is at steady state, the boundary condition of Tank 2 will not be steady resulting in further evolution. With reference to the adaptive time stepping, it is important to select the fastest evolving parameter for use in the time step adjustment algorithm. For example if one were to choose the solute concentration (or m_3) then the time step size would be too large to maintain the other moments or outputs within the specified bounds.

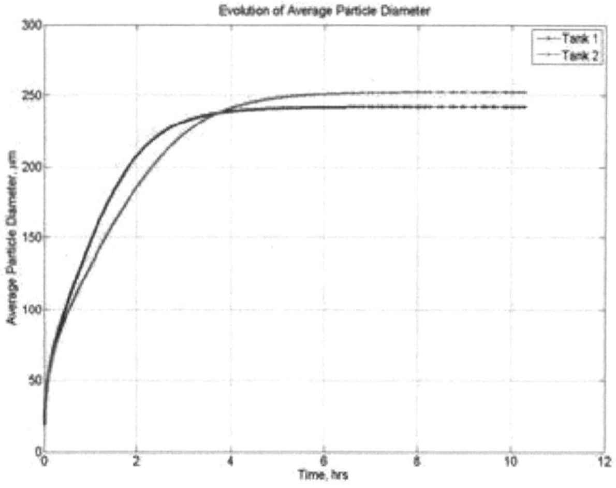

Figure 3. Evolution of the Average Particle Diameter.

The evolution of the average particle diameter in both tanks may be seen in Fig. 3. It can be seen that the average particle diameter does not vary much between tanks at steady state. This increase in particle size is coupled with an increase in the recovery of solute as is evidenced by the larger solids mass fraction seen in tank 2.

Outlined in this section is an analysis of the steady state operating outputs of this system achieved under different operating conditions. The input variables of interest here are the inlet mass flow rate, feed and anti-solvent splits. Investigated is an analysis of the effects of distributing the feed and anti-solvent streams between the two tanks.

Decoupled Physics

In designing a process, questions that may arise are : How do Process Conditions and Configuration effect Product? Can we control particle size? Can we determine the underlying driving mechanisms for particle dynamics? Can we determine the relative contributions of each of the physical phenomena? Can we improve the yield?

Figure 4 shows the impact of growth and nucleation on the average particle diameter in tank 1. (Note the negative axis in the right hand side plot). As would be expected, as the feed mass flow rate is increased, both the growth and nucleation contributions increase. This is due to the decreased residence time and increased supersaturation in each tank. It can be seen in that the trends for aggregation follow (roughly) those of the growth rate. This is due to the fact that an aggregation efficiency terms is used in Eqns. 1.7 and 1.8 which is a function of growth rate. It can be seen that as the mass flow rate is increased the flow tries to drive down the average diameter by introducing particles at a higher rate. At higher flow rates, the average particle size in tank 1 is reduced, resulting in an increase in the flow of small particles into tank 2. This results in an increase in the negative impact of the flow on the average particle diameter in tank 2. i.e. a larger number of smaller crystals are coming at a greater rate at higher flows.

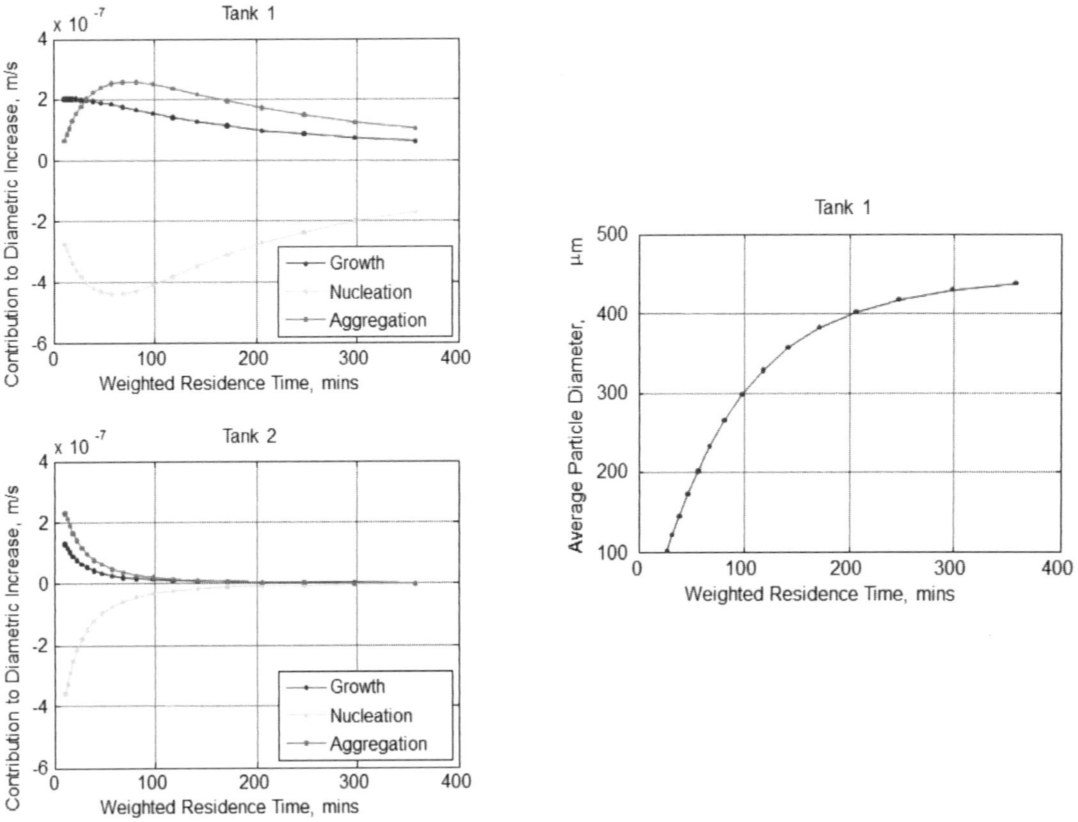

Figure 4. Effect of Feed Flow on Underlying Physics – Growth, Agglomeration and Nucleation.

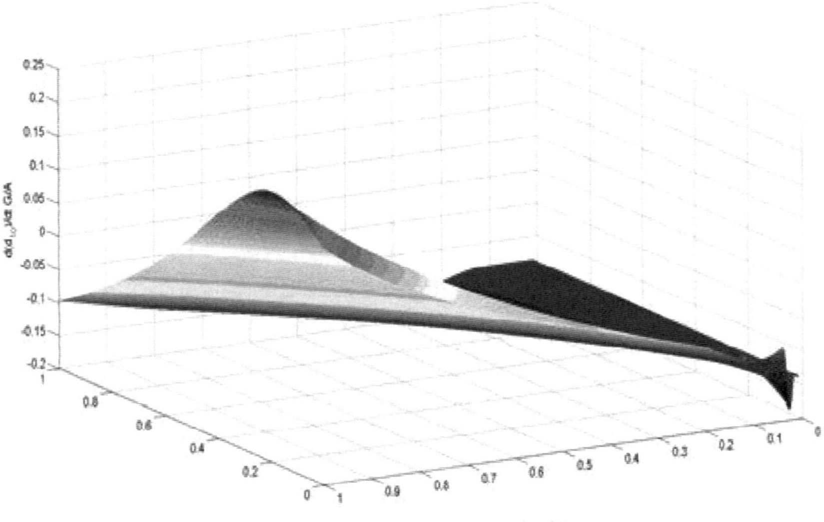

Figure 5. Effect on Underlying Physics - Growth / Aggregation.

Figure 5 illustrates the relative impacts of growth and aggregation on the particle diameter. For ratios less than 0, aggregation is the dominant particle size increase mechanism. For ratios greater than 0 the opposite is true. This may be of importance where weak aggregates are formed or where solvent entrainment is a concern.

Effect of Feed & Anti-Solvent Splitting

In this section the impact of distributing the fed and anti-solvent streams across the two tanks is investigated. The aim of this section is to illustrate that for a given physical cascade set-up there is still some selectivity of average diameters or recovery achieved through simple flow distribution.

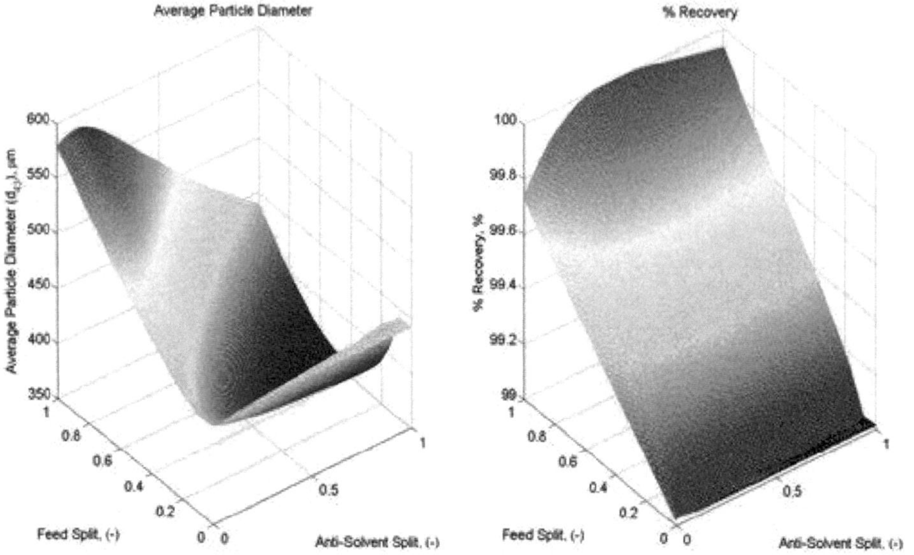

Figure 6. Effect of Feed and Anti-Solvent Splitting on Average Particle Size and Recovery Mass Flow = 0.0125 kg/s.

As can be seen in Figure 6, the impact of the feed and anti-solvent splitting is quite dramatic. There are several reasons that would lead to such a variation. The main reason is a variation in the residence times in each tank with variation of the splitting. For example, a case where both splits are equal to 1, all flow is sent into tank one. If a 50/50 split on each stream is used then the flow to tank 1 is halved resulting in a doubling of that residence time. At the same time the residence time in the second tank is not affected as all flow must still flow through tank 2 before exiting the system. As a result of this the supersaturations experienced in each tank will be quite different leading to large variations in output particle size. As was done previously, the underlying physics may be decoupled and will be done here for illustrative purposes.

One point of interest to note is that on each plot there exists both a maximum d_{43} and maximum recovery point on the surface which typically are not coincident. This leads to the need for the determination of the splits that afford the maximum attainable average particle size for each set of conditions as well as the corresponding recovery. Also required is the average particle size at maximum recovery and the conditions that give this output.

Maximum Recovery

To maximise recovery in all cases, the feed stream must be passed into tank 1 while, depending on the mass flow and volume ratio the anti-solvent stream must be split between the two tanks. As was seen previously, there exists a range of anti-solvent splits that will give maximum recovery (approx.. 30% in Tank 1, 70% in Tank 2 in this case). Shown here are the lower optimum anti-solvent splits. In reality, an optimum split volume may be more appropriate than a surface but the surface is shown here for clarity Fig. 7. It must be noted that for most operating conditions this range is very narrow meaning the "split volume" representation would approximate a surface across the operating space.

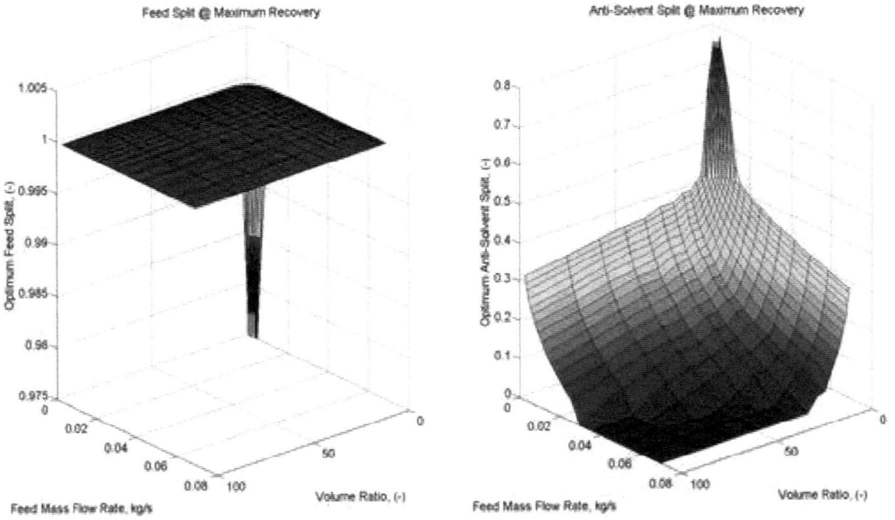

Figure 7. Optimum Feed and Anti-Solvent Splits for Maximum Recovery.

Conclusions

Presented in this text has been an overview of the computational model developed to investigate the behaviour of a two-tank cooling-anti-solvent crystallization of Paracetamol in methanol and water. The model has the capability of fully analysing the overall behaviour of the system as well as affording a near complete understanding into the underlying physics and dynamics of such a system. It must be noted that the results shown here are for a single feed composition with no recycle and at a fixed stage temperature. Further analysis of the effects of composition, stage temperature and recycle is possible without further development.

In conclusion, a powerful computational model has been developed which, if coupled with experimental campaigns and data will afford a very deep understanding of the dynamics of continuous crystallization. This depth of understanding can feed directly into the design of new processes or the optimisation of the operation of existing processes. It is now possible to determine the relative contributions of each of the physical phenomena thus giving the capacity to design/modify continuous systems to maximise/minimise any one or more of the mechanisms of crystallization.

References

1. Zhang, H., Quon, J.,A, Alvarez, A.J., Evans, J., Myerson, A.S., Trout, B. (2012), 'Development of Continuous Anti-Solvent/Cooling Crystallization Process using Cascaded Mixed Suspension, Mixed Product Removal Crystallizers', *Organic Process Reseach & Development,* 16, pp. 915-924.
2. Callahan, C.J., Ni, X.W. (2012), 'Probing into Nucleation Mechanisms of Cooling Crystallization of Sodium Chlorate in a Stirred Tank Crystallizer and an Oscillatory Baffled Crystallizer', *Crystal Growth & Design,* 12 (5), pp. 2525–2532.
3. Brown C.J. and Ni, X.W. (2011), 'Online evaluation of paracetamol antisolvent crystallisation growth rate with video imaging in an oscillatory baffled crystalliser', *Crystal Growth & Design,* 11, pp. 719-725.
4. Lawson, S., Steele, G., Shering, P., Laird, I., Zhao L. and Ni, X.W. (2009), 'Continuous crystallisation of pharmaceuticals using a continuous oscillatory baffled crystalliser', *Organic Process Research & Development,* 13, pp. 1357-1363.
5. Nagy, Z.K., Fevotte, G., Kramer, H. and Simon L.L. (2013), Recent advances in the monitoring, modelling and control of crystallization systems. Chemical Engineering Research and Design, 91(10), pp.1903-1922
6. Wong, S.Y., Cui, Y., Myerson, A.S. (2013), 'Contact Secondary Nucleation as a Means of Creating Seeds for Continuous Tubular Crystallizers', *Crystal Growth & Design,* 13, pp.2514-2521
7. Qamar, S., Elsner, M.P., Hussain, I., Seidel-Morgenstern, A. (2012), 'Seeding strategies and residence time characteristics of continuous preferential crystallization', *Chemical Engineering Science*, 71, pp. 5-17.
8. Ó'Ciardhá, C.T., Hutton, K., Mitchell, N., Frawley, P. (2012), 'Simultaneous parameter estimation and optimisation of a seeded anti-solvent crystallization', *Crystal Growth and Design,* 12 (11), pp. 5247–5261.
9. Ó'Ciardhá, C.T., Frawley, P.J., Mitchell, N.A. (2011), 'Estimation of the nucleation kinetics for the anti-solvent crystallisation of paracetamol in methanol/water solutions', *Journal of Crystal Growth,* 328 (1), pp. 50-57.
10. Randolph, A.D., Larson, M.A. (1998), *Theory of Particulate Processes.* Academic Press Inc., London, U.K.
11. Smoluchowski, M.Z., (1917), 'Versuch Einer Mathematischen Theorie Der Koagulationskinetik Kolloider Losunger', *Zeitschrift fur Physikalische Chemie*, 92, pp. 129-142.
12. Marchisio, D.L., Vigil, R.D., Fox, R.O., (2003), 'Quadrature method of moments for aggregation-breakage processes', *Journal of Colloid and Interface Science*, 258(2) pp. 322-334.
13. Vanni, M. (1999), 'Approximate Population Balance Equations for Aggregation-Breakage Processes', *Journal of Colloid and Interface Science,* 221, pp. 143-160.
14. Batterham, R.J., Hall, J.S., Barton, G. (1981), 'Pelletizing Kinetics and Simulation of Full Scale Balling Circuits', *3rd International Symposium on Agglomeration.* Nurberg, Germany, Paper No. A136.
15. Hounslow, M.J., Ryall, R.I, Marshall, V.R. (1988), 'A Discretised Population Balance for Nucleation, Growth and Aggregation', *AIChE Journal*, 34 pp. 1821-1832.

16. Kumar, S., Ramkrishna, D., (1996), 'On the solution of population balance equations by discretization—II. A moving pivot technique', *Chemical Engineering Science*, 51 pp. 1333-1342.
17. Sotowa, K.I., Kiyoshi, N., Kano, M., Hasebe, S., Hashimoto, I. (2000), 'Application of the method of characteristics to crystallizer simulation and comparison with finite difference for controller performance evaluation', *Journal of Process Control,* 10, pp. 203-208.
18. Fevotte, F., Fevotte, G. (2010), 'A method of characteristics for solving population balance equations (PBE) describing the adsorption of impurities during crystallization processes', *Chemical Engineering Science*, 65, pp. 3191-3198.
19. McGraw, R. (1997), 'Description of Aerosol Dynamics by the Quadrature Method of Moments', *Aerosol Science and Technology,* 27(2), pp. 255-265.
20. Press, W., Teukolski, S., Vetterling, W., Flannery, B. (1992), *Numerical Recipes in C: Second Edition*, Cambridge University Press.
21. Hutton, K., Mitchell, N., Frawley, P. (2012), 'Particle size distribution reconstruction: the moment surface method', *Powder Technology,* 222, pp. 8-14.
22. De Souza, L.G.M., Janiga, G., John, V., Thévenin, D. (2010), 'Reconstruction of a distribution from a finite number of moments with an adaptive spline-based algorithm', *Chemical Engineering Science* 65 (9), pp. 2741–2750
23. John, V., Angelov, I., Öncül, A.A., Thévenin, D. (2007), 'Techniques for the reconstruction of a distribution from a finite number of its moments', *Chemical Engineering Science,* 62 (11), pp. 2890–2904.
24. Bogacki, P., Shampine, L.F. (1989), 'A 3(2) pair of Runge - Kutta Formulas', *Applied Mathematics Letters*, 2(3), pp. 321-325.

Obtaining pure enantiomers and co-crystals using continuous preferential crystallization: A process development investigation using population balance models

T. Vetter[1,2], C.L. Burcham[1], M.F. Doherty[2]

[1]*Eli Lilly & Company, Indianapolis, IN-46281, United States*
[2]*Department of Chemical Engineering, University of California, Santa Barbara, CA-93106, United States*

Keywords: Continuous crystallization, separation of enantiomers, formation of co-crystals, wet milling, process design

Designing processes that yield enantiomerically pure products is of key importance in the pharmaceutical industry where typically only one of the enantiomers shows the desired effect on a patient, while the other enantiomer is either not effective as a drug or even toxic. Since asymmetric synthesis may not yield high enough enantiomeric purity, one has to resort to separation processes that separate enantiomers. For large biomolecules that are administered in liquid form chiral chromatography is the method of choice. However, small molecular drugs are sold as solids that typically originate from a crystallization process. For these products it is therefore interesting to investigate crystallization processes that can separate enantiomers and deliver a solid product in a single process step [1,2].

For conglomerate forming systems of enantiomers, i.e., systems in which crystals are either pure S enantiomer or pure R enantiomer (but no crystals of mixtures of the R and S enantiomers exist), preferential crystallization has been shown to be an attractive processing variant [3,4]. The two key concepts that enable preferential crystallization processes are the dependence of mass uptake of solute into the crystalline phase on the surface area of the crystals, i.e., the amount of solute incorporated per unit time depends linearly on the total surface area of the crystals present in the suspension, and the existence of a metastable zone, i.e., the nucleation of new crystals is an activated process and does not readily happen at low supersaturations. It therefore becomes possible to obtain a larger quantity of the desired enantiomer by seeding a crystallization process with crystals of this enantiomer. As a batch process, this process is best carried out in a polythermal fashion as described by Levilain and Coquerel [4]. Continuous processing concepts that allow preferential crystallization processes to be operated at steady state have also been developed [5-8]. We have recently designed a continuous flowsheet that is capable of separating enantiomers of a conglomerate in a wide range of processing conditions through the use of wet mills for continuous seed generation, recycle streams to obtain high yields and dissolver units to dissolve nuclei of the undesired enantiomer, which was shown to increase the range of processing conditions in which pure enantiomers can be obtained [9]. In that work process simulations were carried out using population balance equation models that allowed identifying key parameters of the process.

In this contribution, we add to these concepts by using this flowsheet to the far broader class of materials that forms racemic compounds. By operating in the three-phase region (solution, crystals of the racemic compound and crystals of the desired enantiomer), the process can guarantee the production of the desired enantiomer in pure from when it is performed with an enriched feed stream. The same concept and flowsheet can be applied to produce co-crystals in a wide region of their phase diagram, which is in contrast to batch processes that are driven to the thermodynamic equilibrium and therefore have to operate in the usually narrow two phase region (co-crystals and solution) of the phase diagram.

Acknowledgments: *This work was supported in part by a Lilly Innovation Fellowship Award to TV from Eli Lilly and Company.*

References:

[1] Jacques, J.; Collet, A.; Wilen, S.H., “Enantiomers, Racemates and Resolutions”, Wiley, NewYork, 1981.
[2] Schroer, J.W.; Wibowo, C.; Ng, K.M., *AIChE J.*, **2001**, *47*, 369-387.
[3] Coquerel, G., “Preferential Crystallization” in “Novel Optical Resolution Technologies”, Springer, Heidelberg, Germany, 2007.
[4] Levilain, G.; Coquerel, G., *CrystEngComm*, **2010**, *12, 1983–1992.*

[5] Levilain, G.; Eicke, M.J.; Seidel-Morgenstern, A., *Cryst. Growth Des.*, **2012**, 12, 5396-5401.
[6] Qamar, S.; Elsner, M.P.; Hussain, I.; Seidel-Morgenstern, A., *Chem. Eng. Sci.*, **2012**, 71, 5-17.
[7] Qamar, S.; Galan, K.; Elsner, M.P.; Seidel-Morgenstern, A., *Chem. Eng. Sci.*, **2013**, 98, 25-39.
[8] Chaaban, J.H.; Dam-Johansen, K.; Skovby, T.; Kiil, S., *Org. Process Res. Dev.*, **2013**, 17, 1010-1020.
[9] Vetter, T.; Burcham, C.L., Doherty, M.F., manuscript in preparation.

378798 Process Modeling, Simulation and Optimization for Continuous Pharmaceutical Manufacturing (CPM) of Ibuprofen

Abstract:

Pharmaceutical corporations face rapidly rising R&D as well as production costs. Batch production processes are dominant in the pharmaceutical industry and have multiple advantages, including equipment flexibility, efficient high-fidelity quality control and the ability to recall specific batches; however, they suffer known disadvantages such as limited heat transfer and mixing scalability, and low operational asset efficiency. Continuous Pharmaceutical Manufacturing (CPM) has a documented promise to suppress costs (Schaber et al., 2011) and foster profitability. Continuous production techniques (Behr et al., 2004) can be easier to scale up and can be designed to be often more efficient in terms of both solvent and energy use (Kockmann et al., 2008). Therefore it is both timely and important to explore the expanding feasibility and applicability limits of this emerging technology, which has been repeatedly demonstrated (Mascia et al., 2013).

The literature has been extensively surveyed in order to identify a series of candidate Active Pharmaceutical Ingredients (API) suitable for CPM production. Ibuprofen (2-(4-isobutylphenyl)propanoic acid), the widely used non-steroidal anti-inflammatory drug, has emerged as an ideal candidate, because it is in high global demand and can provide a market share of appropriate size. Flowsheet synthesis and process modeling are vital for rapidly evaluating R&D potential evaluation. The flowsheet we have considered is based on a published organic synthesis pathway (Bogdan et al., 2009) and produces Ibuprofen using three plug flow reactors (PFRs) in series, followed by a final purification separation. Kinetic and thermodynamic parameter estimation modelling has been employed in order to compute essential data for design and the PFRs have been designed based on reported conversions of feedstock and intermediate organic molecules. The theoretically computed PFR volumes are in good agreement with experimental prototypes constructed for the same organic synthesis reactions.

The development of a continuous final purification step after the third PFR is also essential. The current state of the art relies on a batch process comprising 15 distinct steps in order to obtain Ibuprofen crystals at 99% purity. To fully realise the benefits of CPM, a suitable continuous alternative is necessary and has been designed. Hexane is a promising solvent for separating Ibuprofen, and a high recovery has indeed been achieved. Plantwide economic optimization has thus been considered to explore and ensure the viability and profitability of the proposed design, which is most promising if installed in an existing production facility.

REFERENCES

Behr, A., Brehme, V. a., Ewers, C.L.J., Grön, H., Kimmel, T. et al., 2004. New developments in chemical engineering for the production of drug substances. *Eng. Life Sci.* 4, 15–24. doi:10.1002/elsc.200406127

Bogdan, A.R., Poe, S.L., Kubis, D.C., Broadwater, S.J., McQuade, D.T., 2009. The continuous-flow synthesis of Ibuprofen. *Angew. Chem. Int. Ed.*48(45), 8547–8550. doi:10.1002/anie.200903055

Kockmann, N., Gottsponer, M., Zimmermann, B., Roberge, D.M., 2008. Enabling continuous-flow chemistry in microstructured devices for pharmaceutical and fine-chemical production. *Chem. – Eur. J.* 14, 7470–7477. doi:10.1002/chem.200800707

Mascia, S., Heider, P.L., Zhang, H., Lakerveld, R., Benyahia, B., Barton, P.I., Braatz, R.D., Cooney, C.L., Evans, J.M.B., Jamison, T.F., Jensen, K.F., Myerson, A.S., Trout, B.L., 2013. End-to-End Continuous Manufacturing of Pharmaceuticals: Integrated Synthesis, Purification, and Final Dosage Formation. *Angew. Chem. Int. Ed.* 125(47), 12585–12589. doi:10.1002/ange.201305429

Schaber, S.D., Gerogiorgis, D.I., Ramachandran, R., Evans, J.M.B., Barton, P.I., Trout, B.L., 2011. Economic analysis of integrated continuous and batch pharmaceutical manufacturing: a case study. *Ind. Eng. Chem. Res.* 50, 10083–10092. doi:10.1021/ie2006752

382774 Continuous Purification of Active Pharmaceutical Ingredients Using Multistage Organic Solvent Nanofiltration Membrane Cascade

Abstract:

Generally in the production of pharmaceutical drugs, a large part of the production costs stem from the downstream processing. One important, time consuming and expensive downstream process in pharmaceutical manufacture is Active Pharmaceutical Ingredient (API) purification. To date the processes preferably used for API purification include distillation, adsorption–desorption, solvent extraction, fractional crystallisation and chromatography. For difficult separations where tiny amounts of impurities need to be removed, purification primarily relies on crystallisation or chromatography in their various forms. Crystallisation unit operations are complex processes while process chromatography is expensive and consumes large quantities of solvents, which require further downstream processing. In addition chromatography is poorly suited for large scale and continuous process operations. Thus developing new continuous downstream processing operations is an attractive area for research and development.

Organic solvent nanofiltration (OSN) is an emerging technology for performing membrane separation/purification processes in organic solvents and has been successfully applied for API purification, albeit mainly at laboratory scale. The widely used configuration of constant volume diafiltration usually requires high solvent usage, and, due to imperfect separation (solute rejection <100%) results in significant API losses. These problems could be overcome when using a membrane cascade. Organic solvent nanofiltration performed in a membrane cascade offers an interesting possibility for a continuous downstream processing unit operation.

This work demonstrates continuous purification of Active Pharmaceutical Ingredient (API) Roxithromycin from potential Genotoxic Impurity (GTI) 4-dimethylaminopyridine (DMAP) in a simple and efficient two stage membrane cascade. The presented membrane cascade operates using a single high pressure pump as the primary pressure source, and has no need for a buffer tank between membrane stages which simplifies its control. The cascade configuration is highly flexible in terms of membrane areas and flow rates and can be operated using multiple membrane cells or small membrane modules [1].

Cascade performance is initially evaluated via mathematical simulations, and then validated experimentally. We demonstrated that by careful selection of operating parameters, high purity of the API >99% could be achieved from feed stream purity of 78%. The continuous cascade could be easily coupled with an adsorption column, utilising an inexpensive non-selective adsorbent such as charcoal, as a solvent recovery stage. The batch and the continuous purifications in membrane cascade were evaluated in terms of the environmental impact – E factor and Mass Intensity (MI). With an E factor of ~20 and MI index of 21 the continuous membrane cascade process with a solvent recovery stage falls into the lowest range of the pharmaceutical industry segment [1]. The continuous process combined with a solvent recovery stage generates orders of magnitude less waste than a batch diafiltration which makes it an attractive alternative purification process for pharmaceutical industry.

References

1. L. Peeva, J. da Silva Burgal, I. Valtcheva, A. G. Livingston. Chem. Eng. Sci. (2014) *in Press*.

Drug Encapsulated Aerosolized Microspheres as a Localized, Degradable Cancer Therapy

J. Alaina Floyd[1], Anna Galperin[2], Buddy D. Ratner[1,2]

[1]University of Washington Dept. of Chemical Engineering, [2]UW Dept. of Bioengineering

Malignant tumors of the brain and spinal cord will be diagnosed in over 20,000 patients in the United States in 2014. The American Cancer Society estimates that over half of these will prove fatal. The most common and aggressive type is the glioblastoma multiforme. Ultimately, despite maximal therapies of tumor resection, chemotherapy, and radiation, median survival remains low at approximately one year. This bleak survival rate is in part due to the difficulty of a full tumor resection, the restriction of the blood brain barrier, and the development of tumor drug resistance. Herein, we present the novel formulation of a multi-drug, degradable delivery system that is applied post-resection to significantly increase patient life expectancy.

The system consists of three types of microspheres [poly(lactic acid) (PLA), poly(lactic-co-glycolic acid) (PLGA), and poly(ε-caprolactone) (PCL)] containing encapsulated chemotherapeutics and suspended in a degradable, aqueous poly(N-isopropylacrylamide) (PNIPAM) solution. At room temperature, the PNIPAM solution is a liquid with suspended drug encapsulated microspheres that can be "sprayed on" or injected around a tumor. The increased physiological temperature (37°C) causes the thermoresponsive PNIPAM to phase transition to a gel, entrapping the drug loaded microspheres in intimate contact with the tumor remnants. Over time, PLGA (rapid degradation), PLA (medium speed degradation), and then PCL (slow degradation) microspheres would break down, releasing their drug payloads in a sequential, multi-drug release directly to the tumor site, addressing cancerous regrowth and tumor drug resistance. This work focuses on microsphere formulation and characterization, therapeutic agent encapsulation, *in vitro* release studies, and an *in vivo* subcutaneous tumor pilot study.

Emulsion techniques were used to encapsulate the chemotherapeutic, gefitinib, in PLGA microspheres and the antibody, immunoglobulin G (IgG), in PLA microspheres, producing smooth, nonporous, spherical particles. The encapsulation efficiency was high for both drugs at ~90%. The *in vitro* release of gefitinib from PLGA occurred over 40 days, showing two distinct regimes of release. Through day 20, the drug released at a moderate, continuous rate, attributed to diffusion. After day 20, the release rate increased significantly, suggesting polymer degradation. When entrapped in the PNIPAM carrier, the PLGA-gefitinib release was delayed 4 to 6 days. The release of IgG from PLA had an initial burst release (~20%) followed by minimal release until day 40 and then a steady, continuous release over six months. This delay in release coincides with the end of the gefitinib-PLGA release, demonstrating a sequential, multi-drug delivery. The PLGA-PNIPAM system was also proven capable of an aerosolized delivery and subsequent adhesion on brain tissue preserved at 37°C. After a spray application to the tissue, the PNIPAM phase transitioned to a gel, efficiently entrapping microspheres encapsulating a fluorescent dye and conforming to the convoluted morphology of the brain, providing intimate contact with the tissue (Figure 1). In a pilot animal study, the gefitinib-PLGA-PNIPAM delivery system was tested against subcutaneous, C6 glioma tumors in athymic, nude mice. The drug delivery system was injected around the established tumor and tumor volume growth was measured. Through day seven post treatment, the tumor volume grew at approximately the same rate for both the PLGA-PNIPAM control and the gefitinib-PLGA-PNIPAM treatment. However, between day seven and nine, the drug took effect, reducing the tumor volume growth rate by ~33% until the end point at day 14. A full animal study will be conducted to fully elucidate the efficacy of this drug delivery system against a subcutaneous C6 glioma tumor along with a pilot study to test the extent of diffusion of gefitinib on the brain.

Our drug delivery system, capable of aerosolization or injection, has the potential to overcome tumor drug resistance and tumor regrowth by providing a localized, tunable, sequential drug release of multiple chemotherapeutics to increase patient life expectancy. However, this delivery system is not just limited to gliomas, but can be applicable to most cancer types that require tumor resection. In addition, by using different chemotherapeutic-microsphere combinations, a more personalized approach can be achieved based on tumor characteristics.

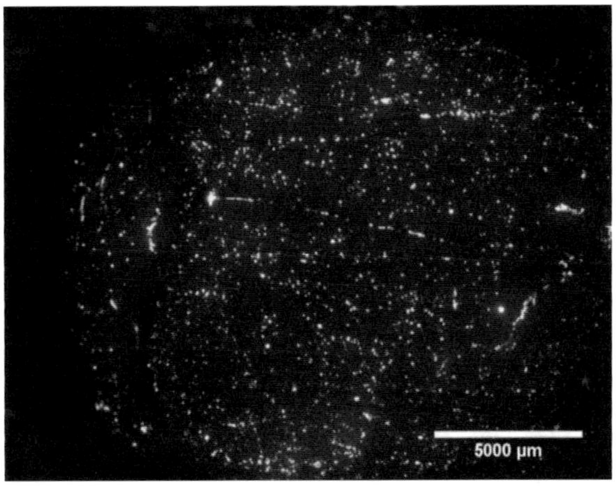

Figure 1: A fluorescent micrograph of an aerosolized delivery of rhodamine B encapsulated PLGA microspheres entrapped in PNIPAM at 37 °C.

Polymer strip films as a robust, surfactant-free platform for delivery of BCS Class II drug nanoparticles

S. Krull, R. Susarla, A. Afolabi, M. Li, E. Bilgili, R. N. Davé
Otto H. York Department of Chemical, Biochemical, and Pharmaceutical Engineering,
New Jersey Institute of Technology
161 Warren St. 150 Tiernan Hall
Newark, New Jersey 07102, USA

Pharmaceutical strip films have garnered significant attention thanks to improved patient compliance, cost-effective scale-up and the potential for continuous manufacturing. The ability to stabilize and release drug nanoparticles makes strip films an ideal platform for the delivery of poorly water-soluble drugs. However, this capability has only been demonstrated for a select few BCS Class II drugs and little is known about how the drug itself impacts film properties and performance. Strip films also offer the potential for a surfactant-free platform for the delivery of poorly water-soluble drugs. In this work, the ability of polymer strip films to physically stabilize and deliver nanoparticles of a variety of BCS Class II drugs with and without the aid of surfactant is investigated. Suspensions containing drug nanoparticles were prepared via wet stirred media milling with and without surfactant, mixed with polymer solution consisting of hydroxypropyl methylcellulose as film-forming agent and glycerin as plasticizer, cast and dried to form strip films loaded with drug nanoparticles. Despite the greater viscosity of mixed film precursor suspensions containing surfactant, this difference did not have a significant impact nanoparticle stability or film uniformity. Nanoparticles of all five model drugs were effectively stabilized throughout the film manufacture process with surfactant while only three of the five drugs were stabilized in the absence of surfactant. Good content uniformity was observed in films for all drugs with effectively stabilized nanoparticles. Films prepared with less water-soluble drugs exhibited greater tensile strength, resulting in slower dissolution relative to the other drugs under investigation. A similar but less prominent trend was observed in films prepared with surfactant which exhibited greater tensile strength and slower dissolution than those without. Raman spectroscopy confirmed that crystallinity was maintained for all drugs upon incorporation into films. Results suggest that polymer strip films can be a reliable and potentially surfactant-free means of delivering a variety of poorly water-soluble drugs.

385049 Enhanced Bioavailability of Poorly Soluble Pharmaceuticals By Means of Free Surface Electrospinning

Abstract:

It is estimated that 90% of active pharmaceutical ingredients (APIs) in research and development are insoluble or partially soluble in water. Due to poor solubility, these APIs exhibit poor bioavailability in solid dosage forms. To improve the release of APIs, we consider free surface electrospinning of microemulsions as a means of producing submicron size domains of API dispersed in an amorphous excipient. Microemulsions containing vitamin E, a poorly solubility API, and Polyvinylpyrrolidone, an excipient, are electrospun to produce a highly porous and high surface area material which promotes rapid drug release. The electrospun materials were characterized by scanning electron microscopy and high performance liquid chromatography to determine the morphology of the fibers and the bioavailability of the final material. As the fiber diameter of the electrospun material decreases, the dissolution rate of the API increases rapidly. In addition to improving the bioavailability of APIs, this technique may be utilized to streamline the downstream processing of pharmaceuticals, resulting in lower operating cost and improved uniformity over current batch manufacturing techniques.

387394 Improved Spray Drying Capability with Affinisol™ High Productivity Polymers

Abstract:

An increasing number of new drug entities suffer from low solubility making the development of a viable deliverable dose challenging. Hypromellose 2910 (HPMC) and hypromellose acetate succinate (HPMCAS) have proven effective at producing stable amorphous solid dispersions for poorly soluble active pharmaceutical ingredients (APIs) that demonstrate enhanced free drug levels in solution and inhibit recrystallization. Currently, there is a variety of HPMC 2910 viscosities available allowing a degree of control over solution viscosity and spray drying throughput; however, there is only a single viscosity of HPMCAS available. The limited viscosity ranges of these polymers impede the ability to maximize spray drying throughput when forming spray dried dispersions.

The present study examines the in vitro drug release rate and crystallization inhibition of amorphous spray dried dispersions formed with poorly soluble model APIs and AFFINISOL™ High Productivity polymers. Novel AFFINISOL™ HPMC High Productivity 2910 and AFFINISOL™ High Productivity HPMCAS are evaluated for physical properties including glass transition temperature and solution viscosity as a function of solids loading. In addition, performance of spray dried dispersions with model APIs will be assessed.

388040 Inhalation Characteristics of Polyanhydride Nanoparticle-Based Pulmonary Delivery Vehicles

Abstract:

Polyanhydrides are biodegradable materials which have been shown to provide sustained delivery of their encapsulated payloads. Copolymers based upon sebacic acid (SA), 1,6-bis-(*p*-carboxyphenoxy)hexane (CPH), and 1,8-bis-(*p*-carboxyphenoxy)-3,6-dioxaocatane (CPTEG) have previously been synthesized into nanovaccines and antibiotic delivery vehicles. There are many advantages in developing a dry powder inhalable nanovaccine including: self-administration, improved patient compliance, and elimination of the cold-chain.

In this work we evaluated the incorporation of nanoparticles formulated from two different candidate chemistries, 20:80 CPH:SA and 20:80 CPTEG:CPH, for utilization as a dry powder inhalable nanovaccine. The role of polyanhydride chemistry on nanoparticle primary size, hydrodynamic diameter, and in vitro aerosol deposition were evaluated. Additionally, the incorporation of these formulations into blister packs and dispersion characteristics from a customized dry powder inhaler were investigated. These studies represent the first steps in using polyanhydride nanovaccines and nanoparticle delivery vehicles in a dry powder inhalation system.

375128 Tablet Design and the Potential Impact on Oral Drug Delivery

Abstract:

The oral administration route is by far the most common way of administering pharmacological substances. In oral drug delivery the challenge is for the drug active to be released in a controlled way from the solid tablet into the blood circulatory system in order to maintain the therapeutic blood plasma concentration levels following a time schedule. Good computational models are needed to predict the drug release and transport into the blood plasma. This would help in reducing the expense, time and effort involved in drug design. Several models are available in the literature describing either the release of drug from the tablet into the gastrointestinal (GI) tract or the transport of drug through GI tract into the blood plasma. We have contributed to the development of a composite computational model for an oral drug delivery system (involving a drug active within a solid polymer matrix), which describes the release as well as the transport of drug from the tablet through GI tract into the blood plasma.

We employ a modified form of the drug release model proposed by Balaji and Peppas [1] based on the following assumptions. As the polymer matrix comes into contact with bodily fluid (modeled as water) inside the digestive system, it penetrates the polymer matrix; the latter then swells forming a gel layer. Drug molecules are released through the gel layer into the digestive system, crossing the enterocyte membrane into the blood stream. In our strategy, we envision two stages in the transport of drug into the blood. Therefore we couple the drug release model with a compartmental absorption and transit model (CAT), see for example Yu and Amidon [2]. The drug release model (Stage 1) is a moving boundary problem consisting of partial differential equations (PDEs) and ordinary differential equations (ODEs) which describe the diffusion of (i) bodily fluid into the drug tablet; (ii) drug out of the tablet; and (iii) polymer chains through a boundary layer. The model accounts for two moving interfaces (namely, tablet/gel interface and gel/bodily fluids interface in the digestive system).

In this presentation we will discuss a strategy for tablet customization, namely designing the dosage form of the tablet based on a desired release profile inspired by a desired plasma concentration profile (therapeutic effect). Our efforts include the systematic identification of tablet geometry and designs that could give specific desired drug release profiles such as a constant release profile or a pulsatile release profile. We next couple the tablet customization with the GI transport model discussed above. We expect the proposed approach to aid in the drug development process and contribute to the development of personalized medicine.

1. Narasimhan, B. and N.A. Peppas, Molecular analysis of drug delivery systems controlled by dissolution of the polymer carrier. Journal of Pharmaceutical Sciences, 1997. 86(3): p. 297-304.

2. Yu, L.X. and G.L. Amidon, A compartmental absorption and transit model for estimating oral drug absorption. International Journal of Pharmaceutics, 1999. 186(2): p. 119-125.

Overview

Twin screw granulation is a continuous wet granulation technique with the advantage of well-controlled, regime-separated growth. A model compound was used to demonstrate the impact of process parameters on granule PSD, flowability, and compactability, and the resulting tablet's strength and dissolution characteristics.

Methodology

After a preliminary ranging study, a comprehensive DOE was designed to deconvolute the relative effects of process parameters on granule responses. The 71-run DOE used factors of:

- GFL: 10-70%
- Screw Speed: 100-500 RPM
- Throughput: 5-100 g/min
- Screw Design: all conveying, 3x30 forwarding elements, 5x90 neutral mixing elements
- Barrel Temperature: 25-60C

The studied responses included:

- Granule exit temperature
- Machine torque
- Power input (W)
- Residence time distribution
- Unmilled granule PSD and bulk density
- Milled granule PSD and bulk/tapped density
- Compaction profile after lubrication
- Dissolution after tableting

Results

A multivariate analysis of the studied responses showed that process parameters had—*in general*—the following relative impact on responses:

GFL>Throughput>Screw Design>Screw Speed>Barrel Temperature

Surprisingly, very few interactions were observed, indicating that the process can be tuned with individual and independent process parameters.

GFL dominated the main responses of PSD and compactability with greater GFLs giving both larger and more compactable granules. That increasing GFL increases compactability is consistent with other HSWG/FBG development. However, the compactability for TSWG was significantly greater than the corresponding HSWG/FBG processes (~5 MPa @ 200MPa vs. 3.5 MPa).

Screw speed showed a linear relationship with power consumption, as expected. Overall power consumption was low compared to HME, with torque never exceeding 20% of machine capability.

Residence time data was fit to a model approximating two CSTRs and a PFR in series, yielding the lag time for the first material to breakthrough (t_{lag}), and the width of the subsequent distribution (τ). The resulting mean residence time was then successfully modeled according to the equation:

$t_{mean} = m/Q + b/SS$ where t_{mean} is mean residence time (equal to $t_{lag} + \tau$), Q is throughput, SS is screw speed, and m and b are experimentally fit constants.

Interestingly, barrel temperature had no significant impact on any response despite expectations that it would decrease liquid penetration time thereby potentially accelerating granulation. That temperature is not significant is a benefit to experimental design since temperature (barrel warming/cooling during long runs) does not need to be considered.

Dissolution was rapid (>85% at 30 min) and does not appear to be affected by process parameters, though further work is warranted.

Conclusions

The twin screw granulation process was demonstrated to be a promising alternative to high shear wet granulation and fluid bed granulation for a 15% drug load formulation. The fundamental impacts of TSWG process parameters on granule attributes were thoroughly characterized over exceptionally wide ranges. The impacts of parameters were rank-ordered to show their relative impacts. Finally, potential failure modes and areas for future development were identified.

Continuous High Shear Wet Granulation: Impact of Process and Design Parameters on Granule Properties

Wei Meng, Xue Liu and Fernando J. Muzzio
Department of Chemical and Biochemical Engineering, Rutgers University, Piscataway, NJ, 08854

Wet granulation (WG) is used broadly in the pharmaceutical industry. This technology has many advantages such as enhancing compression and powder handling, decreasing ingredient segregation and ensuring the content uniformity through size enlargement. A high level of interest exists at the present time in the continuous technology, both by the US FDA, and by large pharmaceutical manufacturers because continuous manufacturing methods can provide significant technical and business advantages relative to batch methods. As has been demonstrated in other process industries, continuous manufacturing methods are more robust and controllable. They achieve the same production rates as batch processes in much smaller and thus less capital-intensive equipment. However, despite these advantages, the pharmaceutical industry still relies almost exclusively on traditional batch methods to produce tablets and capsules. In recent years different continuous WG techniques have been investigated, such as fluid bed agglomeration and twin screw extrusion. However, integration of these unit operations into a complete continuous tablet manufacturing line is still a bottleneck due to high production cost and poor understanding of impact of granule properties.

In this work, a continuous high-shear "one-shaft" wet granulator is examined based on a placebo formulation comprising of 70% ∂-lactose monohydrate and 30% microcrystalline cellulose (Avicel PH101). The process includes two stages. As the powders are fed into the granulator, the blades mounted on the shaft will push the dry powders forward and engender both dispersive axial mixing and convective cross-sectional mixing. This is the mixing stage. Granulation stage occurs once the mixtures reach the nozzle location where a binding liquid (in this formulation water is used as the binding liquid) is fed via a pump. Then the granule growth and decay due to nucleation, coalescence, consolidation and attrition or breakage, can be observed in the process. Currently, two process variables and two design parameters, rotation speed, L/S ratio, blade configuration and nozzle position are selected for the I-optimal design. The collected granules are dried in a hot-air convection oven to a desirable LOD (~ 3%). Granule properties, such as particle size distribution, flow properties, density, porosity, compaction and hydrophobicity are tested and served as responses to generate a design space. We found that the rotation speed and L/S ratio have significant effects on the granule properties. Within the design space, the granules flow behavior is further investigated by measuring the RTD and material hold up. On the next step, an API (APAP) formulation will be examined and the impact of process and design parameters on the drug content uniformity will be characterized based on the VRR of the granulator and RSD of the granules. The influence of different binders and grades of lactose will be also considered in this work.

368641 Inter-Particle Coating Variability in a Continuous Coater

Abstract:

The influence of particle axial motion on inter-particle coating variability is studied in a rotating drum continuous coater. A mathematical framework based on renewal theory is developed and an expression for inter-particle coating variability is obtained that accounts for the variance in the residence time of particles inside the coater. This model makes no assumptions on the nature of the particle axial motion. Discrete element method simulations have shown, however, that the particle axial motion can be accurately modeled by a combination of advective and diffusive motion characterized by an axial Peclet number. Using this advective-diffusive model, it was found that in order to maintain an inter-particle coating variability of less than 1%, typical of what might be needed for functional pharmaceutical tablet coatings, a Peclet number of 20000 is required. Such a large Peclet number would necessitate essentially plug flow for typical continuous coater lengths of 1-2 m, or coater lengths of at least 15 m for typical feed rates and diffusion coefficients.

376905 Evaluation of Different Combinations of Continuous Mixing Operations Across a Range of Formulations

Abstract:

Feeders and mixers are the biggest equipment differences between batch and continuous pharmaceutical processing for the direct compression and roller compaction platforms. For pharmaceuticals, the loss-in-weight feeder design has emerged as the best design for continuously metering powder. However, for mixers there are more design options available. The requirements of the mixing step vary depending on the formula and on whether it is for the short and rapidly changing runs in development or the steady and stable runs in manufacturing. The first requirement is content uniformity. This measures how well an ingredient is spatially distributed. Variation with time is also an issue. The mixing elements must serve as a low pass filter to reduce intrinsic high frequency fluctuations in screw feeders. Therefore, the feeder variability must be known in order to specify the mixer requirements. A surprising amount of mixing also occurs in the tablet press feedframe which serves as an excellent low pass filter for the feeders and reduces the requirement to achieve that result in the mixer itself. With appropriate measurements and modeling, the effects of all these different mixing actions on the subsequent processes can be assessed.

Several negative phenomena can occur in a mixer as well. Examples include accumulation either in the mixer or due to the mixer, particle breakage, and/or distribution of lubricant at a higher level than desired ("over-lubrication"). While sufficient residence time distribution is required to filter the feeders, too much residence time distribution can reduce the system responsiveness. Most of these negative phenomena are very material dependent. The accumulation may be at the same concentration as the in-feed or it may selectively accumulate by species or by size. For the development of a new platform it is required to look at a wide-range of physical properties and formulations to assess the performance, both desired and undesired, of the mixing strategy.

Implementing a Robust PAT Solution and Closed Loop Control on a Continuous Manufacturing Plant for the Production of Pharmaceutical Tables at UPRM.

The US Food Drug Administration (FDA) is encouraging the pharmaceutical industry to innovate their traditional batch operation implementing the continuous manufacturing with in-line sensor to increase the understanding and control of the process. However, the continuous manufacturing in the pharmaceutical context presents several challenges in sensing technology due to the fact the inline sensors must provide efficient and effective acquiring information to control and understand the continuous processes. The FDA have proposed a Process Analytical Technology (PAT) initiative focused on the implementation of inline sensor with the capability of monitoring and controlling in real time any operational conditions to ensure final product quality such as: Active principal ingredient (API) concentration, homogeneity, humidity, stability, etc.

In this work the PAT data management product SynTQ (Synchronized Total Quality) from optimal was used to monitor the critical quality attributes of the continuous process. SynTQ was used in combination with the prediction engine SimcaQP and the multivariate analysis package Simca P+, both of them are from Umetrics. Several of the orchestrations within synTQ were designed and evaluated to achieve a robust PAT solution. The analysis of this tool was essential to create a solid control architecture. The controller used to evaluated the process in closed loop was the Distribute Control System form Emerson process Management DeltaV-S series. The study was focused on the evaluation of the performance of the continuous manufacturing plant for the production of tablets. The plant were carry out for the production of two different product. The Active Principal Ingredients of the tablets were Acetaminophen and Naproxen Nodium. They were manufactured via continuous direct compression method and continuous wet granulation method. The majority of the continuous units integrated on the continuous plant and that were used during the experiments were GMP compliant, they were supplied by recognized manufactures such as: Gericke and Schenck.

A statistical Design of Experiment (DoE) was applied to understand the effect and interaction of operational conditions and process parameters with the critical quality attributes. The DoE allowed us to establish a design space for the implementation of the quality by design concept. Later, some advanced control strategies were implemented to evaluate the performances of the continuous process in closed loop and they were compared with the standard technique of PID control. The aim of this presentation is to show an effective solution to implement a robust PAT and an efficient control strategy in the continuous manufacturing of pharmaceutical tablets.

373791 Model-Aided Development of a Continuous Drug Product Manufacturing Process

Abstract:

This work involves the use of mathematical tools to aid the development of an integrated Powder to Drug Product Continuous Manufacturing Process. Fundamentally driven dynamic models of the different units of the system are the basic frame of a holistic model capable of describing the overall system dynamics. Such a model is used to justify the appropriate usage of process analytical technology (PAT) tools to monitor the critical quality attributes (CQA's) and fundamentally increase the observability of the states of the system. State estimation along with data reconciliation techniques are used to fuse all available measurements and identify the best estimate of the de-noised signals of concentration of active pharmaceutical ingredient (API). Such reconciled values (along with other diagnostics) are then proposed as an integral component of the control strategy.

376470 The Use of CFD Modeling to Understand Variability in Particle Agglomeration in an API Crystallization

Abstract:

Powder property control for Active Pharmaceutical Ingredients (API) is important to ensure efficacy and to enable consistent manufacturing of the drug product. These properties can be controlled through crystallization design, targeting conditions that favor particle growth relative to nucleation and agglomeration. During API crystallization development, the process parameters critical to ensure particle growth are identified, addressing the challenges in scaling from the laboratory to the plant. Here, an API crystallization is discussed where no single mode – nucleation, growth, or agglomeration – is dominant. Initial process scale-up identified a mixing dependence resulting in variable agglomerate formation. Standard mixing correlations (tip speed, energy input) did not provide predictive agglomerate control.

These scale-up results prompted the development team to seek a more predictive mixing model to provide greater insight into agglomerate formation and control. MixIT, a software developed by Tridiagonal Solutions, was used to perform computational fluid dynamics (CFD) analysis of mixing conditions across the pilot scale reactors. For each reactor under consideration, a mesh was generated and the 3D, steady-state Navier-Stokes equations were iteratively solved. As an output of the analysis, important mixing parameters such as tip speed, average strain rate, and impeller power number were predicted. Further, distribution curves (e.g. volume percent vs. strain rate) provided a richer understanding of the strain rate experienced by particles in different pilot scale reactors. These curves provided insight into the different mixing conditions which may influence agglomerate formation. The mixing simulations were then applied to the proposed manufacturing scale crystallizer and process parameters were selected to match the desired strain rate distribution. The resulting batches demonstrated consistent powder property control in the larger scale API crystallizer.

381219 Large Bioreactor Simulation with Graphic Processors

Abstract:

So far, the engineering process of the reactor design was mostly driven by empirical knowledge. The effectiveness of the bioreactor and thus the biopharmaceutical manufacturing process could be enhanced by combining the state of the art empirical knowledge with a multiphase and multiscale simulation. Simulation results could then be used for the development and the design of new reactor generations. However, despite the recent improvement of computational capabilities, only a few seconds of real time operation of an industrial scale reactor takes months to simulate..

Thus, the goal of this study is to use graphic cards to speed up this simulation. The Compute Unified Device Architecture (CUDA) technology of nVidia has made the computational power of graphic processing units (GPUs) available for scientific calculations [1]. To take full advantage of the processor's parallelism, localized calculation algorithms are needed. Therefore the lattice Boltzmann method is used for modelling the fluid flow field, which is agitated by a Rushton turbine. The method is based on the lattice gas automata [2] and uses a regular grid with evenly distributed nodes. The static and moving boundaries are modelled with the modified bounce back algorithm.

The calculation of the bubble movement is done by solving the Newton's equation of motion. The sum of the forces acting on each bubble, i.e. the drag, the buoyancy, the lift force, the history force, the added mass effect and gravity, is used to determine the acceleration of the particle. The acceleration and the time step length give the velocity and the position change at the end of the time step. For coalescence and breakup stochastic models are used. The phases are coupled with a two way approach.

As the simulation results are produced on the graphic card, intermediate data can be visualized while the calculation is running. To simulate large reactors, the code has a multi GPU functionality, hence it can distribute the workload of the simulation on several graphic processors.

Validation is done with literature data as well as with results from holdup measurements of a 150l laboratory reactor.

References

[1] http://www.nvidia.com/cuda

[2] Sukop, M.: Lattice Boltzmann Modeling, Springer, Berlin, 2007.

383506 On the Use of Periodically Inverted Assay Tubes As a Screening Platform for Recombinant Escherichia coli Cultures

Abstract:

The screening of strains and culture conditions is a routine operation in biotechnology settings. Here we illustrate the use of commercial Vacutainer™ assay tubes agitated by successive periodic inversions as an alternative platform to conduct early development culture experiments of two recombinant strains of *Escherichia coli*. As a model system, we use two *E. coli* strains of biopharmaceutical relevance: (a) a C41 derived strain producer of a monoclonal antibody fragment (Fab) and a C41 strain producer of a dimer of the globular region of an influenza antigen. Tubes (effective volume of 3 mL) were gently agitated by periodic inversions in a commercial rotational mixing system (ATR Rotamix, USA). Experiments were conducted using three growth media formulations for *E. coli* with and without replacement of the air in the tube head space. We compared the culture performance of the periodically inverting assay tube mixing system (pITMS for short) with that of shake flasks (150 mL of effective volume) and fully instrumented bioreactors (1.3 L of effective volume) for the same strains and culture conditions. Results produced in the pITMS exhibit high reproducibility and low variance among different culture experiments. Even though the growth rate (μ in h^{-1}) was different in each type of reactor, the general growth behavior of the bacterial culture and the levels of protein expression in assay tubes was practically the same as in the shake flasks and in bioreactor, demonstrating the predictive potential of the pITMS.

Our results suggest that the pITMS is a practical alternative to conduct parallel screening experiments that can resemble the behavior of aerobic bacterial cultures performed in flasks and instrumented bioreactors.

387636 Characterization of Oxygen Availability Measured As Oxygen Transfer RATE and Dissolved Oxygen Tension in Shake Flasks for the Production of O-Mannosylated Proteins in Streptomyces Lividans

Abstract:

Introduction. Shake flask are the most frequently used reaction vessels in biotechnology (Buchs, 2001). In spite of their large practical importance, very little is known about the characteristic properties of shaken cultures from an engineering point of view. Oxygen transfer rate (OTR) is one of the most suitable measurable parameter to quantify the physiological state of an aerobic culture, since most metabolic activities depend on oxygen consumption (Buchs, 2001). The effect of three different shear and aeration conditions using conventional normal (NF), baffled (BF) and coiled (CF) shake flasks on *Streptomyces lividans* cultures on OTR was evaluated online. Also, the dissolved oxygen tension (DOT) was measured on the three flasks designs. Previous reports of our group indicates that different flask designs in *S. lividans* cultures leads to changes in *O*-mannosylation and production of a recombinant protein (Gamboa-Suasnavart et al., 2011), suggesting that oxygen transfer rates are also affected.

Methods. *S.lividans* cultures were performed at 30ºC, 150 rpm, 60 h in 34% sucrose Luria- Bertani medium (Gamboa-Suasnavart et al., 2011). *Corynobacterium glutamicum* cultures in same conditions were carried as a higher respiration activity microorganism for comparison. OTR was measured using the RAMOS (Anderlei and Buchs, 2004). This device analyzes the depletion of oxygen concentration in the gas headspace of the shake flask with an oxygen sensor. The slope of this depletion corresponds to the OTR (Anderlei and Buchs, 2004). Carbon transfer rate (CTR) was calculated as the carbon dioxide formation, measured with a sensor in the flask headspace. Three coiled, three baffled and two conventional flasks were used each time. RQ was calculated as the ratio between CTR/OTR. DOT was measured by a luminescent patch (Pst3) using the system Fibox3 by PreSens.

Results and Discussion. Cultures carried out in CF and BF reached a maximum OTR of 9.16 ± 0.15 and 9.36 ± 0.28 mmol/lh respectively after 40 hrs of culture. In contrast, a significantly lower OTR was reached (0.66 ± 0.48 mmol/lh) in those cultures in NF. *C. glutamicum* cultures presented the same trend, achieving in CF, BF, and NF, 10 ± 0.22, 9 ± 0.31 and 2 ± 0.17 mmol/lh, respectively. RQ values was closer to one in BF and CF, indicating an efficient aerobic metabolism. But in NF, RQ values fluctuates between one and two, suggesting an inefficient carbon oxidation and probably used in other metabolism pathways.

DOT in BF and CF decrease up to 50% at the end of the culture but in NF a total depletion of oxygen was observed. This might be due to the poor OTR observed in NF cultures compared to BF and CF. On the other hand, the pellet sizes of *S. lividans* in NF are seven times (1.57 ± 0.41) higher than those in BF and CF (0.23 ± 0.06 and 0.16 ± 0.05) (Gamboa-S et al. 2011). By using the equation reported by Cui et al. (1997), an oxygen transfer limitation can be assumed by the higher dissolved oxygen tension needed to diffuse to the center of the pellet (170% DOT). In the case of the pellets from BF and CF, just 1.76% of DOT is required, demonstrating that in those geometries no oxygen limitations can occur.

Conclusions. Flask design modifies the OTR in microorganism cultures with low and high respiration activity. Higher OTR values are reached when baffles or a spring were used, due to the increase in mass transfer. OTR changes lead to a differences in metabolism as is shown in RQ values. DOT in conventional flask is insufficient to achieve a properly diffusion into pellets in almost whole culture. These OTR differences could lead to changes in recombinant glycoprotein production in *S. lividans* cultures. The knowledge about the oxygen transfer aspects inside flasks can provide new tools for the development and scale-up of biotechnology processes, mainly in those producing bacterial recombinant glycoproteins.

Acknowledgements. CONACYT-INNOVAPYME 181895, CONACYT 178528, 104951-Z y PAPPIT-UNAM IN-209113, IN-210013

References.

1. Büchs J, **Introduction to advantages and problems of shaken cultures**, *Biochem Eng J*, 2001, **7**:91-98.

2. Anderlei T, Büchs J, **Device for sterile online measurement of the oxygen transfer rate in shaking flasks**, *Biochem Eng J*, 2001, **2**:157-162.

3. Gamboa-Suasnavart RA, Valdez-Cruz NA, Córdova-Dávalos LE, Martínez-Sotelo JA, Servín-González L, Espitia C, Trujillo-Roldán MA, **The *O*-mannosylation and production of recombinant APA (45/47 KDa) protein from *Mycobacterium tuberculosis* in *Streptomyces lividans* is affected by culture conditions in shake flask**, *Microbial Cell Factories*, 2011, 10:110

4. Cui YQ, van der Lans RG, Luyben KC, **Effects of dissolved oxygen tension and mechanical forces on fungal morphology in submerged fermentation**. *Biotechnol Bioeng,* 1997, **57**:409-19.

377912 Preparation and Preliminary Antitumor Activity Evaluation of a New Targeted Nanoparticle System

Abstract:

PREPARATION AND PRELIMINARY ANTITUMOR ACTIVITY EVALUATION OF A NEW TARGETED NANOPARTICLE SYSTEM

, Eva M. Martin Del Valle*, José V. Román, Jesús A. Rodríguez-Rodríguez, Miguel A. Galán

Department of Chemical Engineering, University of Salamanca, Spain

*Corresponding author: emvalle@usal.es

Cancer is one of the main causes of human death. Most of usual antitumor treatments use chemotherapy agents and it involves damages in other organs because of their side effects.

In this respect, the aim of this work has been the design of a new targeted drug delivery nanoparticles system and its support with in vitro and in vivo tests for human lung cancer therapy.

The targeted nanoparticles were produced by electrostatic interactions under acid aqueous conditions. This nanoparticle was used to fix the antitumor drug and identification agent of tumoral cells on its surface.

The in vitro trials' validation has been done by MTT assay in order to check the viability of tumor cells. Besides the in vivo effectiveness is controlled by animal experiments with mice. Mice tumors have been controlled though daily checks and subsequently analyzed by testing pathological anatomy.

Cell experiments have been developed with the H460 line cell and the effect of targeted drug nanoparticles displays a better behavior compared to the current chemotherapy.

Animal experiments present encouraging results for mice treated with nanoparticles and a longer life in these diseased animals compared to non-treated animals and even to animals treated with a conventional chemotherapy.

Therefore, the current study describes how manufacture an advanced targeted drug delivery nanoparticle system and its evaluation in vitro and in vivo experiments.

Osteotropic therapy via Layer-by-Layer Nanoparticles

Stephen W. Morton, Nisarg J. Shah, Mohiddin A. Quadir, Zhou J. Deng, Zhiyong Poon, Paula T. Hammond

Koch Institute for Integrative Cancer Research, Massachusetts Institute of Technology, Cambridge, MA 02139

Current treatment options for debilitating bone diseases such as osteosarcoma, osteoporosis, and bone metastatic cancer are suboptimal and have low efficacy. New treatment options for these pathologies require targeted therapy that maximizes exposure to the diseased tissue and minimizes off-target side effects. This work investigates an approach for generating functional and targeted drug carriers specifically for treating primary osteosarcoma, a disease in which recurrence is common and the cure rate has remained around 20%. This approach utilizes the modularity of Layer-by-Layer (LbL) assembly to generate tissue-specific drug carriers for systemic administration. This is accomplished via surface modification of drug-loaded nanoparticles with an aqueous polyelectrolyte, poly(acrylic acid) (PAA), side-chain functionalized with alendronate, a potent clinically used bisphosphonate. This work represents a tunable approach towards the synthesis of drug carriers, in which LbL enables surface modification of nanoparticles for tissue-specific targeting and treatment.

References: S.W. Morton et al. *Advanced Healthcare Materials*, early view: published online Oct. 9, 2013; doi: 10.1002/adhm.201300465.

387922 Biodegradable Nanoparticle-Based Delivery System for Enhanced Blood-Brain Barrier Transport Utilizing Macrophage Carriage

Abstract:

A major barrier in the treatment of many neurological or neurodegenerative diseases, such as Parkinson's disease and Alzheimer's disease, is the inability of therapeutics to cross the blood-brain barrier (BBB). The prevalence of these diseases is increasing with the median age of our population. Among several strategies to improve therapeutic transport across the BBB is the use of a carriage system based on macrophages. Using this method, improved drug delivery into the central nervous system can occur through ingress of drug-loaded macrophages across the BBB, or via macrophages and brain endothelial cells cell-to-cell contact and drug transfer. Polyanhydride nanoparticle-based delivery systems have been designed to elicit unique cellular responses such as particle internalization and activation of antigen presenting cells such as dendritic cells and macrophages. Through encapsulation of the payload within the polyanhydride particles, sustained release of these compounds can be achieved. Altering the polymer chemistry, functionalizing the particle surface, and/or controlling particle size enables the release of payload to targeted regions within the body.

In this work we investigate the ability of various formulations of biodegradable polyanhydride nanoparticles to enhance the delivery of therapeutic compounds across the BBB. These formulations showed limited toxicity to primary human monocytes, macrophages and brain endothelial cells. Cellular uptake of particles within monocytes and macrophages and the subsequent transport of internalized particles (loaded with a fluorescent imaging agent) to brain endothelial cells were evaluated by confocal microscopy. Quantification of transport of the various nanoparticle formulations was accomplished using flow cytometry. The results suggest that polyanhydride chemistry and surface functionalization of the nanoparticles can be used to enhance the transport and uptake of these particles by primary human brain endothelial cells when macrophages are used for carriage.

392224 Stabilities of exosomes derived from mammalian cells: nanoparticle morphology and functional RNAs

Abstract:

Currently, it has been widely accepted that most mammalian cells release nanoscaled membrane vesicles called exosomes, transferring functional molecules of proteins and RNAs derived from the parent cells. The specific function of exosomes in regulating immunity and promoting oncogenes remains to be investigated. This study assessed the stabilities of exosomes and the functional RNA molecules enclosed. Scanning electron microscope (SEM) was used to inspect the morphology of exosomes, assessing the changes in exosomes due to different production and storage conditions. Optimal SEM sample processing method was investigated for fast and accurate analysis of intact exosome particles using SEM. As a comparison to exosomes, apoptotic vesicles were also inspected and analyzed using SEM. The quality and stability of RNAs contained within the exosomes was characterized using Agilent Bioanalyzer, showing the changes of RNA molecules enclosed in exosomes and stored for days, months, and up to 2 years under -80 degree C. Exosomal RNAs isolated from exosomes that were produced from serum-containing or serum-free conditions, respectively, and at medium incubation time 12 to 48 hours, were analyzed and showed stable and characteristic spectrum for various exosomal nucleotides. Collectively the results suggested improved methods for processing and storage of exosomes that preserve morphological and biochemical characteristics. In order to reproducibly investigate exosome functions, it is essential to maintain a near native state of exosomes and preserve their functional molecules during sample handling, analyzing and storing.

390054 Engineering an Effective Immune Adjuvant By Designed Control of Shape and Crystallinity of Aluminum Oxyhydroxide Nanoparticles

Abstract:

Adjuvants based on aluminum salts (Alum) are commonly used in vaccines to boost the immune response against infectious agents. However, the detailed mechanism of how Alum enhances adaptive immunity and exerts its adjuvant immune effect remains unclear. Other than being comprised of micron-sized aggregates that include nanoscale particulates, Alum lacks specific physicochemical properties to explain activation of the innate immune system, including the mechanism by which aluminum-based adjuvants engage the NLRP3 inflammasome and IL-1β production. This is putatively one of the major mechanisms required for an adjuvant effect. Because we know that long aspect ratio nanomaterials trigger the NLRP3 inflammasome, we synthesized a library of aluminum oxyhydroxide (AlOOH) nanorods to determine whether control of the material shape and crystalline properties could be used to quantitatively assess NLRP3 inflammasome activation and linkage of the cellular response to the material's adjuvant activities *in vivo*. Using comparison to commercial Alum, we demonstrate that the crystallinity and surface hydroxyl group display of AlOOH nanoparticles quantitatively impact the activation of the NLRP3 inflammasome in human THP-1 myeloid cells or murine bone marrow-derived dendritic cells (BMDCs). Moreover, these *in vitro* effects were correlated with the immunopotentiation capabilities of the AlOOH nanorods in a murine OVA immunization model. These results demonstrate that shape, crystallinity and hydroxyl content play an important role in NLRP3 inflammasome activation and are therefore useful for quantitative boosting of antigen-specific immune responses. These results show that the engineered design of aluminum-based adjuvants in combination with dendritic cell property-activity analysis can be used to design more potent aluminum-based adjuvants.

388607 Novel Chemoradiotherapeutic Magnetic Nanoparticles for Targeted Treatment of Non-Small Cell Lung Cancer

Abstract:

Lung cancer is the leading cause of cancer-related death in the United States and 85% of all lung cancers are classified as non-small cell (NSCLC). Unfortunately, NSCLC is extremely difficult to treat and the survival rates are low; in 2012, the five year survival rate was only 17%. An innovative approach that may ultimately allow for the clinician to target and aggressively reduce tumor burden in patients with NSCLC will be presented. Platinum (Pt)-based radiosensitizers; cisplatin, carboplatin, oxaliplatin and holmium-166 radionuclide (166Ho; $E_{\beta\text{-max}}$ = 1.84 MeV; $t^{1/2}$= 26.8 h) have been incorporated in a garnet magnetic nanoparticle (Pt-HoIG) for selective delivery to tumors using an external magnet. The Pt anticancer drugs can be used as radiosensitizers, and the ^{166}Ho emits beta particles and gamma photons, which can be used for therapy and imaging, respectively. A major obstacle in therapy is the non-specificity of current treatments. Nonspecific anticancer drugs administered via intravenous injection are distributed throughout the entire body and are harmful to healthy tissues as well as tumor sites, leading to significant side effects. The nanoparticles reported here could potentially deliver their chemoradiotherapeutic 'cargo' directly to tumors in a patient, guided by a magnet. The HoIG and Pt-HoIG nanoparticles were characterized using PXRD, SEM, EDX, TEM, ICP-MS, and Zeta potential measurements. Neutron activated cisplatin-HoIG and oxaliplatin-HoIG showed greater toxicity than HoIG, free cisplatin and oxaliplatin toward NSCLC A549 cells. Thus, cisplatin and oxaliplatin act synergistically with the radioactive Ho making nontoxic doses of radiation extremely effective.

383300 Uptake, Co-Localization, Cytotoxicity and Transepithelial Transport of Acid-Labile Doxorubicin-Dendrimer Conjugates in an in Vitro Model of the Lung Adenocarcinoma Epithelium

Abstract:

Lung cancer is the leading cause of cancer death among both men and women in the United States. Adenocarcinoma accounts for more than 40% of all lung cancer cases. Doxorubicin (DOX) is an FDA-approved chemotherapeutic that has been widely used as primary anticancer drug in the treatment of a variety of cancers. However, rapid elimination, uncontrolled release, systemic distribution, and life-threatening cardiotoxicity, has hindered the applicability of DOX and other potent anti-cancer agents. The development of polymeric carriers for the modulation of transport, targeting and controlled release of potent anti-cancer agents and their formulation for local lung delivery is of great relevance in the treatment of lung adenocarcinoma.

In this work, a series of PEGylated poly(amidoamine) dendrimer nanocarriers with acid-labile and acid-nonlabile DOX conjugates were synthesized and characterized. We employed a two-step PEGylation strategy to increase the payload of the hydrophobic DOX. We investigated the impact of pH (neutral and acidic pH), PEGylation density (low, medium and high) and number of DOX conjugates (low and medium) on the release of DOX from the dendrimer nanocarrier, the kinetics of carrier uptake, intracellular release kinetics of DOX from the nanocarrier, toxicity in an alveolar adenocarcinoma cell line (A549) and effect of PEGylation degree on transport of dendrimer nanocarriers across lung epithelial cell line (calu-3). PEGylation retards the release of DOX in acidic medium, and also intracellularly, as determined by co-localization studies with confocal microscopy. We also observed that the kinetics of cellular entry of the nanocarrier with DOX increased significantly compared to free DOX, as determined by flow cytometry. At the highest PEGylation density, the rate of internalization of the nanocarrier containing DOX was even higher than that of free DOX. PEGylation density also affects cytotoxicity as seen by an increase in IC_{50} for DOX-conjugated dendrimer compared to free DOX, due to the controlled release of the therapeutic. Mucus layer secreted by airway epithelial cells (calu-3) retained most of dendrimer carriers on apical side. However, high PEGylation degree significantly helped dendrimer nancarriers escape mucus trapping compared to bare dendrimer, aiding their transport to the basolateral side. Reduced transepithelial electrical resistance (TEER) and limited cellular internalization demonstrated paracellular transport plays a vital role in transporting dendrimer nanocarriers across airway epithelium. Dendrimer-DOX conjugates were formulated in the portable metered-dose inhalers using a new particle engineering strategy, and their aerosol characteristics were observed to be conducive to deep lung deposition.

Key words: Polyamidoamine dendrimer (PAMAM), doxorubicin, lung cancer, pressurized metered-dose inhalers (pMDI), controlled release.

389575 Crystallinity Detection Methods for Polymer-Stabilized Solid Dispersion Formulations

Abstract:

Process technologies such as hot melt extrusion and spray drying can significantly enhance the bioavailability of low solubility drugs through creation of polymer-stabilized amorphous solid solutions. On the other hand, the inherent thermodynamic instability of the saturated amorphous state must be managed over the shelf-life of the product. Formulations, processes, and packaging configurations are typically designed to avoid conditions leading to recrystallization, but an accurate evaluation of these conditions depends on the availability of analytical techniques of sufficient sensitivity. Traditional approaches to detection of crystallinity in amorphous materials include XRPD and ssNMR, which typically have sensitivity ~1% crystallinity by weight in the drug product. After dilution with excipients, this may translate to far lower sensitivities for detection in the drug product on an API basis. This is perceived to be an area of concern for demonstration of adequate process control to Health Authorities.

We describe methods to address these regulatory concerns by characterization of products in development. Examples of quantitative limit methods for crystallinity detection via XRPD and FT-Raman are provided to show how sensitivity can be improved over typical levels by an order of magnitude in some cases. The enhanced sensitivity methods have a line of sight to potential use beyond development as validated release tests at commercial facilities, and compliment novel research techniques. Case studies illustrate contrasting characterization strategies for active-polymer mixtures with low and high crystallization tendency.

378935 In Vitro and in Vivo Validation of a New Drug Delivery System for Human LUNG Cancer

Abstract:

IN VITRO AND IN VIVO VALIDATION OF A NEW DRUG DELIVERY SYSTEM FOR HUMAN LUNG CANCER

Eva M. Martín del Valle*, José V. Román, Miguel A. Galán

Department of Chemical Engineering, University of Salamanca, SPAIN

*Corresponding author: emvalle@usal.es

Human lung cancer is the main cancer cause of death in men and the third cancer cause in women. As it is well-known, current antitumor treatments involve chemotherapy which in most cases entails aggressive side effects.

In this respect, in this work we will present a new targeted drug delivery system, based on double functionalized microcapsules and its validation in vitro an in animal models for human lung cancer treatment. These microcapsules, produced using a fan jet nozzle, could be used in order to avoid the side effects of chemotherapy. It was shown that these microcapsules, used in situ tumor place, are biodegradable and biocompatible.

In order to validate the formulation, the in vitro effectiveness is tested by the MTT assay which measures the decrease of lived cells. The in vivo effectiveness is controlled by animal models with mice. Their tumor size is checked every day by palpation and possible damages in internal organs are analyzed by autopsy after the death of the animals. Autopsies and pathological anatomy tests were not shown metastasis in mice.

Cell experiments are carried out with the H460 line cell and the activity of bioconjugated antitumoral generally shows a faster effect of cell mortality and better behavior compared to the free administration solution of the drug.

The animal experiments present a good survival for mice with microcapsules therapy and an important advantage compared to non treated animals and even to animals treated with a free administration solution of the drug.

As a result, this study reports the preparation and in vitro/in vivo validation of an advanced drug delivery system, which in the future will enable clinical trials in humans.

369696 Screening Microneedle Formulations for Influenza Vaccine Stabilization

Abstract:

Influenza is a potentially life-threatening infectious disease that causes up to 5 million cases of severe illness annually during seasonal epidemics in the US alone. Microneedle (MN) patches are designed to deliver vaccines, such as influenza vaccine, to the skin where there is an abundance of antigen-presenting cells, thus eliciting similar or improved immune response when compared to conventional intramuscular injection.

Dissolving polymer MN are projections that measure hundreds of microns in length, which are composed of water-soluble polymer with a vaccine payload embedded along with formulation excipients such as stabilizers and dissolution enhancers. These MN dissolve within minutes upon insertion into skin, thereby releasing their vaccine payload. However, influenza vaccine can be damaged due to the drying process during manufacturing and instability during storage. The excipient formulation is critical to maintaining vaccine antigenicity during the drying and storage of MN patches. This study addresses stabilization of influenza vaccine in microneedle patches during storage at elevated temperature, which will facilitate mass distribution for annual and pandemic influenza vaccination campaigns.

In this study, we have developed a high-throughput screening method to optimize microneedle formulations for influenza vaccine stability. Producing MN arrays is time consuming, thus requiring a faster method for screening formulations. We have addressed this bottleneck by using flat, 6 mm-wide, polydimethylsiloxane (PDMS) "chips" as a surrogate for the molds used to form MN. Test formulations are dried and stored on the surface of these chips and then tested for vaccine activity. Our method for assessing vaccine activity is an ELISA, used to quantify the amount of antigenic protein, hemagglutinin, still in the native conformation.

A large number of compounds were screened against their ability to retain influenza vaccine activity. Through this study, several promising formulations were found to maintain vaccine activity for at least 3 months at 40°C. These were comprised of suitable buffers and stabilizing compounds including sugars, amino acids, and long-chain carbohydrates. The best formulations thus far utilized a combination of stabilizers where the combination performed better than the constituent single stabilizers. Long-term stability experiments in MN patches are underway and will be reported in this presentation. In addition, experiments to elucidate the mechanism of activity loss and retention are being conducted.

372754 Tablet Press Weight Control Model

Abstract:

Tablet weight control is a key requirement when producing tablets for pharmaceutical products. Currently the control systems used to control tablet weights by a particular manufacturer can potentially lead to significant variability in the tablet weights. There has not been much attention given to the details of the equipment control strategies used because the tablet production equipment is bought as a packaged item from vendors. This presentation discusses the detailed control system analysis used to control tablet weights and discusses a model that was built to reflect this new understanding. In particular, the analysis and modeling has shown that the Compression Force and Weight Control Loops have opportunities for improvement and that there is an advantage to having a unique control strategy for each product. Results from this modeling effort have been used to identify changes that have been implemented to reduced tablet weight variation.

388949 Quality-By-Design Evaluation of an Immediate Release Tablet Platform

Abstract:

Excipient variability, and its impact on drug product performance, has been previously examined with risk-based, experiment-based, and data-based approaches. However, there has been very little investigation into understanding how much variability in excipients impacts drug product performance relative to variability in API properties and processing parameters or method. Therefore, the present study provides an example assessment of the relative impact of variability in API, excipients, and the manufacturing process on final performance of two, model, immediate release, drug products to aid those that pursue Quality-by-Design drug product development. Ibuprofen and Theophylline served as the model APIs, microcrystalline cellulose (MCC) served as the ductile diluent, lactose (spray dried) served as a brittle diluent, croscarmellose sodium served as the disintegrant, and magnesium stearate served as the lubricant. The impact of filler:lubricant particle size ratio variation (3.4 – 41.6) on the attributes of an immediate release tablet were compared to the impacts of the manufacturing method used (direct compression or dry granulation) and drug loading (1%, 5%, and 25%), particle size (D[4,3]: 8–114 microns), and drug type (Theophylline or Ibuprofen). These factors were evaluated using a 2^{5-1} fractional factorial statistical design. 5-kg batches of the various formulations were manufactured using suitably-scaled equipment. All batches were successfully manufactured, except for the direct compression of 25% drug loading of 8 micron (D[4,3]) drug, which exhibited very poor flow properties. All manufactured tablets possessed adequate quality attributes: tablet weight uniformity < 4% RSD, tablet potency: 94-105%, content uniformity < 6% RSD, acceptance value < 15, solid fraction: 0.82-0.86, tensile strength > 1 MPa, friability < 0.2% weight loss, and disintegration time < 4 minutes. The filler:lubricant particle size ratio exhibited the greatest impact on blend and granulation particle size and granulation flow, while drug property variation dominated blend flow, ribbon solid fraction and tablet quality attributes. Although statistically significant effects were observed, the results of this study suggest that the manufacturability and performance of this immediate release tablet formulation is robust to a broad range of variation in drug properties, both within-grade and extra-grade excipient particle size variations, and the choice of manufacturing method.

361247 Excipient-Process Interactions and Their Effects on Tablet Compaction and Film Coating

Abstract:

When formulating drugs with low aqueous solubility and high sticking tendency, it is a common practice to include sodium lauryl sulfate (surfactant) and magnesium stearate (lubricant) in the formulation. Even though sodium lauryl sulfate has some lubricant properties, it is not as effective as magnesium stearate and therefore these excipients may be present together in a given formulation. The objective of this study was to establish the main effects and interactions of sodium lauryl sulfate, and magnesium stearate, and the manufacturing process on final blend compaction properties and the performance of the resultant tablets during film coating. A full factorial study (12 batches) was conducted using 2 levels each of sodium lauryl sulfate (0, 1% w/w) and magnesium stearate (0.5, 1.75% w/w), along with three different manufacturing processes (direct compression, high-shear wet granulation, and dry granulation). Compaction profiles were generated using Stylcam® compaction simulator. At a tablet solid fraction of 0.9 the resulting tablet hardness values were found to vary in the range of 13-42 SCU, highlighting their compactability differences. Both sodium lauryl sulfate and magnesium stearate had a significant negative effect on compaction likely due to the lubricant properties of both these excipients. Additionally, during film coating of these batches logo bridging tablet defects were observed for cases when either sodium lauryl sulfate and/or magnesium stearate was present at the higher concentrations. The negative effects on compaction and coating performance were magnified for the dry granulation process likely due to the overall increased shear experienced by these excipients during the dry granulation process, which in turn increases the 'effective concentration' of these excipients. A low level of MgSt (0.5% w/w) with no SLS present showed good compaction and no logo bridging across all manufacturing processes. The findings from this study highlight the importance of the manufacturing process when considering the use-level of excipients such as sodium lauryl sulfate and magnesium stearate in the formulation.

Investigation of the Effect of Ethylcellulose Viscosity Variation on Slightly Soluble and Freely Soluble Drugs from Extended Release Multiparticulates

Cody Schoener[†], Shari Workentine[†], Raxit Y. Mehta[‡], Jason Teckoe[‡], and Ali R. Rajabi-Siahboomi[‡]
[†]The Dow Chemical Company, Midland, MI 48674 USA; [‡]Colorcon Inc., Harleysville, PA 19438, USA

Purpose: Ethylcellulose is a commonly used polymer for extended release (ER) multiparticulate formulations due to its excellent barrier properties, safety, and global acceptance. The critical material attributes of ethylcellulose which may influence drug release are viscosity and ethoxyl substitution. These attributes must be evaluated to aide in developing robust ER multiparticulate formulations and fulfill the Quality by Design (QbD) precedence being set forth by the FDA. The current study investigated the effect of viscosity variation for ETHOCEL™ Std 10 cP, 20 cP, 100 cP Premium product grades on the extended drug release from multiparticulates containing slightly soluble (acetaminophen, APAP) or freely soluble (metoprolol tartrate, MT) drug compounds.

Methods: APAP or MT was spray coated onto sugar spheres (Suglets® #18 – 20) using Opadry® 02A (Colorcon Inc., USA) as binder. A drug to binder ratio of 50:50 w/w was maintained for APAP and 70:30 w/w for MT. ETHOCEL™ QbD samples were identified and represent the viscosity specification range for each ETHOCEL™ product grade (Table 1). Ethylcellulose and dibutyl sebacate (9:1 w/w) were dissolved in isopropyl alcohol and water (9:1 w/w) and spray coated onto the drug layered beads using a wurster fluid bed coater (Glatt Air Techniques Inc. GPCG-2) at 5, 10, and 15% weight gain. USP Apparatus I (baskets) was utilized for *in-vitro* dissolution studies and drug release data was compared using similarity factor (f_2) analysis.

Table 1: ETHOCEL™ QbD Samples

Viscosity Grade	Viscosity Specification cP	Viscosity of QbD Samples cP
ETHOCEL™ 10cP	9-11	9, 10, 11
ETHOCEL™ 20cP	18-22	18, 20, 22
ETHOCEL™ 100cP	90-110	90, 100, 110

Results: APAP release profiles from the ER multiparticulates demonstrated minimal variation for each ETHOCEL™ 10 cP, 20 cP, 100 cP QbD samples. For example the 9 cP, 10 cP, and 11 cP (viscosity specification for ETHOCEL™ Std 10 cP) coating at 5, 10, and 15% weight gain had similarity factor (f_2) values of >75, >74, and >83 respectively (Figure 1, A – C). Similar f_2 values for APAP ER MP were computed for the remaining QbD samples.

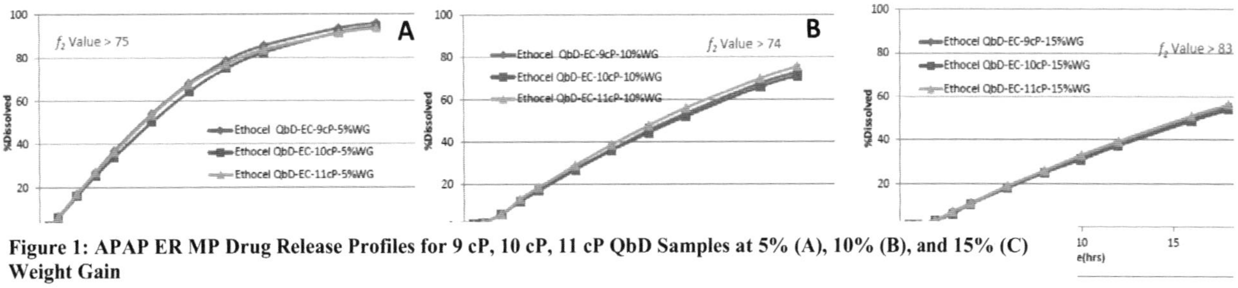

Figure 1: APAP ER MP Drug Release Profiles for 9 cP, 10 cP, 11 cP QbD Samples at 5% (A), 10% (B), and 15% (C) Weight Gain

Metoprolol tartrate release profiles from the ER multiparticulates had similar computed f_2 values to that of APAP ER MP for ETHOCEL™ QbD 10 cP and 20 cP samples (Figure 2, A – C). However, at higher weight gains (10 and 15%) of 90 cP, 100 cP, and 110 cP QbD samples, f_2 values were lower. More variability in the MT ER MP drug release profiles as compared to APAP ER MP is attributed to MT's higher solubility in water.

*** Formatting Issue - Best Available Paper/Graphic ***

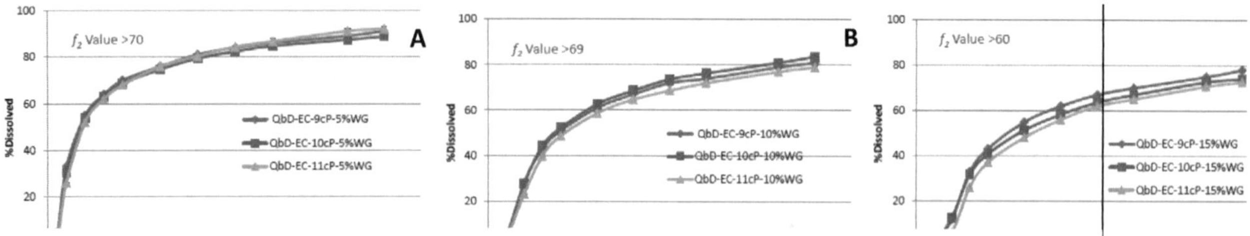

Figure 2: MT ER MP Drug Release Profiles for 9 cP, 10 cP, 11 cP QbD Samples at 5% (A), 10% (B), and 15% (C) Weight Gain

Conclusion: These studies show that viscosity variation within the specification of ETHOCEL™ 10 cP, 20 cP, and 100 cP has minimal impact on drug release for APAP ER MP. Viscosity variation within the specification of ETHOCEL™ 10 cP and 20 cP had minimal impact on drug release for MT ER MP. ETHOCEL™ 100 cP QbD at high weight gain on MT ER MP demonstrated low f_2 values. Overall, these results highlight the consistency of the ETHOCEL™ product and utility of ETHOCEL™ QbD samples as a means to assess the robustness of ER multiparticulates dosage forms during product development.

364276 The Virtual Product-Process Design Lab

Abstract:

The Virtual Product-Process Design Laboratory (VPPD-Lab) has been developed as a computer aided tool for product-process development. Rather than design chemicals based products through an experimental-based trial and error approach, the powerful methodologies and tools in VPPD-Lab help to replace some of the experimental steps by performing virtual experiments using model-based techniques. VPPD-Lab offers a quick search of active ingredients, solvents, and additives to determine a set of promising formulations that match the desired (target) product properties (functions). This is followed by experimental-based validation, and evaluation of the most promising candidates for eventual selection as the final product, reducing thereby, valuable time and resources at the preliminary design stage.

The earlier version of VPPD-Lab consisted of templates for the design of microcapsules, pesticide uptake, and homogeneous liquid formulated products such as paint formulation, sunscreen lotion, and insect repellent lotion. The algorithms in VPPD-Lab can perform ingredients selection and testing, solubility tests, and property measurements using its available library of property models and chemicals databases. VPPD-Lab is also able to directly access ICAS (Integrated Computer Aided System) as a support tool to (i) consult several other chemicals databases, (ii) predict property values which are missing in the databases predicted through the ICAS property prediction tool (ProPred), and (iii) perform process design using process models available in ICAS MOT (Modeling testbed) to design the processes. However, the main limitations of property models available in VPPD-Lab is that the application range of the models is not large enough to include a wider spectrum of chemicals based products, such as, emulsified formations, blended synthetic fuels and lubricants, and many more. Covering a wider spectrum of products requires the use and management of a wide range of property models, calculation procedures and a very large amount of data. A systematic framework is needed to manage the data and models are necessary as well as the modules for a better communication between VPPD-Lab and support tools.

This work proposes a systematic framework to improve and extend the application range of VPPD-Lab. New property prediction models have been developed and integrated to the different steps of the product design algorithms. The chemicals databases have been extended and integrated, and property models needed for each type of products are managed through a knowledge base representation system (ontology) enabling a flexible and efficient communication between the product-design templates and supporting tools (databases, models, and connected software). The new version of VPPD-Lab has included new templates for the design of blended and emulsified products with the new user interface of a generic workflow to design products and processes. Therefore, VPPD-Lab is now generic, user friendly, and widely applicable with options for a wide range of chemicals based liquid products.

Two case studies dealing with tailor-made design of jet-fuels and an emulsified formulation product is highlighted to illustrate the new features and options available in VPPD-Lab.

368170 Oscillatory Amplitudes Regulation Method in 1,3-PD Production By Klebsiella Pneumoniae

Abstract:

Oscillatory amplitudes regulation method in 1,3-PD production by *Klebsiella pneumoniae*

Hangzhou Wang[a], Tong Qiu[b], Jinsong Zhao[b],Bingzhen Chen[b*]

[a]Memorial University, St. John's, N.L., Canada

[b]Tsinghua University, Beijing, China

1,3-propanediol (1,3-PD) is an important chemical raw material, and it can be used to produce polyesters, polyether and polyurethanes with excellent properties. Microbial fermentation production has many advantages [1, 2] over chemical synthesis of 1,3-PD. But oscillatory behavior was observed in both experiments and mathematical simulation during the continuous fermentation of *Klebsiella pneumoniae*, in which biomass, product and substrate oscillations exist under certain operating conditions. It was reported that these oscillations were caused by the existence of Hopf bifurcations in the system[3]. To design a more stable process and achieve high product quality, these operating condition points corresponding to the Hopf singularity ones should be avoided. And in some particular situations, the singularity operation point is located exactly at the maximum concentration point; the oscillation is unavoidable or is prone to generate. In these situations, it is a better way to regulate the oscillatory phenomenon, that is, to reduce the oscillatory amplitude so as to increase process stability to some extent. In this paper, a regulation method of oscillatory amplitudes is proposed so that the oscillatory amplitude can be attenuated when the oscillation itself is unavoidable. The detailed steps for this regulation method of oscillatory amplitudes as follows:

1. Calculate all steady state solutions for the dynamic system[4].
2. Identify all Hopf singularity points in the steady state curve[5, 6].
3. Simulate the oscillatory phenomenon and find the amplitudes under different operation conditions.
4. Establish the relationship model between the amplitudes and operation conditions(such as feed dilution rate, initial concentration of substrate).
5. Use the relationship model to regulate oscillatory amplitudes in process design.

According to the proposed method, with the results of oscillation simulation around the singularity operation condition points, the product concentration oscillatory amplitudes under different operation conditions can be obtained. These calculated amplitude data can be regressed to a smooth amplitude curve. The curve fits the varying tendency of the process, and can represent the relationship between amplitude and operation conditions. As a result, the characteristics of the oscillatory behavior in dynamic process can be studied quantitatively The oscillation behavior can be predicted and regulated so that a process with better stability can be designed.

Keywords: 1,3-PD, Hopf Singularity points, Oscillatory Amplitudes, Process Design and Regulations

References

1. Wong, C.-L.; Huang, C.-C.; Chen, W.-M.; Chang, J.-S., Converting crude glycerol to 1,3-propandiol using resting and immobilized Klebsiella sp. HE-2 cells. *Biochemical Engineering Journal* **2011,**58–59, (0), 177-183.

2. Menzel, K.; Zeng, A. P.; Deckwer, W. D., High concentration and productivity of 1,3-propanediol from continuous fermentation of glycerol by Klebsiella pneumoniae. *Enzyme and Microbial Technology* **1997,**20, (2), 82-86.

3. Sridhar, L. N., Elimination of Oscillations in Fermentation Processes. *AIChE Journal* **2011,**57, (9), 2397-2405.

4. Wang, H.; Chen, B.; He, X.; Zhao, J.; Qiu, T., Numerical Analysis Tool for Obtaining Steady-State Solutions and Analyzing Their Stability Characteristics for Nonlinear Dynamic Systems. *Journal of Chemical Engineering of Japan* **2010,**43, (4), 394-400.

5. Wang, H.; Zhang, N.; Qiu, T.; Zhao, J.; He, X.; Chen, B., Analysis of Hopf Points for a Zymomonas mobilis Continuous Fermentation Process Producing Ethanol. *Industrial & Engineering Chemistry Research* **2012,**52, (4), 1645-1655.

6. Wang, H.; Zhang, N.; Qiu, T.; Zhao, J.; He, X.; Chen, B., A process design framework for considering the stability of steady state operating points and Hopf singularity points in chemical processes. *Chemical Engineering Science* **2013,**99, (0), 252-264.

---

* Corresponding author. Tel.: +86 10 62781499;fax: +86 10 62770304.

E-mail address: dcecbz@tsinghua.edu.cn (Bingzhen Chen)

388102 Molecular Design of Optimum CO2 Capture Solvents: From Conceptual Screening to SAFT-Based Validation

Abstract:

The wide adoption of chemical absorption/desorption systems for post-combustion CO_2 capture in industry is currently challenged by the high energy penalty in solvent regeneration and the environmental impacts associated with solvents and their derivatives. Intense research efforts reported in recent years are predominantly based on lab and pilot-scale experiments to select solvents which may potentially improve the overall performance of absorption/desorption CO_2 capture. However, this is very challenging due to a) the highly non-ideal solvent-CO_2-water chemical interactions, b) the countless combinations of potential capture solvent and blend candidates and c) the need for combined consideration of numerous thermodynamic, kinetic and sustainability properties as performance criteria prior to selecting solvents with optimum capture features. Computer-aided molecular design methods (CAMD) can help address these challenges and have been successful in supporting the synthesis of molecules with desired physical, chemical and environmental properties in non-CO_2 separations [1]. Despite extensive developments in CAMD methods, few recent works have reported their utilization in the design of CO_2 capture solvents or mixtures for chemical and physical absorption using the Statistical Associating Fluid Theory with potentials of Variable Range (SAFT-VR) [2, 3] and for physical absorption using the Perturbed Chain Polar Statistical Associating Fluid Theory (PCP-SAFT) [4]. While these approaches enable an accurate and reliable determination of solvent-process vapour-liquid equilibria, the set of few solvents screened to date will be further expanded as research efforts extend the rigorous predictive capabilities of SAFT-based models towards additional molecular structures. On the other hand, few CAMD approaches have also been reported [5, 6, 7] that use less accurate, but well-established group contribution methods to screen a much wider set of molecular structures by approximating as solvent performance criteria mainly properties relevant to thermodynamics and toxicity.

In our current work, we propose the use of an optimization-based CAMD method to select post-combustion CO_2 capture solvents of optimum performance in molecular and mixture properties associated with thermodynamics, kinetics and sustainability. The problem is first approached in a fast screening stage where solvent structures are evaluated based on the simultaneous consideration of important pure component properties. For the first time numerous properties are considered as performance criteria reflecting solvent characteristics based on thermodynamic (e.g., vapour pressure, CO_2 solubility etc.), kinetic (solvent basicity, steric hindrance etc.) and sustainability (e.g., health and safety hazard, life cycle assessment etc.) behaviour. The simultaneous consideration of properties selected to capture the molecular chemistry effects on the absorption/desorption process compensates for the initial utilization of simpler models and ensures the selection of fewer but more effective solvents. A few highly-performing solvents are further evaluated using the SAFT-VR and SAFT-γ equations of state to predict accurately the very non-ideal solvent-water-CO_2 mixture vapour-liquid equilibrium behaviour. Different functionalities of the employed CAMD method are used both to design optimum, novel molecular structures and to screen a dataset of commercially available amine solvents suitable for CO_2 capture. The obtained results reveal interesting structure-property trade-offs and point to commercial molecules, which have very recently been considered or have yet to be employed as capture solvents.

Acknowledgements

Funding from the European Commission under grant FP7-ENERGY-2011-1-282789-CAPSOL is gratefully acknowledged.

Cited References

[1] Adjiman, C.S., Galindo, A., 2010, Process Systems Engineering: Volume 6: Molecular Systems Engineering, Pistikopoulos, E.N., Georgiadis, M.C., Dua, V. (Eds).

[2] Mac Dowell N., Pereira F.E., Llovell F., Blas F.J., Adjiman C.S., Jackson G., Galindo A., 2011, Transferable SAFT-VR models for the calculation of the fluid phase equilibria in reactive mixtures of carbon dioxide, water, and n-alkylamines in the context of carbon capture, Journal of Physical Chemistry B, 115, 8155-8168.

[3] Pereira, F.E., Keskes, E., Galindo, A., Jackson, G., Adjiman, C.S., 2011, Integrated solvent and process design using a SAFT-VR thermodynamic description: High-pressure separation of carbon dioxide and methane, Computers and Chemical Engineering, 35 (3), 474-491.

[4] Bardow, A., Steur, K., Gross, J., 2010. Continuous-molecular targeting for integrated solvent and process design. Ind. Eng. Chem. Res. 49, 2834–2840.

[5] Juan Salazar, Urmila Diwekar, Kevin Joback, Adam H. Berger, Abhoyjit S. Bhown, Solvent Selection for Post-Combustion CO2 Capture, Energy Procedia, Volume 37, 2013, Pages 257-264

[6] Bommareddy, S., Chemmangattuvalappil, N.G., Solvason, C.C., Eden, M.R. Simultaneous solution of process and molecular design problems using an algebraic approach (2010) Computers and Chemical Engineering, 34 (9), pp. 1481-1486.

[7] Chemmangattuvalappil, N.G., Eden, M.R. A novel methodology for property-based molecular design using multiple topological indices (2013) Industrial and Engineering Chemistry Research, 52 (22), pp. 7090-7103.

389735 Structure Generation of Optimal Reactants and Products Using Signature Descriptors

Abstract:

Selection of optimal chemicals and maximization of process efficiency has successfully been carried out by utilizing process design techniques that integrate product design considerations in a reverse problem formulation. The combined process-product design problem is solved for design targets in terms of properties that are later used to design optimal products. However, most of these techniques have been implemented in non-reactive systems. Different efforts to address the paucity of techniques for reactive systems have been restricted to single unknown reactants. Also, utilization of linear property models and solution schemes that do not have a provision for treating property constraints are among some of the other restrictions. In this work an algorithm has been developed for the design of optimal reactants and products that is not restricted by the number of unknown reactants and products. Nonlinear property models can be incorporated in the algorithm and the solution scheme considers property constraints. The products of the reaction(s) are constrained by property bounds arising from the process to ensure the optimality of the process. An optimization problem is formulated for each of the products in order to select the best chemical structures that satisfy the property bounds and have the best dominant property. The recently developed molecular signature descriptors have been utilized to perform molecular design. The signature descriptor is a systematic codification system over an alphabet of atoms describing the extended valence of the atoms of a molecule. The signature of a molecule can be expressed as a linear combination of its atomic signatures. Topological indices of molecules that form various property models can be linearly represented in terms of molecular signature descriptors [1]. Different quantitative structure activity/property relationships (QSARs/QSPRs) have been applied in the developed algorithm to estimate properties from molecular structures. QSARs/QSPRs can be expressed in terms of different topological indices. Thus the target properties can be expressed in terms of signatures. In the developed method, the problem is formulated in terms of the number of occurrences of each signature, i.e. the number of times each signature appears in the molecular structure. The dominant property to be optimized and the property constraints can thus be ultimately expressed in terms of the number of occurrences of the signatures in the product molecule. After identifying the occurrence values from each of the formulated problems, the molecular structure of products is generated by using an algorithm developed by Chemmangattuvalappil et al. [2]. Next, using the chemical equation to track the migration of groups from reactants to products, the structure of the reactants is determined by considering the appropriate addition and/or removal and/or rearrangement of groups. Thus, candidate reactant structures can be generated irrespective of the type and number of reactions as the introduced algorithm focuses on the design of products first and then uses the chemical equations to identify the reactant structures. Graph theory principles have been utilized to track the signatures and to avoid the generation of infeasible molecular structures. To solve the resulting nonlinear integer optimization problem(s), MI-LXPM, a real coded genetic algorithm (RCGA) developed by Deep et al. [3] has been incorporated in the solution scheme. RCGAs are general purpose population based search techniques that mimic the principles of natural selection and natural genetics laid down by Charles Darwin and encode the solution (represented as chromosomes) in real numbers. Property functions that may be discontinuous, non-differentiable and/or highly non-linear can be handled by genetic algorithms (GAs). Real coding is generally more efficent than binary and gray coding and it helps avoid the "Hamming-Cliff" problem. MI-LXPM is a robust and efficient GA for solving nonlinear integer programming problems. It utilizes the recently developed and self adaptive Laplace crossover (LX) operator and a tunable power mutation (PM) operator [4]. A special truncation procedure is utilized to handle integer restrictions on variables instead of rounding off carried out in previous algorithms. One may not begin with an array of feasible points since a parameter free constraint handling procedure dependent on tournament selection is used. Thus search of feasible points is not a constraining factor. This contribution will illustrate the developed methods and highlight their use through a case study.

[1] J.L. Faulon, D.P. Visco Jr., R.S. Pophale, 2003, Journal of Chemical Information and Computer Sciences, 43, 707-720

[2] N.G. Chemmangattuvalappil, M.R. Eden, 2013, Industrial & Engineering Chemistry Research, 52, 7090-7103

[3] K. Deep, K.P. Singh, M.L. Kansal, C. Mohan, 2009, Applied Mathematics and Computation, 212, 505-518

[4] K. Deep, M. Thakur, 2007, Applied Mathematics and Computing, 193, 211-230

390512 An Application of Computer-Aided Molecular Design (CAMD) Using the Signature Molecular Descriptor: Designing New Water-Reducing Admixtures for Cement through Cement Paste Rheology Evaluation

Abstract:

Concrete is a material that forms the literal foundation of our modern society. Concrete is no longer simply cement, water and aggregates, but rather, it also typically contains a complex array of chemical admixtures that modify its many properties to produce a cost effective, durable, formable and mechanically sound material. Water-reducing admixtures (WRAs) have been developed to reduce the required water content of concrete for a given workability or increase the workability for a given water content.

The development of new chemical admixtures for concrete is normally an experimental approach wherein one uses prior knowledge to suggest potential effective compounds. This approach is time-consuming, incremental and typically expensive. To accelerate the process, we propose a computer aided molecular design (CAMD) approch using the Signature molecular descriptor as a powerful alternative. CAMD is the application of computer-implemented algorithms that are utilized to identify and design new molecules with optimally predicted properties such that they can be tested and evaluated for efficacy. In this work, we are interested in the flow properties of cement paste, described by its rheological properties. Accordingly, a parallel-plate rheometer was used to measure the effect of eighteen low molecular weight organic chemicals belonging to different classes (sugars, sugar alcohols, and sugar acids) on rheology by directly obtaining yield stress and viscosity data. Workability, measured by a simple industry standard flow (mini-slump cone) test, was also obtained. Setting time, the time required for cement paste to stiffen, was measured since a higher workability is normally accompanied by an extended setting time, which is an undesired side-effect in some applications. After creating and refining four different quantitative structure–property relationship (QSPR) models for our four properties of interest (yield stress, plastic viscosity, workability, and setting time), a structure enumeration algorithm wass employed to generate structures outside of the original training set that have optimally predicted properties such as high workability, low yield stress and viscosity, and short setting time. Over 670 structures were newly generated and screened against a non-commercial chemical database, ZINC, to identify commercially available compounds. As a result, 6 compounds were purchased and tested for their effect on the flowability (i.e. workability, yield stress, and viscosity) and the setting time of cement paste for evaluation. Experimental results validated the previously developed QSPR models. Results confirm that the CAMD strategy is a powerful tool for generating novel and non-intuitive structures with the ability to increase flowability of cement paste with controlled setting time.

Graph Based Database Development for Gasoil Reactive Systems

AIChE Abstract
Charles Solvason and Jorge Martinis

Sessions:

- 10A03 Product and Molecular Design

In the petroleum refining industry, the accurate design and optimization of hydrocarbon conversion processes for gasoil feedstocks demand a growing effort in developing rigorous and reliable kinetic models (Van Geem *et al.*, 2002). Existing techniques rely on the approximation of gasoil stream compositions using a set of pseudo-components called "oil cuts" characterized by a set of lumped kinetic models. Although the lumped models are easy to construct and implement, they lack molecular information which makes them feedstock dependent and of limited use in process simulation and design. Developments by Neurock *et al.* (1994), Trauth *et al.* (1994), and Stark *et al.* (1995) address this limitation using molecular reconstruction techniques to approximate the individual components of the gasoil streams via stochastic optimization methods coupled with Monte Carlo kinetic models. The application of stochastic methods is, however, limited by the computational intensity imposed by the large sampling frequency needed to achieve statistical significance (Van Geem 2007). For specific classes of gasoils, Hudebine and Verstraete (2011) utilize a hybrid stochastic method that maximizes an entropy function with Lagrangian parameters associated with analytical constraints in order to reduce the computational burden. However, the method strongly depends on the initial set of molecules which must be chosen and built in relation to the type of gasoil studied.

An alternative, deterministic approach to molecular reconstruction has been developed by Bryan Research & Engineering (BRE) for the process simulator, ProMax®. The approach replaces the stochastic methods with a parallelized SQL call to a molecular database that contains all chemical constituents important to gasoil products and streams. The method is dependent on two competing factors, the completeness of the database and the size of the database. Unfortunately, most gasoil reconstructions require representations of chemical constituents of up to 40 atoms in a hydrogen suppressed, molecular graph. Enumerating all chemical constituents up to this size is cumbersome for even state-of-the-art graph generation algorithms and would result in a database too large to be efficiently searched (Peironcely *et al.*, 2012).

To address these limitations, a hybrid method is proposed that limits the size of the graph database using a set of constraints provided by molecular group and point group theory. In the first step of the method, a modified set of molecular functional groups described by Joback and Reid (1987) are used as building blocks to construct a complete set of vertex subgraphs that contain all double bonds, aromatic ring structures, and non-aromatic ring structures necessary to model gasoil systems. Next, these vertices are perturbed using point group symmetry operations to create a set of isomeric vertex graphs. Of these graphs, constitutional isomeric vertices are selected and separated into terminal and internal vertex sets. The internal vertices are constructed using graph generation algorithms and the terminal vertices are added at the final step to create complete molecular graphs. By utilizing coarse-grained sub-graphs of vertices, specific isomer classes common to gasoil systems can be created, without overloading the database, thus ensuring the capability of the SQL call to search the database in a timely manner. In addition, the database generation routing is constructed such that it is expandable by toggling various stereoisomer point groups and vertex rotations. The molecular library will be utilized to illustrate the database generation method as it is applied to a diesel hydrotreater case study.

Van Geem, K.M., Hudebine D., Reyniers M.F., Wahl F., Verstraete J.J., Marin, G.B. (2007): "Molecular reconstruction of naphtha steam cracking feedstocks based on commercial indices." *Computers & Chemical Engineering,* 31, pp. 1020-1034.

Hudebine D. and Verstraete J.J. (2011): "Reconstruction of Petroleum Feedstocks by Entropy Maximization. Application to FCC Gasolines." *Chemical Reaction Modelling of Refining Processes,* 66(3), pp. 461-477.

Joback K.G. and Reid R.C. (1987): "Estimation of pure-component properties from group contributions." *Chemical Engineering Communications*, 57(16), pp. 233-243.

Peironcely, J.E., Rojas-Chertó M., Fichera D., Reijmers T., Coulier L., Faulon, J. and Hankemeier T. (2012): "OMG: Open Molecule Generator." *Journal of Cheminformatics* 4(21), pp. 1-13.

Neurock, M., Nigam, A., Trauth, D., and Klein, M.T. (1994): "Molecular representation of complex hydrocarbon feedstocks through efficient characterization and stochastic algorithms." *Chemical Engineering Science*, 49, pp. 4153-4177.

Stark S.M., Neurock M., Klein M.T. (1995): "Comparison of MIMD and SIMD Strategies for Monte Carlo Modeling of kinetically coupled reactions." *Computers& Chemical Engineering*, 19(6), pp. 719 -742.

Trauth D.M., Stark S.M., Petti T.F., Neurock M., and Klein, M.T. (1994): "Representation of the molecular structure of petroleum resid through characterization and Monte Carlo modeling. *Energy & Fuels*, 8(3), pp. 576-580.

386157 Determining Optimal Groups for Group Contribution Methods

Abstract:

Given the sheer number of chemical compounds in use today, determining a new or untested compound's properties is a problem beyond current laboratory resources. This necessitates the need for quantitative structure activity relationships (QSARs), which are models used to predict certain properties based solely on the chemical structure of a compound. Often, QSARs are in the form of group contribution methods, which assume that a property of a compound can be modelled based on the number of occurrences of molecular subunits, or "groups". Many group contribution methods are derived from regression over large datasets, but many of these regression problems require first specifying the identity of these groups, a requirement which introduces some bias in the model building. While many group contribution methods have seen a substantial degree of success [1], many new methods are still based to some degree on the original UNIFAC groups [2]. Some recent approaches have applied best subset selection methods to alleviate this bias [3,4], but these methods still must specify a larger set from which to select the best components. We propose a method that aims to investigate the entire space of subgroups and select the set of groups which provides the best predictive accuracy from a given dataset. This is essentially optimal group selection for a given model. This method has promise in improving group contribution methods for properties that have historically been difficult to predict and design for as well as determining important substructures for more complicated properties like toxicity or bioactivity.

[1] Group-contribution based estimation of pure component properties. Jorge Marrero and Rafiqul Gani.Fluid Phase Equilibria. 2001. 183, 183-208.
[2] Group-contribution estimation of activity coefficients in nonideal liquid mixtures. Aage Fredenslund, Russel L. Jones, and John M. Prausnitz. AIChE Journal. 1975. 21(6), 1086-99.
[3] Choosing Feature Selection and Learning Algorithms in QSAR. Martin Eklund, Ulf Norinder, Scott Boyer, and Lars Carlsson. Journal of Chemical Information and Modeling 2014 54 (3), 837-843
[4] Towards Optimal Descriptor Subset Selection with Support Vector Machines in Classification and Regression. Holger Fröhlich, Jörg K. Wegner and Andreas Zell. QSAR & Combinatorial Science. 2004. 23(5), 311-318

377386 Mathematical Modeling of Macromolecular Release of Stimuli-Responsive Polyelectrolyte Multilayer Films

Abstract:

Polyelectrolyte multilayers (PEMs) prepared via the layer-by-layer (LbL) assembly technique have gained attention in many research fields due to their simplicity, versatility, and wide applicability [1]. The LbL technique is based on the alternating deposition of oppositely charged polyelectrolytes or any other multivalent charged species. The design of weak PEMs for delivering functional agents (e.g., antibiotics, growth factors) and in response to stimuli (e.g., changes in solution pH) is of interest for surface delivery applications, with the kinetics being dependent on molecular weight of the polyelectrolytes and differences between assembly pH conditions and the pH at which release takes place.

Despite the great potential for the use of weak PEM films as a stimuli-responsive system with desired release characteristics, a lack of understanding of the fundamental mechanisms underlying the behavior of these films has slowed development. This presentation presents a mathematical model suitable for describing and predicting the release behaviors of weak polyelectrolytes in response to external pH variations and swelling. The mathematical model also estimates the total fraction of polyanions released by the pH-induced charge imbalance within the film. To test the accuracy of the models, the theoretical results are compared to published experimental data.

More specifically, a mathematical model was developed to describe the overall release behaviors of polyelectrolyte chains based on theories of sticky gel electrophoresis, biased reptation, and polymer film swelling. Theoretical expressions for the degree of ionization of 20% quaternized poly-4-vinylpyridine (Q20) and poly-methacrylic acid (PMAA) in a weak PEM film showed good agreement with published experimental data [2]. Comparisons with experimental release profiles indicated the need for the inclusion of the effects of swelling and rearrangement into the mathematical model. This mathematical model predicted an induced electric field within the film of about 0.001 V/nm, and provided physical insights into chain configuration, extent of chain entanglement, and diffusion coefficient in the initial state. The model also demonstrated very close agreement with experimental release profiles [2] for weak PEM films over the full range of polyanion molecular weights M_w of 6, 22, 72, and 350 kDa. A quantitative comparison of the model with the experimental data [2] indicated a high accuracy and correlative power of the model. Furthermore, the mathematical model accurately predicted release profiles for a factor of fifty reduction in the molecular weight of the polyanion.

The comparison of model predictions with experimental data suggested that the swelling model describes the dynamic effects of diffusional resistance on the mobility of long chains in short times. The mathematical model with swelling successfully accounted for the dependency of the chain release time on the molecular weight of polyanion and the time-dependent entanglement length. The predictive capability of the mathematical model suggests the potential for its use in tuning stimuli-responsive release of macromolecules from weak PEM films and to provide kinetic information for designing weak PEM films with predictable release characteristics for many surface delivery applications such as therapeutic release of macromolecular components such as proteins, DNA, and enzymes.

References:

1. Paula T. Hammond, Form and function in multilayer assembly: New applications at the nanoscale. *Advanced Materials*, 2004. **16**(15):1271-1293.

2. Eugenia Kharlampieva, John F. Ankner, Michael Rubinstein, and Svetlana A. Sukhishvili, pH-induced release of polyanions from multilayer films. *Physical Review Letters*, 2008. **100**(12):128303.

378315 Formulation Design Evaluation for a Single Layer Osmotic Capsule

Abstract:

A single layer osmotic capsule for controlled-release has been examined to provide an easy to manufacture capsule analog to single layer osmotic tablet delivery systems. The impact of the single layer capsule core contents on dissolution performance was evaluated. Factors included: 1) drug type and level – acetaminophen (high solubility), caffeine (high solubility), and carbamazepine (low solubility); 2) grade, particle size, and level of the viscosifying polymer (hydroxyethylcellulose); 3) type and level of binder (copovidone and hydroxypropylcellulose); 4) solid fraction of the capsule cores; and 5) type and level of the osmogen (sorbitol and sodium chloride). Single layer capsule cores containing various amounts of the preceding materials were encapsulated in a semi-permeable osmotic capsule shell comprised of 60% cellulose acetate and 40% PEG 3350. Screening studies were performed to evaluate the impact of hydroxyethylcellulose viscosity and particle size, binder type, capsule core solid fraction, and drug solubility on the lag time and the initial, linear release rate from the osmotic capsule. The screening studies indicated that drug solubility had an impact on the lag time and the initial, linear release rate, while the other factors showed minimal impact on the drug release from the single layer osmotic capsule. A subsequent statistical design of experiments (DOE) study examined the formulation space for the single layer osmotic capsule using acetaminophen. The DOE study indicated that formulation changes to the single layer core could result in T_{50} (time until 50% of the drug is dissolved) values ranging from 5 hours to greater than 24 hours. Further, drug was released faster from capsules with higher osmogen to drug loading ratios. For lower osmogen to drug loading ratios, the amount of hydroxyethylcellulose and the ratio of sorbitol to sodium chloride had a significant effect on the rate of release from the osmotic capsule, with higher levels of sodium chloride and hydroxyethylcellulose increasing the release rate. In summary, the design of the single layer capsule core formulation can provide a drug product formulator with a broad range of flexibility in order to achieve varying release rates from an osmotic capsule.

Process Control of Modified-Release Matrix Tablet Manufacturing Operations and Key Excipient Enablers

True L. Rogers (TLRogers@Dow.com), Kathryn O. Hewlett (KOHewlett@Dow.com), Karen M. Balwinski (KMBalwinski@Dow.com), Jörg Theuerkauf (JTheuerkauf@Dow.com), Louis F. Ruocco (Ruocco@Dow.com)

Purpose: Modified-release (MR) matrix tablets are typically produced using hydroxypropyl methylcellulose (HPMC) as the rate-modifying polymer. Several ingredients in the matrix tablet formulation, particularly the active pharmaceutical ingredient (API) and HPMC, may be irregularly shaped and prone to poor flow and segregation. Although direct compression would be the most streamlined and cost-efficient manufacturing process, inclusion of granulation steps is often necessary to resolve flow and segregation issues. In this study, HPMC matrix tablets were produced via direct compression of a powder blend or by including intermediate roller compaction or high-shear wet granulation steps prior to tabletting. Figure 1 compares the three manufacturing processes investigated. The purpose of this talk is to compare process control, productivity, and achievability of desired end-product attributes by the manufacturing process employed and also to understand how HPMC morphology impacts formulation processability.

Figure 1. Process flow diagrams comparing wet granulation, dry granulation, and direct compression.

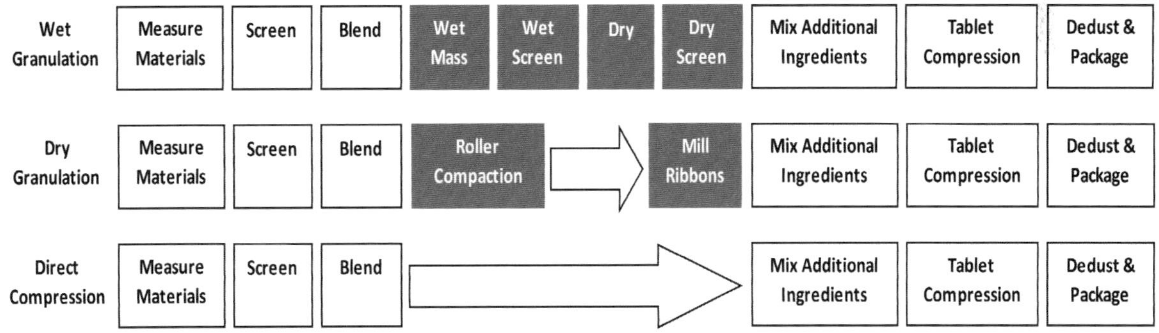

Methods: The matrix tablet model formulation contained 50% metformin HCl, 30% HPMC, 19% microcrystalline cellulose (MCC), 0.5% talc, and 0.5% magnesium stearate. The targeted tablet weight was 1000 mg, so the target API dose per tablet was 500 mg. The standard CR grade of HPMC was used to make tablets via intermediate wet granulation processing steps, a typical practice of matrix tablet manufacture. A new direct compression grade of HPMC, METHOCEL™ DC2, was used to make tablets via direct compression or by using intermediate roller compaction processing steps prior to tabletting. For direct compression, the API, HPMC, MCC, and talc were blended together for 10 minutes. Magnesium stearate was then blended into the powder mixture for one additional minute, and the powder blend was subsequently compressed to caplet-shaped matrix tablets (9.09 x 18.92 mm) using a 26.7-kN compression force. For wet granulation, the API, HPMC, and MCC were blended together, high-shear granulated to desired granularity, and passed through a 20-mesh screen. The granules were oven-dried at 45 °C overnight and passed again through a 20-mesh screen. The sieved granules were blended with talc for 1 minute, and then blended with magnesium stearate for one additional minute before being tabletted. For dry granulation, the API, HPMC, MCC, and talc were roller compacted under 18-kN of force. The roller compacted ribbons were then milled to granules, which were subsequently blended with magnesium stearate and compressed to tablets. Tablets were collected every 3 minutes during each tabletting trial, so run charts could be generated to gauge process control over time.

Results: It is typical to include wet granulation processing steps when manufacturing matrix tablets containing irregularly shaped, poorly flowing ingredients, such as METHOCEL CR, the established standard of HPMC. METHOCEL DC2 and METHOCEL CR are chemically identical; but DC2 exhibits a spheroidal morphology to enable flow in dry processing techniques, such as direct compression and roller compaction. Figure 2 shows the morphological differences between DC2 and CR.

Figure 2. Morphological differences between METHOCEL DC2 and METHOCEL CR.

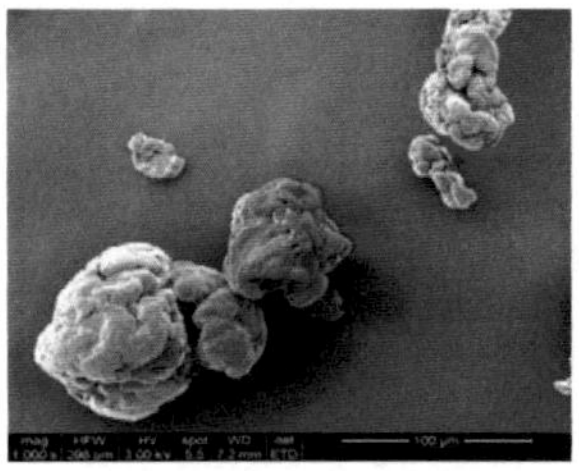

METHOCEL DC2

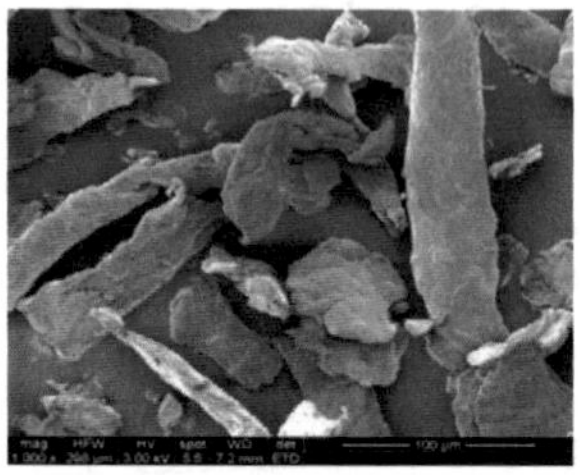

METHOCEL CR

Tablet weight, breaking force, API unit-dose uniformity, and dissolution performance were compared across each tablet manufacturing process employed. DC2 matrix tablets produced via direct compression weighed 971 ± 23 mg, reflecting satisfactory process control (see Figure 3a). DC2 matrix tablets produced using roller-compaction intermediate steps weighed 995 ± 10 mg, reflecting exceptional process control (See Figure 3b). CR matrix tablets produced using wet granulation intermediate steps weighed 838 ± 14 mg, well below the 1000-mg target (see Figure 3c). A control tabletting trial was conducted using direct compression to make CR matrix tablets (see Figure 3d). Like the CR tablets made from wet granulation, the CR control tablets from direct compression were below the 1000-mg target weight, and the CR control tablets reflected significantly inferior process control compared to DC2 tablets produced either by direct compression or using roller compaction.

Figure 3. Process run charts for individual tablet weight measurements collected throughout the tabletting trials for a) DC2 matrix tablets made by direct compression, b) DC2 tablets made using roller compaction, c) CR tablets made using wet granulation, and d) CR control tablets made by direct compression.

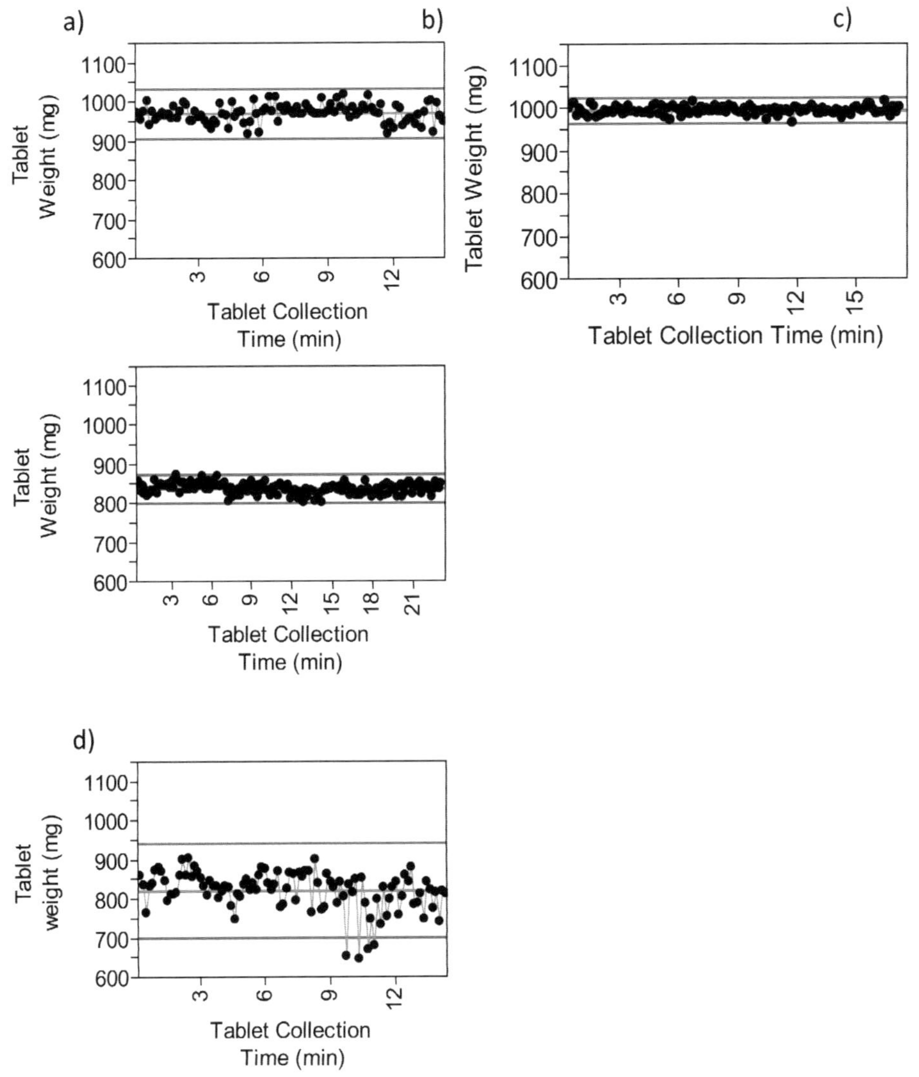

DC2 matrix tablets produced via direct compression or using roller compaction exhibited breaking forces of 112 ± 22 and 108 ± 12 N, respectively. CR tablets produced using wet granulation exhibited a breaking force of 69 ± 4 N. Tablets produced via direct compression or using roller compaction exhibited satisfactory unit-dose uniformity within 90-110% of the labeled claim of 500 mg API per tablet. Expected MR performance was achieved in all cases.

Conclusions: Direct compression is well known as the most streamlined of the tabletting processes but often is not a feasible option due to flow or segregation issues. A key learning from this study is that matrix tablets produced by direct compression required the use of METHOCEL DC2 in order to achieve the desired end-product attributes and sufficient process control. DC2 tablets produced using roller compaction exhibited more reproducible attributes than DC2 tablets produced via direct compression, but did require three additional processing steps. DC2 matrix tablets produced via both manufacturing processes met the desirable end-product attributes. Streamlined manufacturing coupled with sufficient process control rendered direct compression a viable matrix tablet manufacturing option as long as the more flowable METHOCEL DC2 excipient was utilized.

385705 QbD in Development of Film Coated Tablets: A Risk Assessment Based Doe Towards a Robust Product and Process

Abstract:

The principles of Quality by Design (QbD) were applied during the formulation development of a film-coated tablet. A risk based approach was used to identify the impact of formulation components and process parameters on the critical quality attributes of the coated tablets. Formulation composition and manufacturing process parameters were defined by carefully constructed design of experiments with focus on impact to critical quality attributes of the final dosage form at release and on shelf-life. The case study will also highlight the integration of process analytical tools (PAT) to monitor in-process material attributes together with parametric controls to ensure a robust and cost effective strategy to transfer and scale-up at the commercial manufacturing site.

Wavenumber Selection for NIR Calibration Modeling: Application to Water and Drug Content Estimation

Taku Uchimaru[1], Takuya Miyano[2], Koichi Fujiwara[1], Manabu Kano[1], Hideaki Tanabe[2], Hiroshi Nakagawa[2], Tomoyuki Watanabe[2], Naoki Wakiyama[3]

1 Department of Systems Science, Kyoto University, Kyoto, Japan

2 Pharmaceutical Technology Division, Daiichi Sankyo Co., Ltd., Hiratsuka, Japan

3 Daiichi Sankyo Propharma Co., Ltd., Hiratsuka, Japan

Abstract:

To improve understanding of manufacturing processes, process analytical technology (PAT) has played an important role. Although a large number of successful applications of PAT have been reported, it is still needed to answer the question: how to develop an accurate and robust calibration model based on near infrared (NIR) spectroscopy. In general, one has to carefully select data for modeling and validation, preprocessing methods, input variables, modeling methods, and so on to build an accurate and robust model. In calibration modeling, partial least squares (PLS) are usually adopted to cope with collineartiry among input variables, i.e., absorbance or reflectance at several hundred or thousand wavenumbers. Not all the input variables are necessary to estimate an output variable. In fact, to achieve highly accurate estimation, input variable selection is crucial.

In the present work, new efficient wavenumber selection methods are proposed to achieve higher estimation accuracy than conventional methods and also to significantly reduce the computational load required by conventional wavenumber selection methods such as interval PLS. In the first method, referred to as R+VIP, two conventional indexes for input variable selection, i.e., the correlation coefficients R and variable influence on projection (VIP) are combined into one index. As a result, R+VIP has advantages of both indexes and functions effectively. The second method is named as spectral fluctuation dividing (SFD). In this method, a whole spectrum is divided into multiple spectral intervals at local minimum points of the spectral fluctuation profile, which consists of the standard deviation of absorbance at each wavenumber in a calibration set, and the spectral intervals of significant relationships with a target response are selected.

The usefulness of the proposed methods in terms of both improving the estimation accuracy and reducing the computational load is demonstrated through its application to the problems of estimating water and drug contents in granules.

Granules containing a drug substance (Daiichi-Sankyo, Japan) were used as analyte. In granulation process, real-time NIR measurement was performed using a Fourier-transform NIR spectrometer MPA (Bruker GmbH, Germany) and equivalent Matrix-F (Bruker GmbH, Germany) through a fiber-optic probe mounted in the fluid bed granulator. The NIR spectra were obtained at every one minute during the granulation process. For blended granules, NIR measurement was performed using a Fourier-transform NIR spectrometer MPA (Bruker GmbH, Germany). In addition, granules were sampled from the fluid bed granulator at every ten minutes during granulation and at the end of both the spraying and the drying. Water content of the sampled granules was measured by the loss on drying (LOD) method, and drug content in the blended granules was measured by the high performance liquid chromatography (HPLC) method.

In this work, the following wavenumber selection indexes/methods were compared.

1) Correlation coefficients (R)
2) PLS-beta
3) Variable influence on projection (VIP)
4) R+VIP
5) Interval PLS (iPLS)
6) Nearest Correlation Spectral Clustering (NCSC)
7) Spectral fluctuation dividing (SFD)
8) Moving averaged SFD (MASFD)

The results demonstrated the usefulness of R+VIP and SFD. In fact, the calibration models developed by using R+VIP were more accurate and robust than the models developed by using the conventional indexes. In addition, SFD improved estimation accuracy with less computational load compared with the conventional methods. Furthermore, MASFD, which combines SFD and the moving average method, reduced the number of spectral intervals without deteriorating the estimation accuracy. Hence, the proposed methods are superior to the conventional methods, and will be widely applicable to calibration modeling in PAT.

373600 Improvement of Iterative Optimization Technology (Calibration-Free/ Minimum Approach) with Dimensionality Reduction of Spectra

Abstract:

Improvement of Iterative Optimization Technology (Calibration-Free/ Minimum Approach) with Dimensionality Reduction of Spectra

Hiromasa Kaneko[1], Koji Muteki[2], Daniel O. Blackwood[2], Yang A. Liu[2] , Sonja S.Sekulic[2], Kimito Funatsu[1]

[1]Department of Chemical System Engineering, The University of Tokyo, 7-3-1 Hongo, Bunkyo-ku, Tokyo 113-8656, Japan

[2]Pfizer Worldwide Research & Development, Eastern Point Rd, Groton, Connecticut 06340, United States

Keywords: Process analytical technology, Iterative optimization technology, Dimensionality reduction

Process analytical technology plays an important role in the pharmaceutical industry. PAT is used extensively in process development, process understanding and process control. Often, quantitative measurements are desired/required and a calibrated model will have to be developed and implemented. The development, implementation and maintenance of these quantitative models are both resource and time intensive.

Muteki et al (2013)[1]. previously proposed calibration-free/ minimum approach, iterative optimization technology (IOT), which is used to predict the composition of a mixture while maintaining a similar predictability to calibration models (e.g. Partial Least Squares). It involves using only pure component spectra and mixture component spectra (without a calibration data set). Linear and non-linear IOT have been successfully applied (by offline/online) to some practical pharmaceutical process (e.g. blending, feed frame, solvent mixture, reaction, etc). However, a key (remaining) question was how to deal with a dependency (multi-collinear relationship) among pure component spectra during the optimization computation. For the mixture case which includes similar structured materials, it would be essentially difficult to provide good prediction on mixture component ratio. This problem has often been an important obstacle when applying IOT to mixture cases having more than several components.

This study presents a calibration method which can improve the prediction ability of IOT through reducing dimensionality of spectra with optimal selection of wavelength. It involves using Latent Variable Models (e.g. PCA, PLS) for dimensionality reduction of spectra in IOT and genetic-algorithm-based wavelength selection (GAWLS)[2] for optimal wavelength selection. The proposed methods achieved higher predictive accuracy compared to the traditional IOT. The effectiveness of the proposed methods is demonstrated using simulation data where pure spectra have strong correlation and real industrial data of multicomponent mixtures.

1. Muteki K, Blackwood DO, Maranzano B, Zhou Y, Liu YA, Leeman KR, Reid GL. Mixture Component Prediction Using Iterative Optimization Technology (Calibration-Free/Minimum Approach). Ind. Eng. Chem. Res. 2013;52:12258– 12268.
2. Arakawa M, Yamashita Y, Funatsu K, Genetic algorithm-based wavelength selection method for spectral calibration. J. Chemom. 2011;25:10–19.

Application of Multivariate data analysis to Cell Culture Development

John S. Bowers, Senior Principal Scientist, BPD
Balrina Gupta, Associate Principal Scientist, BPD
T. Craig Seamans, Principal Scientist, BPD
Louis Obando, Principal Scientist, Chemistry, MMD PAT.

A large amount of data on a number of variables are collected to understand the behavior CHO cell culture processes, like the process for the manufacture of monoclonal antibodies. In the early development environment, statistical analysis based on set points, averages, or visual review of trends, have typically been used to understand results. This approach does not fully utilize the hundreds of data points taken per batch. In order to gain better understanding of a monoclonal antibody cell culture process, Multivariate data analysis (MDVA), using SIMCA from Umetrics, has been applied to the initial process ranging studies performed in BioProcess Development in New Jersey.

MDVA evaluations on control runs confirmed process robustness. Variations in parameters during control runs had weak effects on performance and quality attributes. Evaluations on DOE experiments using pH, temperature, feed timing and amount, and elevated pCO_2 as the perturbed parameters confirmed conclusions from average based analysis, but MDVA highlighted when the variations on parameters were most important. Conditions when most of the antibody was produced, during the production phase, controlled quality far more than conditions in the cell growth phase. Conversely, variations in the growth phase impacted performance more. Short transients, less than 1 day, of operating parameters had very little impact on quality. MDVA analysis was better at describing continuous behavior, such as cell growth and death during the batch, than average measurements since the entire time based behavior is included in the analysis.

Model Development for NIR-Based Real-Time Monitoring of Ingredient Concentration

Hiroshi Nakagawa[1], Manabu Kano[2], Shinji Hasebe[2],
Takuya Miyano[1], Tomoyuki Watanabe[1], Naoki Wakiyama[3]
1 Pharmaceutical Technology Division, Daiichi Sankyo Co., Ltd., Hiratsuka, Japan
2 Department of Systems Science, Kyoto University, Kyoto, Japan
3 Department of Chemical Engineering, Kyoto University, Kyoto, Japan
4 Daiichi Sankyo Propharma Co., Ltd., Hiratsuka, Japan

Abstract:

There has been considerable research on process analytical technology (PAT) and real-time monitoring based on NIR, but the model development is still an important issue and persons in charge have difficulty in building good models. In the present work, to verify model development techniques in their applications to real-time monitoring of ingredient concentration with NIR during blending, we evaluated the effect of calibration sets, spectral analysis techniques, modeling parameters and manufacturing conditions on the prediction. Then, we proposed a model development procedure on the basis of the evaluation results. The analytical performance of the developed models under various conditions was evaluated on the basis of the fitting performance to the calibration set, the prediction error for data sets acquired with various blending equipment, and the analytical validation according to the United States pharmacopeia (USP <1119>) and draft guideline on the use of NIR.

Granules including the API (Daiichi-Sankyo, Japan), which was under development, were used in this work. The granules were manufactured with fluid-bed granulators and diffusion mixers with V-blender and bin blenders. The samples with nine different API concentrations (70%, 75%, 85%, 90%, 100%, 110%, 115%, 125%, and 130%) were manufactured with various blenders of scale 1 kg, 50 kg, 200 kg, 400 kg, and 500 kg. Corona (Type: Remote NIR-HR, Carl Zeiss, Germany), which has a diode array type NIR spectrometer with 3 nm resolution, was used for real-time monitoring (dynamic measurement). Blending experiments with the various blenders, scales, and manufacturing conditions were conducted to acquire spectra by dynamic measurement based on the diffuse reflectance method. Static measurement with Corona was also conducted under the same measurement conditions as dynamic measurement.

To build a model that can accurately predict the API concentration, a calibration set, a spectral analysis technique, and modeling parameters need to be selected appropriately. Here, modeling parameters represent spectral preprocessing techniques and wavelengths. Conventional partial least squares (PLS) and locally weighted PLS (LW-PLS) were used as spectral analysis techniques. In addition, six spectral preprocessing techniques were evaluated: 1) no preprocessing, 2) standard normal variate (SNV), 3) 1st derivation, 4) 2nd derivation, 5) 1st derivation and SNV, and 6) 2nd derivation and SNV.

The results demonstrated that it is crucially important to include the dynamic measurement spectra acquired with the target blender in a calibration set. In addition, it was found that the prediction error of LW-PLS models was smaller than or at least similar to that of PLS models, through the model development study and the analytical validation. The present work also clarified that the modeling parameters, i.e., spectral preprocessing and wavelengths, affected the prediction error interactively. In conclusion, we propose to include the dynamic measurement spectra in a calibration set, to apply LW-PLS as spectral analysis technique, and to select modeling parameters by taking into account the interaction in order to construct a model with small prediction error.

374197 Detection of Process Fingerprints Applying Multivariate Analysis Methods to Monoclonal Antibody Cell Culture Data

Abstract:

This work presents a sequential data analysis path, which was successfully applied to identify process characteristics in fed-batch cultivation data. The analyzed data set incorporates 116 batch experiments of a fermentation process for the production of a Fc-Fusion protein at 3.5 liter scale. Starting with the univariate analysis of the investigated process and response variables as well as the effect of experimentally manipulated parameters a basic pre-characterization of process is performed. Consequently, the multivariate analysis mainly based on principal component analysis (PCA) and partial least squares regression (PLSR) is used for further investigation of the dynamic system. Various models are developed and evaluated using different techniques for data-pretreatment, data-recovery, variable transformation, data unfolding and grouping of the experiments. Based on the elaborated models characteristic patterns in the system, process fingerprints, are concluded. Those reveal an understanding for the dynamic importance of the variables for process modeling as well as an understanding for the dynamic predictability pattern of the process based on a tunable incorporation of its history. The obtained results are compared to an analogous analysis for a similar large scale (5000 liter) process data set and multivariate model-based scale-up techniques are discussed. The performance of the linear-based models is compared to alternative statistical techniques such as random forest revealing their opportunities and challenges. Finally, the elaborated fingerprints are interpreted regarding their supportive and simplifying applicability to bioprocess monitoring, analytics and control as well as design of experiments.

383038 In-Process Particle Size Analysis for Commercial Pharmaceutical Manufacturing

Abstract:

Particle sizes of drug substances, excipients, and intermediates are of great importance for most of solid oral dosage forms and are traditionally monitored and controlled by laboratory (off-line) analysis. Owing to advances in particle sizing techniques as well as implementation of process analytical technology (PAT) and quality by design (QbD), in-process particle size analyzers have been used in a variety of pharmaceutical applications recently. In-line or on-line particle size analysis has been demonstrated to be a useful tool to provide continuous real-time information, enhance product and process understanding, and ensure final drug product quality. However most of these applications focus on product and process development stages and little information is available on how to implement in-process particle size analysis for large-scale commercial manufacturing processes. The objective of this presentation is to discuss the current state of art regarding use of in-process particle size analysis for routine commercial manufacturing processes as well as its benefits and challenges. Some regulatory considerations with regard to submission and evaluation of in-process particle size analysis at commercial pharmaceutical manufacturing processes will also be discussed in this presentation.

On-line Reaction Profiling in Pharmaceutical Process Development

Authors: Sarah Pulicare, Elie Chaaya, Paul Oram, Zhihao Lin, Ralph Calabria, Sherry Song, Jeonghan Park, Feng Xu, and Azzeddine Lekhal, Merck & Co. Inc., Rahway, NJ

The first GMP step in the production of a desired Active Pharmaceutical Ingredient (API) involves a cyclization reaction that was shown to be oxygen sensitive. Despite consistent degassing procedures in the lab and plant, key oxidative impurities were present in all reaction streams at varying levels. It was unclear whether this variability should be attributed to processing conditions or oxidation during sampling and subsequent analysis. In order to differentiate between these instances, on-line HPLC and FTIR were used throughout this study. These two techniques were particularly used to track the consumption of the starting material, formation of the product, and generation of the oxidative impurity throughout the course of the reaction. In addition, a dissolved oxygen (DO) probe was implemented to monitor the concentration of oxygen in the headspace and solution.

Prior to proceeding with the reaction, different degassing techniques were evaluated to ensure consistency among runs. In parallel, some work was also done to define a sampling and analytical strategy that would guarantee sample stability. Despite these measures, high levels of oxidation were still observed in the lab runs. An initial analysis showed that this result was directly related to the solvent system in use, THF/DMAC. In particular, it was found that the level of the oxidative impurity was strongly correlated to the level of anti-oxidant, i.e. Butylated hydroxytoluene- BHT, usually present in solvents such as THF. To confirm this hypothesis, spiking studies were carried out and a significant improvement in product stability was demonstrated for high levels of this additive. This result was further confirmed after reassessing the reaction streams of the scaled-up batches run in our pilot plant. It was clearly demonstrated that batches with the highest level of oxidative impurities had non-detectable levels of BHT in the end of reaction streams. With adequate oxygen control strategies and with a better understanding of the mechanism of formation of the oxidative impurity, the experimental plan was further expanded to encompass range-finding studies of key process parameters (e.g. temperature, solvent composition, reagent equivalence, addition rates) to pave the way for a full QbD development strategy.

During this work, it was clearly demonstrated that the use of PAT reaction profiling tools helped in (1) streamlining the developmental work, (2) identifying critical experiments for downstream processing, and (3) ensuring a high probability of success upon scale-up by developing appropriate control strategies. Most importantly, it was also found that in comparison to traditional off-line techniques, a large body of data can be generated over a relatively short period of time and subsequently available for post reaction analysis.

368612 Applying PAT for Pharmaceutical Powder Blending Process: Challenges and Opportunities

Abstract:

Pharmaceutical powder blending presents challenges for high risk category products such as highly potent drugs, low dose drugs, drugs with narrow therapeutic indices (NTI), or drugs that have strong segregation or cohesion tendencies. Depending on the goals, applying PAT for pharmaceutical powder blending can help to achieve different levels of process understanding, such as real-time process monitoring and fault detection, process control strategies to ensure final product quality, or multivariate calibration model development and verification for real time testing and release, etc. When applying PAT for pharmaceutical powder blending, certain unique challenges are to be addressed from a multi-disciplinary or integrated approach. For example, i) materials aspect: inherent heterogeneity of pure components or mixture; ii) sampling aspect: sampling strategy, sample size, and scale of scrutiny; iii) measurement aspect: experimental errors associated with sampling practice, dynamic process environment, dynamic measurement environment; and iv) analytical aspect: appropriate reference method establishment. Unprecedented opportunities exist for innovation and process understanding. For example, particles with inherent heterogeneity (due to for example morphology difference) or strong segregation tendency (due to for example particle size difference) may call for particle engineering strategies (such as milling), novel formulation strategies, or innovative equipment/process design strategies such as continuous processing. In this presentation, case studies based on FDA research results will be presented to highlight several technical issues, such as dynamic measurement errors, sampling strategy (sample location and sample size), sampling bias, analytical verification, general linear modeling for process design space development, and scale-up effect. The effect of process variables on the dynamic measurement errors, design space, and scale-up will be discussed. Finally, a science- and risk-based approach for pharmaceutical powder blending will be discussed from a regulatory science perspective.

PAT Implementation and QbD Control Strategy for Commercial Production of a Drug Substance

(For presentation at AIChe 2014)

Elie Chaaya, Tseng-En Hu, Zhihao Lin, Lei Sun
Busolo Wabuyele, George Zhou, Cindy Starbuck
Merck Sharp & Dohme Corp.

Process analytical technologies (PAT) play an important role in process development and control throughout the pharmaceutical industry. In this presentation, established PAT tools such as mass spectrometry (MS), Fourier transform infrared (FTIR) and near-infrared (NIR) spectroscopy have been successfully used in the process development of a drug substance, enabling a Quality-by-Design (QbD) control strategy of an innovative manufacturing process resulting in greater process stability, portability, and flexibility. To realize this QbD benefit at the production site, online FTIR, NIR and MS, have been installed at different phases of this production process. First, at the crude step, an online FTIR has been used to monitor important process parameters of an enzymatic reaction and to ensure optimal operation conditions. The analyzer has also been used to detect the end-point of reaction to minimize process time and formation of impurities. Second, in the purification step, , an online NIR has been used to measure batch composition after solvent switch. By adjusting the batch composition across the design space, the optimum seeding temperature can be determined, ensuring a smooth process downstream to achieve the desired quality of the API. Not only does NIR play a role in feedback to adjust the solution composition, but also it helps to determine the seeding-point temperature in feed-forward. Third, in the final drying step, an online MS has been used for tracking the key solvent evaporating from the batch for better process understanding and robustness. Drying of the API is critical as any residual solvent left behind after drying could results in potential agglomeration during storage, impacting the quality and performance of API. Online MS helps to address this potential agglomeration by ensuring complete removal of unbounded water from the API solid.

Implementation of PAT methods has enabled the manufacturing site to realize QbD control strategy for the production process of this drug substance This allows for close monitoring of key process properties and ensures the process being operated within the predefined QbD control space. Any deviation from the control space will be detected by the PAT methods and correction procedures will be taken accordingly. In addition, PAT has enabled significant cycle time reduction by eliminating waiting time for offline lab results, bringing cost savings to the production site.

Always the Same? Location of Segregation via Spectroscopy

R. Hohl[1], O. Scheibelhofer[1], P. Wahl[1], Stephan Sacher[1], Johannes G. Khinast[1,2]

[1] Research Center Pharmaceutical Engineering GmbH, Graz, Austria
[2] Institute for Process and Particle Engineering, Graz University of Technology, Graz, Austria

Meeting the Quality Target Product Profile by means of a robust manufacturing process and proper in-line monitoring techniques is the target of state of the art pharmaceutical processing. Especially the correct quantity of active ingredient in the final dosage form is crucial. Considering tablet manufacturing as exemplary system, the shaping of the final dosing form happens in the tablet press, while the API concentration is determined in connected upstream blending steps. The API concentration is supposed to be invariant throughout the tablet press. However, an unavoidable phenomenon might spoil this: segregation.

In the recent years the application of near-infrared spectroscopy for testing the API-content has become widespread, and indeed offers a non-destructive method, that is fast enough to enable 100% control. Additionally, the application of hyperspectral imaging even allows to asses not only inter-tablet, but also intra-tablet homogeneity.

During this work, we created a method to follow a volume of material throughout the tableting process. This is achieved using the combination of a new multifiber spectrometer, which enables access of different measurement points at the tablet press. In addition, a hyperspectral imaging unit is used to assess the quality of the final tablets. A raw material combination promoting segregation effects is used as model formulation. By determining the concentration along the material stream inside the press as well as concentration fluctuations in the final tablets the kind and extent of segregation inside the press can be determined.

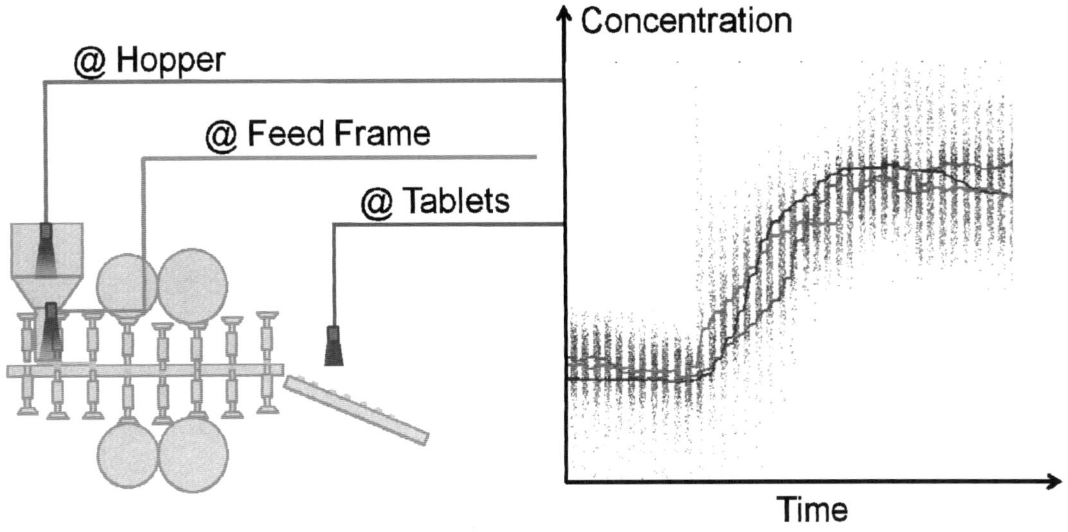

Figure 1: Scheme of the investigational set-up. Simultaneous measurements at several positions allow the location of segregation occurrence.

357614 PAT Data Management As a Tool to Support Continuous Production in Secondary Manufacturing

Abstract:

PAT data management as a tool to support continuous production in secondary manufacturing

Pamela Docherty, Siemens, Vertical Pharma

Kjell Francois, Siemens NV, Vertical Pharma

Pharmaceutical secondary manufacturing has long stood in vast contrast to the drug discovery end of the business, as well as to other sectors, when it comes to innovation such as continuous manufacturing. Is that about to change? We argue that the coming decade could see secondary manufacturing not just catch up but leapfrog other sectors but that mindset will be a crucial factor.

Secondary manufacturing has been for a long time an inefficient and time consuming part of the production cycle. Now, however, the traditional business model is breaking down with consequent pressures on all parts of the pharma value chain. Manufacturing's contribution to improving yield, reducing time, cost and waste is increasingly critical. Regulation which previously had insisted on batch testing is now moving to be much more supportive of real time product release and process analysis, heralding a future where the validation and establishment of continuous manufacturing will be easier.

Currently, more and more companies are making the move towards Continuous manufacturing. Smart data solutions must be provided on order to streamline the dataflow and get the essential information out of the big data stream. In this presentation we are diving in the mechanism of continuous manufacturing, leaving the isolated 'islands of automation' and working towards a fully integrated approach. Based on a couple of real established use cases we will explain how a PAT approach can be established in a continuous manufacturing area.

The Use of Single Use Mobius® MIX Vessels with Protein Solutions

Impact of Impeller Flow and Particle Shedding on Protein Turbidity and Aggregation

Dana Kinzlmaier and Elizabeth Goodrich, EMD Millipore
AIChE Conference, November 2014

Overview

1	Background
2	Evaluation of Mixing Action on Protein Quality
3	Evaluation of Particle Shedding during Mixing
4	Evaluation of Particle Presence on Protein Quality
5	Questions

Uses for Single Use Mobius® MIX Systems with Protein

- Process Intermediates
  - Product pool concentration typically changes over the course of a unit operation
- Pool Conditioning
  - pH and conductivity adjustments between unit operation
- TFF Recycle Vessel
  - Protein concentration and buffer composition change throughout UF/DF

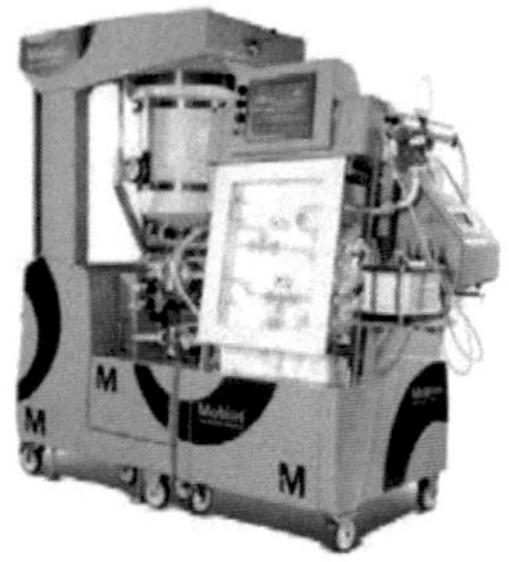

Potential Mobius® MIX Impacts on Protein Quality

- **Fluid Shear?**
  - Off Impeller tip

Effect of Shear on Proteins in Solution, C.R. Thomas and D. Greer, Biotechnol Letters, 2011

The paradigm that emerges from most studies is that shear in the fluid mechanical sense is unlikely by itself to damage most proteins and that interfacial phenomena are critically important.

Potential Mobius® MIX Impacts on Protein Quality

- Fluid Shear?
  - Off Impeller tip

- **Air-Liquid Interfaces?**
  - Turbulence at liquid surface
  - Vortexing

Characterization of Tank Mixing Regimes

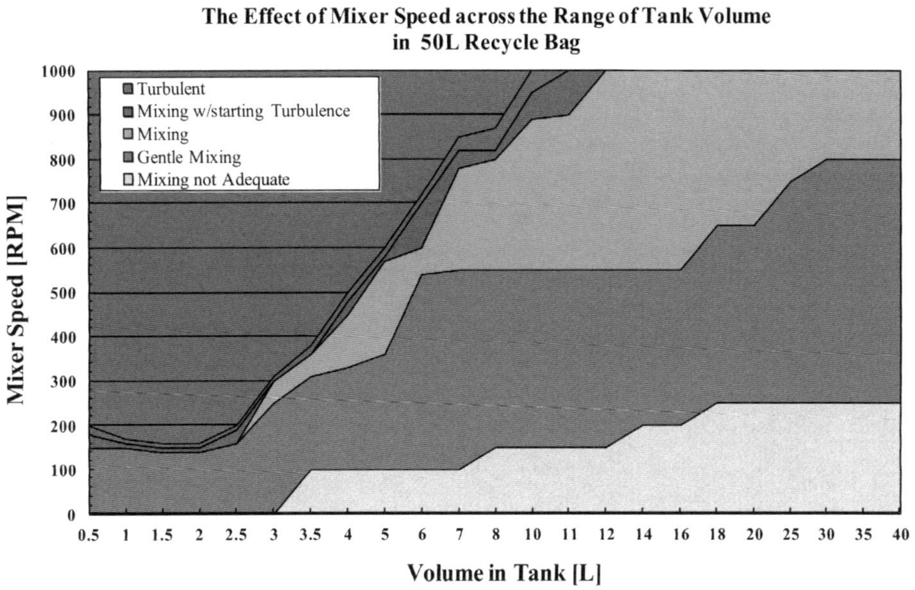

Potential Mobius® MIX Impacts on Protein Quality

- Fluid Shear?
  - Off Impeller tip
- Air-Liquid Interfaces?
  - Turbulence at liquid surface
  - Vortexing
- **Solid-Liquid Interfaces?**
  - "Grinding" action in impeller cup
  - Exposure to shed particulates
    - IgG Particle Formation During Filling Pump Operation, T. Randolph, Wiley Interscience, May 2008 (positive displacement piston pump used)

Evaluation of Mixing Action on Protein Quality

Design of the Mixing Study

- Model Feedstock
  - Polyclonal IgG in 25mM Sodium Acetate pH 5
  - 0.22μm filtered just prior to use
- Mixing Vessel Used:
  - Mobius® FlexReady Solution for TFF with 50L Recycle Bag
- Parameters evaluated:
  - Impeller speed – 100, 550, 750, 1000 rpm
  - IgG concentration – 1, 5, 10 g/L
  - IgG volume in the recycle bag – 2, 7, 12 L
- Monitored Turbidity vs Time

Use of Turnover Volume to Compare All Test Results

- Turnover Volume: ratio of liquid flowing through the openings present in the impeller to the total liquid volume in the recycle bag
- Fluid is pulled from tank through impeller holes and into the cup by the "pumping" action created by the rotating bottom surface of the impeller, the narrow gap, and the static base of the cup
- Flow throughout the rest of the tank is not considered (rotor shape is not a factor)

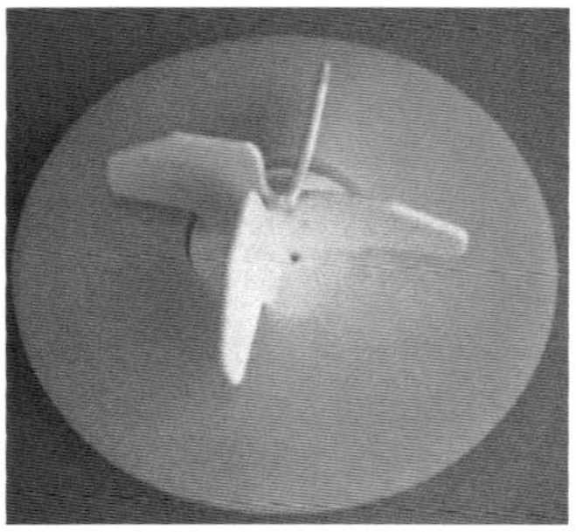

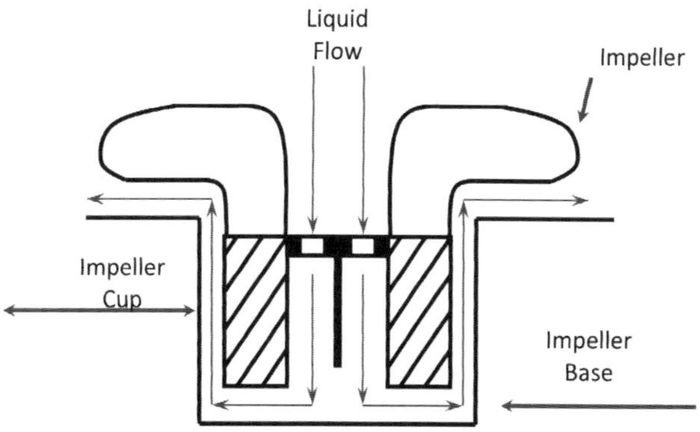

Comparison of Turnover Volume in 50L, 100/200L and 500/1000L Recycle Bags

Calculations are based on worst case scenario:

- o Minimum volume of solution in a recycle bag (10% of total recycle bag volume)
- o Impeller speed: 20% of full scale of the mixer

	Mobius® TF-2 System	Mobius® MIX		Mobius® MIX 500/1000	
	(50L)	100L	200L	500L	1000L
Turnover Volume/Hour	0.97	0.87	0.43	0.21	0.11

Turbidity versus Tank Turnover Volume

Turbidity as a Function of Turnover Volume using IgG Solution.

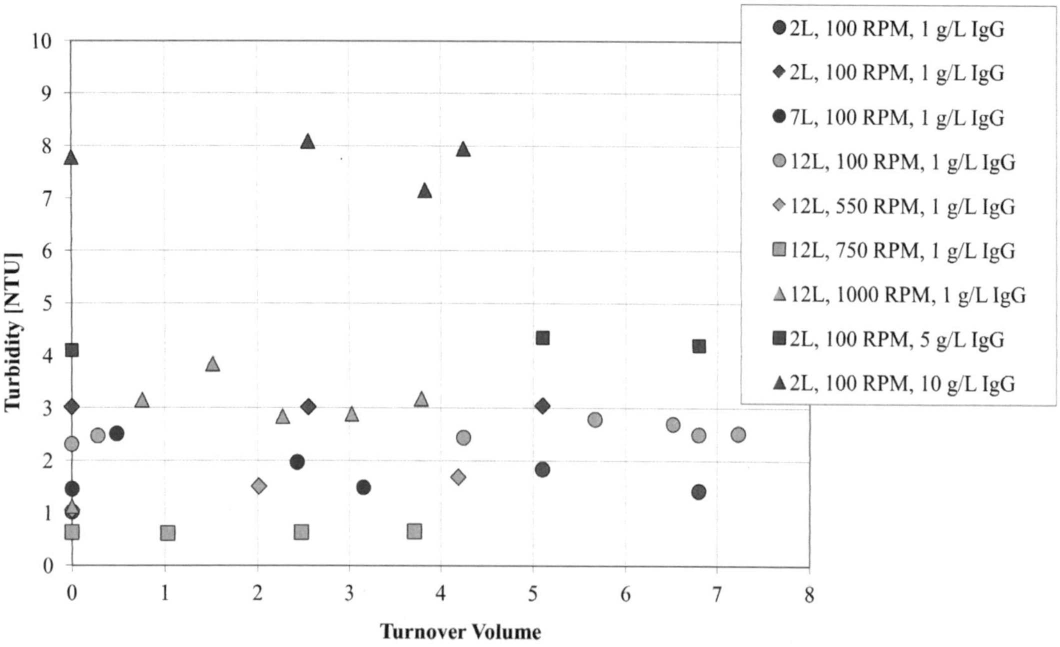

Mixing Action Conclusions

Impeller Speed

- No effect observed during 8 hours of process

Protein Concentration

- No significant effect of concentration on turbidity rise with mixing
- Initial turbidity is higher when concentration is higher

Initial Volume

- Calculated turnover for Mobius® Single-use Mixing Solution is < 1 per hour of process time
- Mixing does not have any effect on turbidity over 8 hour process time

Evaluation of Particle Shedding during Mixing

Particle Shedding Evaluation

Sample

- 12 L of filtered water
- 50L Recycle Bag, mixed for 48 h at 1000 RPM
- 0.22 μm capture and re-suspension

Control 1

- 12 L of filtered water
- no 50L Recycle Bag, no mixing
- 0.22 μm capture and re-suspension

Control 2

- 12 L of filtered water
- 50L Recycle Bag, no mixing
- 0.22 μm capture and re-suspension

Particle size analysis

- Liquilaz® Particle Measuring System: 0.2 μm to 2 μm detection range
- Mastersizer® System: 0.05 μm to 800 μm detection range

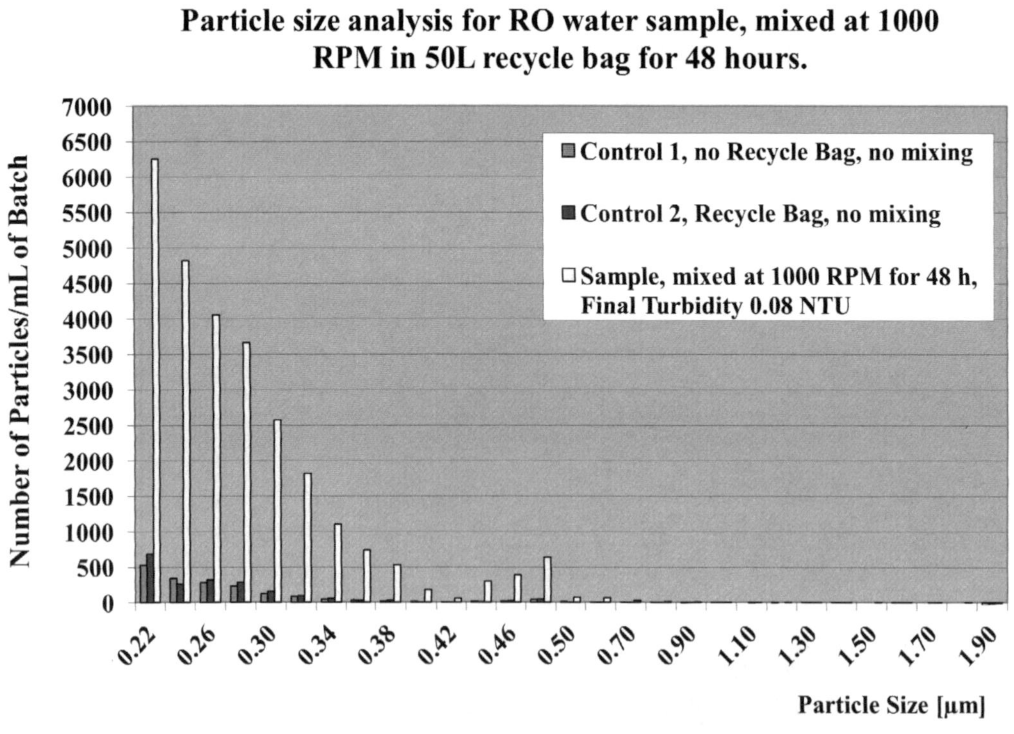

Evaluation of Particle Presence on Protein Quality

Particle Shedding Impact Procedure #1

- 10 g/L IgG solution diluted 1:1 with collected Particle Shedding sample and controls

- Particle concentration is 24X higher in diluted samples than in actual recycle bag
  - 12L of particles resuspended in 250mL, diluted 1:1 with protein

- SEC HPLC analysis performed for monomer/dimer %
  - t = 0, 3, 24 hours after particle addition

HPLC Results

Time	t = 0 hrs			t = 3 hrs			t = 24 hrs		
Sample	Monomer Peak Area	HMW Peak Area	Monomer %	Monomer Peak Area	HMW Peak Area	Monomer %	Monomer Peak Area	HMW Peak Area	Monomer %
Control 1	143.1	14.5	**90.8**	144.0	14.2	**91.0**	144.7	13.0	**91.8**
Control 2	144.4	14.6	**90.8**	145.0	14.3	**91.0**	146.3	13.2	**91.7**
Sample	146.2	14.7	**90.9**	146.1	14.2	**91.1**	146.4	13.1	**91.8**

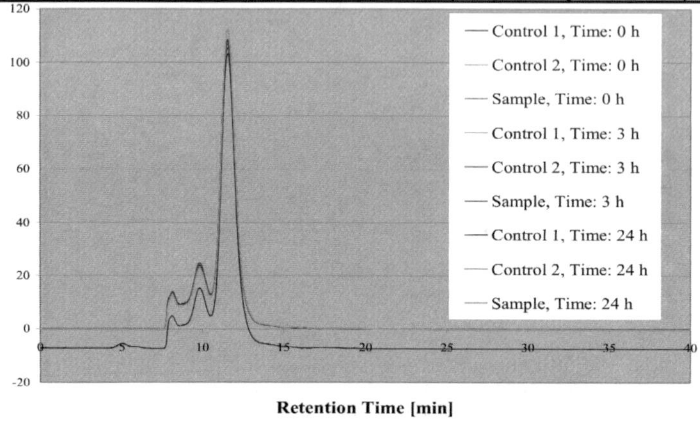

Particle Shedding Impact Procedure #2

12L of 1 g/L IgG Solution (0.22 μm filtered just prior to use)

Mixed at 750 RPM in 50L Recycle Bag for 48 hours

Sample 1: Control, no mixing, 198 hour RT hold

Samples 2 & 4: 48 hours mixing @ 750 RPM, 150 hour RT hold

Sample 3: 48 hours mixing @ 750 RPM, 150 hour 4 ° C hold

Sample	Monomer [%]	Dimer [%]
Sample 1	95.7	4.3
Sample 2	95.6	4.4
Sample 3	95.6	4.4
Sample 4	95.7	4.3

Particle Shedding Conclusions

- Particle analysis using the Liquilaz® Particle Measuring System, shows the presence of particles smaller than 0.8 µm
- Control samples indicate that mixing is the primary cause of particulates
- HPLC analysis confirmed that particles generated by the impeller used in single use recycle bags don't act as nucleation sites for protein to aggregate in 24 hour time frame
- 48 hours of active mixing does not significantly impact protein aggregation over an approximately 1 week duration

Acknowledgements

Herb Lutz

Tina Carbrello

Nick Harrington

Venkatesh Natarajan

Thank You!

375141 Practical Implementation of Single Use Filtration Assemblies in the Final Filling Operation

Abstract:

Single use assemblies have gained broader acceptance over the past decade with more recent focus further downstream in the formulation and filling area. However, installing single use filling lines can also prove challenging for operators who are more familiar with stainless steel systems. The handling and the operation of a single use assembly is different compared to a stainless steel filtration train. This paper will describe the practical experiences of implementing a single use final filtration assembly from the design stage through to the day to day operation. It will cover such topics as process design considerations, unforeseen problems with the implementation of the assembly, and the operational steps involved to conduct a pre-use, post-sterilization integrity test.

Selecting Single-Use Components for the Fill/Finish Operation

The selection of components in a Single-Use assembly is contingent on defining clear objectives and the boundary where Single-Use products are to be applied. This presentation focuses on the fill/finish steps in a biopharmaceutical operation. These steps make up a critical phase of the production process. Single-Use components applied here must meet the strictest regulatory scrutiny.

We rationalize the qualification of components that are used in fill/finish assemblies and define how these components are selected for optimum operation. We review the major areas of consideration that make up the typical Single-Use fill finish assembly including:

1. Therapeutics supply vessel
2. Pumping section
3. Area classification and transfer into restricted access area
4. Filling needles
5. Automated filling line

We also present a case study where needle revalidation caused a scramble to finalize Single-Use assembly qualification/production. It's the little things that unexpectedly arise to create setbacks. We highlight how these "little things" can be addressed in the implementation project plan.

David Richardson
presentation ID# 362186:
Clarification of CHO Bioreactor (2-6 lpm) with Single Use Centrifuge-Unifuge®

Session: Single Use Systems: A Revolution in Biotechnology Processing scheduled for Wednesday, November 19, 2014: 03:15 PM - 05:45 PM, Hilton, 206

The large scale economic collection of proteins is increasingly an important problem for the Biotechnology industry. Separation of the desired protein from the Bioreactor broth poses a formidable challenge.
Advances in cell culture techniques have greatly increased the pcv and cell populations per ml. and increased titers of the target proteins in culture fluid. This increase in upstream efficiency has led to a separation challenges in downstream processing at the clarification stage.
Clarification of the harvested cell culture fluid is an important process for protein yield. When the target protein is extracellular, cell clarification techniques can be used to optimize the separation of the product from the cells and cellular debris.

Pneumatic Scale Angelus

barrywehmiller bw betterworld

UniFuge ®
Introduction

UniFuge ® -Single Use Centrifuge

- UniFuge is a Single-use tube bowl centrifuge based on proven platforms. In one application, 1200 hours or 320 runs without a single production batch loss. Available since 2010.
- **Mammalian** Applications: CHO, MD, HAL2, MRC5, SF9,HEK, PERC6
  - No shear at inlet and no shear at discharge
  - Clarification, separation, collection
  - Sterile cell harvest, stem cell and cell banking
  - vaccines
- 10-1000 liter bioreactor clarification: 200-4000 g
  - Process Feed rates 0.25 – 6 L/min (to 360 L/hr)

Pneumatic Scale Angelus

barrywehmiller bw betterworld

UniFuge®
Process Visualization

CARR UniFuge®

Sterile Separation Module in Single Use Technology

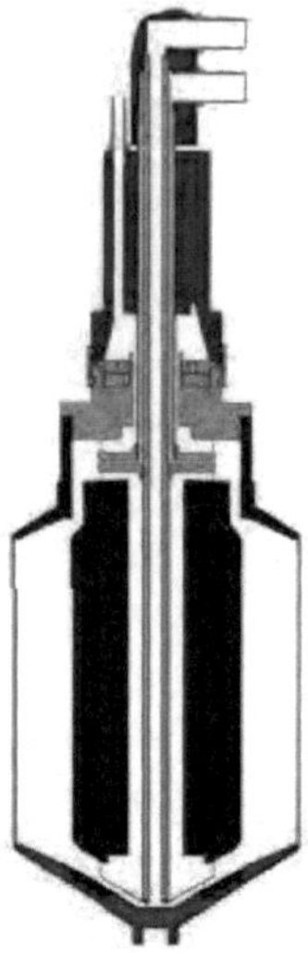

CARR Centritech SEPARATION SYSTEMS

pneumatic scale ANGELUS

Pneumatic Scale Angelus

barrywehmiller bw betterworld

UniFuge® Pilot Layout

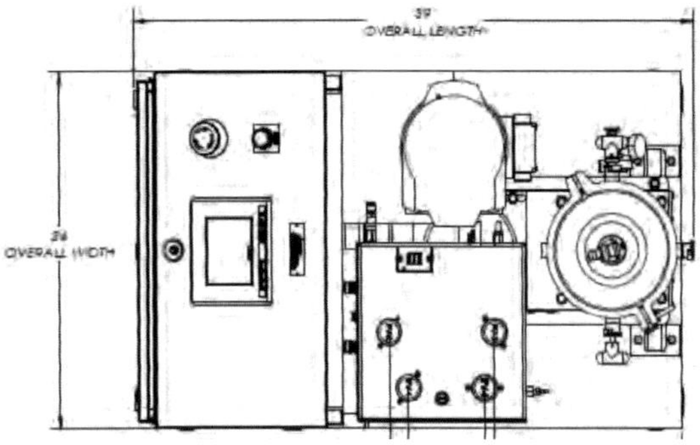

- Electrical: 120/230VAC, 1 phase, 50/60Hz, 15/7.5A
- Pneumatic: 80 PSI, 10 SCFH
- Feed: 200 ml/min to 6L/min, ½" tubing
- Cooling jacket-1L/min; max pressure: 90 psi

Pneumatic Scale Angelus

barrywehmiller bw betterworld

CHO 1-6 LPM NTU/Vmax MA CMO

G force	Feed rate lpm	Feed volume liters	Ntu	Vmax at 20 psi
3000	2	3	87	356 Vs 156 d.s.
3000	3	3	94	217
3000	4	3	111	
3000	5	3	120	
3000	6	3	144	
3000	1	3	77	
4000	2	3	80	322
4000	4	3	102	
4000	6	3	118	80
4000	1	3	72	
4000	5	3	115	
4000	4	22	117	
Control 3500 rpm			61	
Feed			1260	

15E6 cells/ml 70% viability

Pneumatic Scale Angelus

barrywehmiller bw betterworld

UniFuge® Loading and Pressure

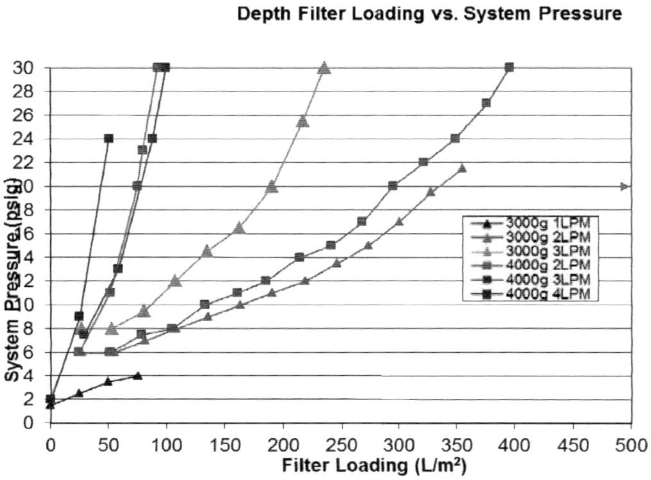

Millipore X0HC grade depth filter followed in series by a Sartorius Sartopore 2 0.45/0.2um filter

Pneumatic Scale Angelus

barrywehmiller bw betterworld

27 % viability CHO 12E6 JP

Parameter set Result Table 1:	Feed rate lpm	Feed volume liters	Ntu	Discharg e volume	G force	Time at NTU sample	Avg. NTU pooled (2 cycles)	Pmax rating
1	1	2	39		4000	0		#1
1	1	7	44		4000	5		
1	1	12	45		4000	10		
1	1	17	44		4000	15		
1	1	22	51		4000	20		
1	1	27	48		4000	25		
1	1	33	50	1.4/1.8	4000	33	51	
2	3.4	2	159		4000	0		#3
2	3.4	8.8	56		4000	2		
2	3.4	15.6	74		4000	4		
2	3.4	22.4	74		4000	6		
2	3.4	29.2	72		4000	8		
2	3.4	33	72	1.4/1.7	4000	10	86	

Pneumatic Scale Angelus

barrywehmiller bw betterworld

27% Viability CHO 12E6 JP

Parameter set	Feed rate lpm	Feed volume liters	Ntu	Discharg e volume	G force	Time at ntu sample	Avg ntu of pooled 2 cycles	Pmax rating
3	3.4	2	192		3000	0		#4
3	3.4	8.8	63		3000	2		
3	3.4	15.6	66		3000	4		
3	3.4	22.4	65		3000	6		
3	3.4	29.2	63		3000	8		
3	3.4	33	64	1.3/1.6	3000	10	85	
4	6.0	2	124		4000	0		#2
4	6.0	8	90		4000	1		
4	6.0	14	83		4000	2		
4	6.0	20	79		4000	3		
4	6.0	26	80		4000	4		
4	6.0	30	81		4000	5	88 one cycle	

Pneumatic Scale Angelus

barrywehmiller bw betterworld

Why UniFuge® ?

- 85-90 % filter reduction
- 7-30% titer increase
- Mass balance increases 7% vs. centrifuge,15% vs. depth filter,
- Easy to use, 15 minute turn around time, no CIP or SIP.
- FDA friendly- closed process , no cross contamination
- Small foot print.
- No WFI flushing
- Module weighs 1 kg vs. heavy filtration cartridge

Pneumatic Scale Angelus

barrywehmiller bw betterworld

382787 A Method for Pre-Use Integrity Assurance with Large Volume Single-Use Bin and Mixer Bags

Abstract:

Before committing high value products or feed streams into single-use bags, it is a wise business decision to test the bag for integrity following installation into the carrier. Despite rigorous quality control measures, assembly release integrity testing, and advanced packaging methods for transportation from the vendor's site, film bags may still incur damage during unpacking, handling and installation if robust handling practices are not in place. As more biotech firms incorporate single-use bags into their manufacturing work flows, the field continues to lack a recommended method for integrity assurance testing by the end-user. This work describes such a method for large volume bags, adapted from one vendor's release testing criteria, and attempts to quantify performance around some common breaches of integrity. The advantages of this method is ease of implementation, rapid execution, simple interpretation of results, and a defect detection as sensitive as the vendor testing. Furthermore, a correlation is drawn with liquid seepage from the man-made defects to assess the resolution of the integrity assurance method to detect a potential leak prior to use.

373244 A Mathematical Model for Drug Delivery and Dose Optimization for Solid Tumors

Abstract:

Recent advancements in computing power has allowed for the mathematical simulation of systems that are otherwise too complicated to solve analytically. This work will focus on the mathematical modeling and simulation of the solid tumor microenvironment and how various doses and infusion times will affect the solid tumor.

The first part of this work presents a brief literature review on the mathematical modeling of drug release from diffusion-controlled drug release devices. These are drug systems that are comprised of a polymer that encapsulates the drug in some form, and dictates the diffusion and release of drug. In many cases, the polymer can swell due to the increase in the volume of material due to absorption of solvent. Therefore, an investigation on the behavior of drug diffusion in a polymer-swelling system was performed. Sensitivity analysis revealed that the diffusion coefficients in played a major role in the drug release profiles in such systems.

The second part of this work models the diffusion of drug in solid tumors. Specifically, a mathematical model of drug behavior in solid tumors is performed in order to characterize the mass transport of drug in the solid tumor microenvironment. The equations used to model the system are based on the convection-diffusion equation. Clinical parameters such as dose and infusion time were varied to determine optimal tumor regression, and thus, optimal dose treatments. It was found that for drug efficacy could be maintained while reducing cardiotoxicity by two methods: (1) increase infusion times for large doses (e.g. 40 min or greater for doses of 100 mg/m^2 or more), or (2) Fractionate a standard dose (approx. 40 – 100 mg/m^2) into several smaller doses (approx. 15 – 25 mg/m^2 each) which favor quicker infusion times.

Based on the modeling framework presented in the second part, a mathematical model accounting for more complex biochemical phenomena, such as intracellular signaling, is presented. Intracellular signaling plays a major role in various pathways that regulate a cell's metabolism, gene expression, or the ability to undergo cell division. Clinically relevant parameters, such as dose and infusion time, were varied in order to characterize what affect various treatment regiments have on the biochemical response of solid tumor systems. It was found that higher infusion times had the most significant impact on tumor regression than any other parameter.

Characterization of a Continuous Granulator

Madeline Candelaria, Carlos Velázquez
Department of Chemical Engineering, University of Puerto Rico-Mayagüez
Center for Pharmaceutical Engineering Development and Learning (CPEDaL)

ABSTRACT: Inefficient energy consumption, enormous facility requirements, lack of flexibility of variations in batch size and difficulty for scale-up have been some of the reasons that motivate the pharmaceutical industry to move from batch processes to continuous processes. Currently, the granulation process is still considered one of the most unrepeatable processes in the pharmaceutical industry being carried-out in a batch-wise manner. Continuous granulation has been identified as an advantageous solution, providing the implementation of continuous monitoring and control, assuring the quality of the product; reducing the human intervention so that human mistakes are reduced, among other substantial benefits. Granules are currently obtained from two types of processes: dry and wet granulation and as recent studies have demonstrated, most of the proposed continuous granulation techniques (spray drying, fluid bed agglomeration, roller compaction, twin screw granulation) are in certain cases not suitable for the pharmaceutical industry or even more, those processes are not always fully understood. It is therefore that this project wants to develop a granulator that could produce granules continuously, validate it and investigate the effects of process parameters as well as material properties in the granule growth. The main objective of this research is to understand the fundamentals behind the continuous granulation process by means of an experimental design and by using the population balance equation to model the agglomeration mechanism of the particles.

The experimental design consists in carry-out granulations at different granulator paddle speeds using Anhydrous Lactose, Monohydrate Lactose and Povidone as a binder. The speed will be varied from 150 rpm's to 500 rpm's. The analysis will be supported using a Laser Diffraction Method and conventional sieving test to determine the particle size distributions (PSD) of the granules and evaluate the PSD against granulator speed.

Keywords: continuous processing, wet granulation, particle size, granule growth, continuous paddle granulator

Preparation of structured meso–macroporous silica materials: Application as delivery system

Santamaría. E, Maestro.A, Gutiérrez, J.M* and González. C.
Chemical Engineering Department, Faculty of Chemistry, University of Barcelona, Martí i Franqués 1-11, Barcelona 08028, Spain
**e-mail: carme.gonzalez@ub.edu*

Meso–macroporous silica materials with a well-ordered array of mesopores were prepared from oil-in-water emulsions. The influence of the following three composition variables on material characteristics was studied: emulsion dispersed phase fraction, concentration of silica and concentration of surfactant. The obtained materials were characterized via small-angle X-ray diffraction scattering, scanning electron microscopy, transmission electron microscopy, Hg intrusion porosimetry and nitrogen adsorption–desorption isotherms. A network of structured mesopores was obtained even when using a highly concentrated emulsion (volume of the dispersed phase, > 0.75). The mesopore network presented a hexagonal arrangement, with mesopore diameters between 4 and 7 nm. Non-ordered macropores, with diameters between 50 nm and 10–15 µm were also present, depending on composition variables. The isotherms were of type IV, typical of mesoporous materials, but at high p/p_0 they were the usual shape for the macroporous materials. The possibility of tailoring mesopore and macropore structures by modifying composition variables could extend the application of these materials.

In figure 1, example of meso-macroporous material with macroporous and mesoporous ordered are presented.

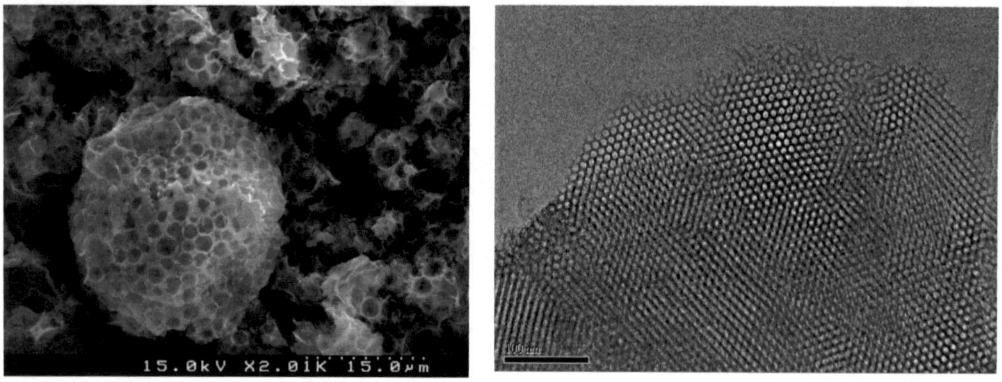

Figure 1. SEM and TEM images showing macroporous and mesoporous structures obtained

The effect of the materials' properties on ibuprofen (IBU) drug loading and release was studied. The results showed that the loading of IBU increases as the macropore presence in the material is increased. The IBU adsorption process followed the Langmuir adsorption isotherm. A two-step release process, consisting of an initial fast release and then a slower release was observed. Macropores enhanced the adsorption capacity of the material; this was probably due to the fact that they allowed the drug to access internal pores. When only mesopores were present, ibuprofen was probably adsorbed on the mesopores close to the surface. Moreover, the more macropore present in the material, the slower the release behavior observed, as the ibuprofen adsorbed in the internal pores had to diffuse along the macropore channels up to the surface of the material. The material obtained from a highly concentrated emulsion was functionalized with amino groups using two methods, the post-grafting mechanism and the co-condensation mechanism. Both routes improve IBU adsorption in the material and show good behavior as a controlled drug delivery system.

Another application of mesoporous materials is in the cosmetic world. Most sunscreen formulations are emulsions or creams. However, formulations based on nanoparticles as the delivery vehicle for the sunscreen compounds potentially have advantages in terms of retention on the skin, lack of penetration across the epidermal layer and UV attenuation by both absorption and scattering. Polyphenols were also adsorbed in the materials in order to obtain an efficient UV filter mixed with the cosmetic cream.

Keywords: Silica, emulsion, meso-macroporous, ordered mesopores, cooperative templating mechanism, ibuprofen, polyphenols, cosmetic cream.
Financial support from CYCYT CTQ2011-29336-C03-02 is gratefully acknowledged

383719 Process Control of the Continuous Tumbler Mixer

Abstract:

The development of new process technologies permits the characterization and innovation in the pharmaceutical industry. An efficient automatic control system is highly desired to satisfy regulatory constraints and market demands. The online monitor and information of the system are necessary factors to reach a continuous process. For the online monitor of the system, process sensors such as Near infrared, camera, temperature, level etc. had been proposed to obtain process information. Today's computational resources allow the design of more complex controllers and algorithms, this controllers can use similar techniques as human learning in order to adapt and maintain the desire parameters.

The aim of this study is to develop an efficient way to control specific process parameters of a continuous tumbler mixer. Throughput, API concentration, blend homogeneity are key factors to maintain in the process. The material throughput across the line is monitor with the feeders feed rate and adjust with a PID controller. The API concentration is monitor with a Near Infrared sensor and adjusts with a PID controller. The blend homogeneity is maintained with a Sony Hdr-Cx110 camera and an algorithm base on Image analysis and Fuzzy Controllers. The integration of the monitored techniques (NIR spectroscopy, feeder feedrate and Sony Hdr-Cx110 camera) with control programs and systems (Matlab and Emerson Delta V) permit to adjust process parameters. These parameters can be monitor and control to produce a robust continuous mixing process. The next step of the investigation is to observe the behavior of the factor interaction in the proposed control algorithm.

Investigation of The Applicability And Properties of Date Stone Extracts As Pharmaceutical Excipients

Abstract

Excipients play an important role in the manufacture of medicines, helping to preserve the safety, stability and effectiveness of active pharmaceutical ingredients (APIs) and ensuring that they provide their promised benefits to the patients. Optimal utilization of such excipients provides pharmaceutical manufacturers with cost-savings in drug development, improved functionality and help in drug formulation novelty [1]. They perform as vehicles which transport the incorporated drug to the site of absorption and are expected to guarantee the stability of the incorporated drug, the precision and accuracy of the dosage, and also increase the organoleptic characteristics of the drugs, where required, in order to enhance patient adherence [2].

The use of natural excipients to deliver the bioactive agents has been hindered by the preference for synthetic materials. Natural excipients have rewards over synthetic ones as they are chemically inactive, biodegradable, nontoxic, cheap and widely available [3]. They can also be modified in different ways to achieve tailor-made materials for drug delivery systems and therefore can compete with the existing synthetic excipients. Starch is one of the most widely used natural polymers in pharmaceuticals due to its nontoxic, biocompatible and biodegradable properties. Furthermore, it is reasonably pure and does not need extreme purification treatment, especially when administered orally. The contribution of bioactive food compounds such as phenolics in several metabolisms and their neutralizations with disease-developing mechanisms has caused an increasing attraction to functional foods and drugs [4]. Date palm pit extract is a rich source of polyphenols that exhibit metal scavengering, anti-mutagenes, and antimicrobial activities. Also, the date stone represent about 13-15% of the total weight of the date fruit and contains approximately 55-70% nitrogen free extract. Almost 55-73% of the nitrogen free extract in the date stone is mainly starch [5]. Date stone starch can be simply isolated by treatment with water under mild conditions, thus, date stones are a great alternative source of starch. The use of natural polyphenols and starch in formulation of drugs instead of synthetic materials can overcome the disadvantages of their instability and alleviate unpleasant tastes, in addition to improving the bioavailability and dissolution profiles.

This proposal aims to carry out a preliminary analysis of a well-known and promising source of starch and polyphenol from date stones, raw and thermally treated, for a possible application as a pharmaceutical excipient through physicochemical morphological and thermal characteristics. Moreover, it is important to search the feasibility to extract starch and polyphenols from thermally treated stones, providing extracts with different properties.

References

[1] E. Sjögren, B. Abrahamsson, P. Augustijns, D. Becker, M.B. Bolger, M. Brewster, J. Brouwers, T. Flanagan, M. Harwood, C. Heinen, R. Holm, H.-P. Juretschke, M. Kubbinga, A. Lindahl, V. Lukacova, U. Münster, S. Neuhoff, M.A. Nguyen, A.v. Peer, C. Reppas, A.R. Hodjegan, C. Tannergren, W. Weitschies, C. Wilson, P. Zane, H. Lennernäs, P. Langguth, In vivo methods for drug absorption – Comparative physiologies, model selection, correlations with in vitro methods (IVIVC), and applications for formulation/API/excipient characterization including food effects, European Journal of Pharmaceutical Sciences, 57 (2014) 99-151.
[2] N.C. Ngwuluka, B.A. Idiakhoa, E.I. Nep, I. Ogaji, I.S. Okafor, Formulation and evaluation of paracetamol tablets manufactured using the dried fruit of Phoenix dactylifera Linn as an excipient, Research In Pharmaceutical Biotechnology, 2 (2010) 25-32.
[3] A. Shirwaikar, A. Shirwaikar, S.L. Prabu, G.A. Kumar, Herbal Excipients in Novel Drug Delivery Systems, Indian Journal of Pharmaceutical Sciences, 70 (2008) 415–422.
[4] M.J. Jivan, M. Yarmand, A. Madadlou, Encapsulation of date palm pit extract via particulation of starch nanocrystals in a microemulsion, International Journal of Food Science and Technology, 49 (2014) 920–923.
[5] O.M. Yousif, M.F. Osman, G.A. Alhadrami, Evaluation of dates and date pits as dietary ingredients in tilapia (Oreochromis aureus) diets differing in protein sources, Bioresource Technology, 57 (1996) 81-85.

384673 CHO Cell Line Stability: Impact of Cell Banking

Abstract:

The CHO cell line is widely used as the workhorse for therapeutic protein production in the Biopharmaceutical industry. Identifying factors that impact the stability of CHO cell lines is crucial in ensuring robustness of cell line selection. To understand the impact of banking (one freeze-thaw cycle for primary seed bank generation) on CHO cell line stability, we studied 24 cell lines, producing 3 monoclonal antibodies The experiments focused on understanding the change in cell culture Key Performance Indicators (KPI), product quality and flow cytometry characteristics after banking. Three rounds of shake flask studies were conducted over the course of three months to compare banked versus unbanked cells in parallel. For the 24 test case cell lines, we observed comparable cell culture KPI profiles, product quality attributes and flow cytometry profiles between banked and unbanked cell lines across the three sets of shake flask experiments. From these experiments we were able to conclude that one round of banking (one freeze-thaw cycle) does not have an impact on stability of the 24 cell lines tested.

386566 A Simple Process Transfer Method for Roller Compaction

Abstract:

A probability based process transfer method for roller compaction, from development scale to commercial scale, was found to accurately describe the mechanism of ribbon formation at both scales. In this method, a statistical relationship between roller compaction parameters and ribbon attributes was first defined using a development Alexanderwerk WP120 roller compactor with an 18-run design of experiment. The statistical relationship was then adapted to a commercial Alexanderwerk WP200 roller compactor with one experimental run on the Alexanderwerk WP200. The experimental run served as a calibration of the statistical model parameters. The proposed transfer method was then confirmed by conducting a mapping study on the Alexanderwerk WP200 using a factorial DoE, which showed a match between the predictions and the verification experiments. The study demonstrates the applicability of the roller compaction transfer method using the statistical model from the development scale calibrated with one experiment point at the commercial scale.

Investigation of impact of roller compaction process parameters on critical product attributes

Rakesh C. Dontireddy[1, 2], Mary E. Crowley[1], Graham O'Mahony[1], Raghu Peddipatla[1], Michael A. McAuliffe[3] and Abina M. Crean*[1,2]
[1]School of Pharmacy, University College Cork, Ireland
[2]Pharmaceutical Technology Manufacturing Centre, Ireland
[3]Centre for Advanced Photonics and Process Analysis, Institute of Technology, Cork, Ireland
*a.crean@ucc.ie

INTRODUCTION

Roller compaction (RC) is a continuous dry granulation technique in which a dry powder blend is compressed between two rotating rolls to form ribbons which can be subsequently milled to form granules. A number of reports have described how granulate and tablet properties depend upon the properties of the ribbon such as solid fraction, ribbon density and strength[1,2,3]. The primary aim of this study is to determine how differences in the API particle size (within the blend to be compacted) impacts on ribbon and subsequent granule properties.

Objectives

- To determine whether ribbon properties can be used as an indicator of granule and ribbon properties for formulations containing API with a range of particle size.
- To determine the effect of (1) process parameters and (2) micronised versus unmicronised API on the quality of ribbons.
- To explore the applicability of PAT tools to measure in-process material attributes during a roller compaction process

RESULTS

Ribbon Density

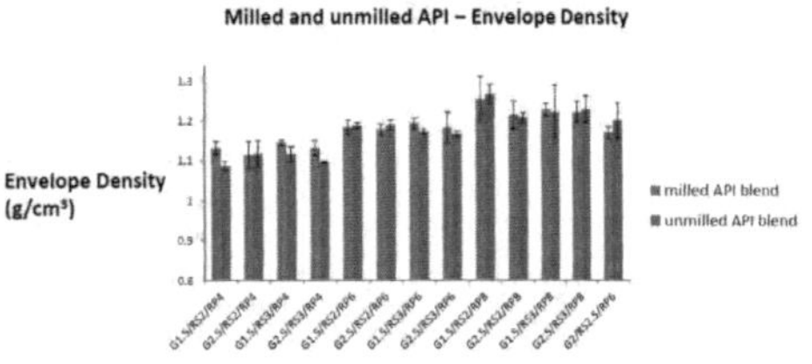

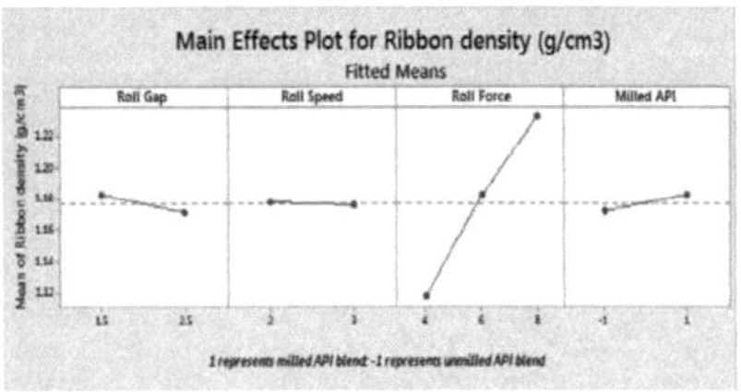

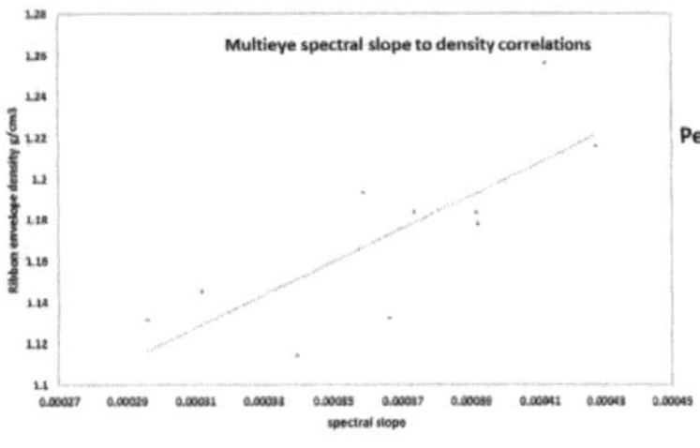

Pearson Coefficient = 0.77, P<0.01

MATERIALS AND METHODS

Formulation

Metoclopramide HCl	49.4% w/w
Microcrystalline Cellulose PH102	24.8% w/w
Lactose Anhydrous	24.8% w/w
Croscarmellose Sodium	0.8% w/w
Mag. Stearate	0.3% w/w

Note: Metoclopramide HCl investigated was micronised and un-micronised

Design of Experiments (Full Factorial+centre point, 13 runs)

Parameter	Levels
Roller Gap	1.5 mm 2.5 mm
Roll Speed	2 RPM 3 RPM
Roll Force	4 kN/cm, 6 kN/cm, 8 kN/cm

Processing Equipment Roller compactor: CCS220 Fitzpatrick

Analysis

Ribbons	Granules
Multieye®: At-line NIR Geopyc 1360/AccuPyc Density	Eyecon® In-line particle size Sieve analysis % fines Compaction studies Piccola Riva with MCC AIM system

RESULTS

Granule Particle size

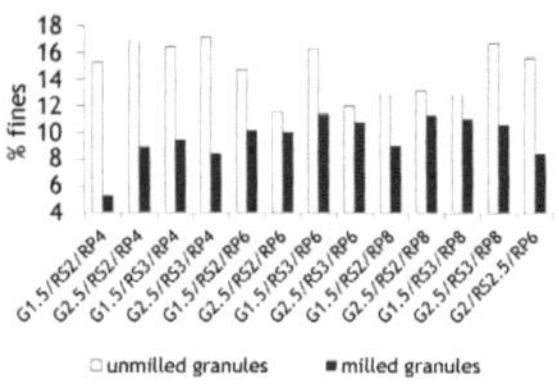

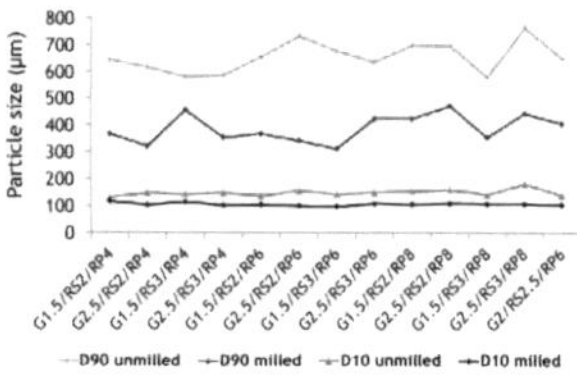

Granule Compaction

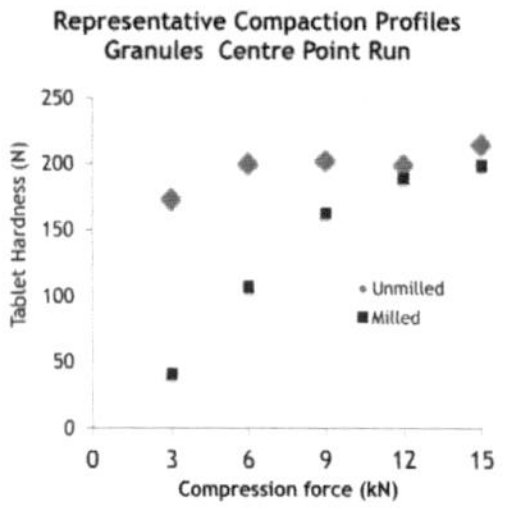

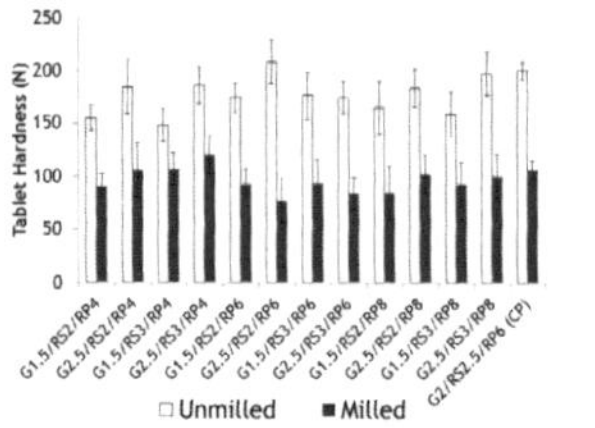

CONCLUSIONS

- Ribbon density did not indicate subsequent granule and tablet properties for blends and process parameter ranges studied.
- API particle size showed the greater influence on granule compactability.
- PAT tools, Multi-eye N-IR at-line and Eyecon in-line showed applicability as tools to measured in-process materials.

ACKNOWLEDGEMENTS

Research is funded by the Pharmaceutical Manufacturing Technology Centre, an Enterprise Ireland, Irish Development Agency funded centre. Roller compactor study supported by the Fitzpatrick Company. Multi-eye® was supplied by Innopharma Labs. Avicel PH102 was donated by FMC. Work is performed in collaboration with SSPC, Ireland.

REFERENCES

[1] Zinchuk, A. V.; Mullarney, M. P.; Hancock, B. C. *Int. J. Pharm.* **2004**, *269*, 403.
[2] Acevedo, D.; Muliadi, A.; Giridhar, A.; Litster, J. D.; Romanach, R. J. *AAPS PharmSciTech* **2012**, *13*, 1005.
[3] Nesarikara, V. V.; Vatsaraja, N.; Patela, C.; Earlya, W.; Pandeya, P.; Sprockela, O.; Gaoa, Z.; Jerzewskia, R.; Millerb, R.; Levin, M. *International Journal of Pharmaceutics 426 (2012) 116– 131* **2012**, *426*, 116.

387244 Novel Antifungal Fusion Proteins

Abstract:

Development of recombinant platform to express novel antifungal peptides with the use of an anchor protein forms the motivation for the present work. Although numerous soluble peptide signaling molecules are known, there are very few options of psuedobiospecific anchor peptides useful for the development of new fusion proteins. OmpL37, a protein found on the surface of leptospire spp. is hypothesized to be useful for the binding of biologics to skin, as it has strong preferential affinity for components of the (epi)dermis while giving a nil immune response.

This poster will discuss the methodology of developing OmpL37-based constructs for the effective delivery of a novel antifungal peptide. The poster describes the cloning strategy of intermediate (GFP-based affinity) and final sequences, defines conditions for fed-batch cultivation (media and inducing strategies), and examines efficacy of the antifungal.

Microfluidic Platform for Crystallization Optimization and Structure Determination of Pharmaceutical Solid Forms

Elizabeth Horstman[1], Sachit Goyal[1], Yuchuan Gong[2], Geoff G.Z. Zhang[2], and Paul J.A. Kenis[1]

[1]Department of Chemical & Biomolecular Engineering, University of Illinois at Urbana-Champaign, 600 South Mathews Avenue, Urbana, Illinois 61801,USA.

[2]Materials Science, Global Pharmaceutical R & D, AbbVie Laboratories, 1401 Sheridan Road, NCR13-317B, North Chicago, Illinois 60064, USA.

Success in finding a crystalline solid form of a pharmaceutical parent compound (PC) with optimized properties using conventional screening procedures is limited by small quantities of PCs available during the early stages of drug development that constrain the screening of experimental conditions. Microfluidic platforms reduce sample volumes of PC and precipitant (salts, cocrystal formers, or antisolvents) solutions [1] thereby allowing for screening a larger number of conditions using the limited amount of PC. High throughput screening utilizing small quantities of PC could enable early identification of possible solid forms of PCs and help reduce the cost in terms of time and money invested in solid form development. To date, we have reported on a polymer-based microfluidic array-type platform to screen for solid forms of PCs that employ free interface diffusion mixing [2], antisolvent addition [3], and solvent evaporation [4].

In this work, we present a microfluidic approach for crystallization optimization of pharmaceutical parent compounds (PCs) and their solid forms (cocrystals) via on-chip crystal seeding and subsequent X-ray data collection from crystals on-chip. The microfluidic platform is comprised of an 8 x 9 well array that enables screening of unique seeding conditions (dilutions) by metering PC solution (or PC/cocrystal former solutions) and seed solution at varying ratios (1:4 to 4:1) across each row. After crystals were observed in at least some wells via optical microscopy, X-ray data was collected from those crystals at room temperature. This on-chip seeding approach aids in crystallization of a specific desired solid form (*e.g.*, a specific polymorph) over a mixture of multiple solid forms when seeds of the desired solid form are utilized. To validate this approach, diffraction quality crystals of three model PCs (piroxicam, piracetam, and carbamazepine) and a cocrystal (carbamazepine 4-hydroxy benzoic acid) were grown on-chip and their structures were determined entirely from on-chip X-ray data. Easy handling, automated seed dilution, high throughput screening while using small quantities of PCs (~5 microgram/well), and on-chip X-ray analysis of multiple crystals makes these platforms attractive for early solid form identification and characterization.

References

[1] C. Hansen, E. Skordalakes, J. Berger, S. Quake, Proc. Nat'l Acad. Sci., 2002, 99, 16531-16536.

[2] M.R. Thorson, S. Goyal, B.R. Schudel, C.F. Zukoski, G.G.Z. Zhang, Y. Gong, P.J.A. Kenis, Lab on a Chip, 2011, 11 (22), 3829-3837.

[3] M.R. Thorson, S. Goyal, Y. Gong, G.G.Z. Zhang, P.J.A. Kenis, CrystEngComm, 2012, 14 (7), 2404-2412.

[4] S. Goyal, M.R. Thorson, C.L. Schneider, G.G.Z. Zhang, Y. Gong, P.J.A. Kenis, Lab on a Chip, 2013, 13 (9), 1708-1723.

389483 Protein Disaggregation By Thermal Cycling

Abstract:

Protein aggregation is inevitable in biopharmaceutical processing. Extensive processing required during synthesis and purification of biopharmaceutical products exposes these protein molecules to various kinds of stresses which destabilize them and cause them to aggregate. Factors that can directly affect protein stability include the solution pH, protein concentration, ionic strength, buffer type, temperature, pressure, exposure to gas-liquid interface, phase change (e.g. freeze-thawing) and presence of chemical additives. The native structure of a protein is only marginally more stable than the unfolded structure and it can therefore be perturbed quite easily. Unfolding and refolding of proteins occur continuously and the partially unfolded intermediates formed during such processes are highly susceptible to forming aggregates. These aggregates are non-native, i.e. there is significant alteration in secondary and tertiary structure and therefore loss in biological activity. Protein aggregates are generally lack therapeutic activity and could potentially be immunogenic, i.e. could elicit severe immunological reactions. We developed a thermal-cycling technique for simultaneous disaggregation of protein oligomers and refolding of the released monomer. Monoclonal antibodies of the immunoglobulin G (or IgG) class form stable non-native oligomers through Fab-Fab interactions. Monoclonal antibody samples containing known amounts of oligomers were exposed to rapid heating and cooling cycles using a laboratory thermal-cycler. The heating phase of the thermal-cycle released the partial unfolded monomer and the rapid cooling phase led to protein refolding and minimized the probability of protein re-aggregation. Orthogonal analytical techniques such as size exclusion chromatography, hydrophobic interaction chromatography, and circular dichroism spectroscopy were used to assess the extent of oligomer disaggregation and refolding.

390133 The Ion Exchange Proteome of Pseudomonas Fluorescens

Abstract:

Pseudomonas fluorescens is gaining popularity as an expression host for a variety of recombinant DNA products that range in application from therapeutics to industrial enzymes. In this poster, we present a characterization of the sub-proteome of *P. fluorescens* that interact with ion exchange chromatography. Batch and fed-batch cultures of *P. fluorescens* that overexpress Green Fluorescent Protein, a model product, provided extracts that were fractionated by DEAE chromatography. The proteins encountered during chromatography were identified, and a strategy was developed to improve GFP recovery (purity and column capacity).

363894 Development of a Reverse Phase HPLC Assay to Characterize a Monoclonal Antibody

Abstract:

Monoclonal antibodies (mAbs) represent the largest class of therapeutic proteins. One major challenge for biopharmaceutical industry is to understand the mAb microheterogeneity which could affect its therapeutic efficacy. mAb microheterogeneity could result from post translational modifications such as deamidation, oxidation and clipping. The reverse-phase HPLC (RP-HPLC) analysis is one of the widely used separation techniques to monitor the purity and stability of mAbs during manufacturing and storage. In this study, a RP-HPLC assay was developed to characterize the purity of a mAb. The mAb was reduced and separated on a RP HPLC column to reveal the Heavy Chain (HC), Light Chain (LC) and impurities. Oxidation and clipping induced impurities were identified by LC-MS. Reduction condition, columns, separation temperature and gradient were optimized to achieve the best separation and minimize assay induced artifacts. The RP-HPLC assay was validated for its accuracy, precision, linearity, specificity and range following FDA and ICH guidelines.

378960 Characterization and Modeling of Continuous Processing Unit Operations with Experimentally Measured Residence Time Distributions (RTD)

Abstract:

Residence time distributions (RTDs) are a valuable predictive tool in continuous processing systems including the continuous powder systems that can be used in pharmaceutical secondary manufacturing. RTDs are commonly investigated in detail for fluid systems, while continuous powder unit operations are relatively unexplored. The RTD of a system mathematically describes how a material travels inside the unit operations of a continuous processing system. This has valuable in raw material traceability, sensing, and control. Utilizing RTD it is possible to trace any upset or process change through the entire system, which can be used to improve quality. Since the RTD is the pulse response of the system, it can also be used to define the frequency of sensing required to detect any instantaneous pulse perturbation.

This work focuses on the experimental trials used for measurement and determination of residence time distributions of all the processing equipment used in a direct compaction (DC) pharmaceutical line. The DC process line consists of: several feeders, a mill, a continuous blender, and a tablet press. Each unit operations is investigated individually as well as in the overall line. The RTDs are determined through a series of experiments where a tracer is added to the system or the individual unit operations in either a pulse or step-change and a NIR system is used to measure the resultant outlet stream.

The RTDs of each unit operation within the DC line are then programmed into a flowsheet modeling software (gPROMS), where disturbances to the system are simulated. A statistical approach is taken to determine adequate sampling strategies for detecting various disturbances within the system. Operational characteristic curves are generated for the various sampling strategies, which are then compared to one another as well as industry standards, such as USP <905>.

Rapid prediction of facility fit and debottlenecking of antibody purification facilities

EPSRC CENTRE for Innovative Manufacturing in Emergent Macromolecular Therapies

Yang Yang[1], Suzanne S. Farid[2], Nina F. Thornhill[1]

[1] Department of Chemical Engineering, Imperial College London, London SW7 2AZ, UK
[2] Department of Biochemical Engineering, University College London, WC1H OAH, UK

UCL
Imperial College London

Introduction

Background

Higher titre and higher concentration formulations can pose facility fit challenges in legacy biopharmaceutical purification suites which originally designed for lower titre processes and lower concentration formulations. Those facility fit bottlenecks may result in unexpected product losses and hence discarding expensive product.

Overall Aims

To develop a data mining and optimisation decisional tool for rapid prediction of facility fit issues and debottlenecking of antibody purification facilities.

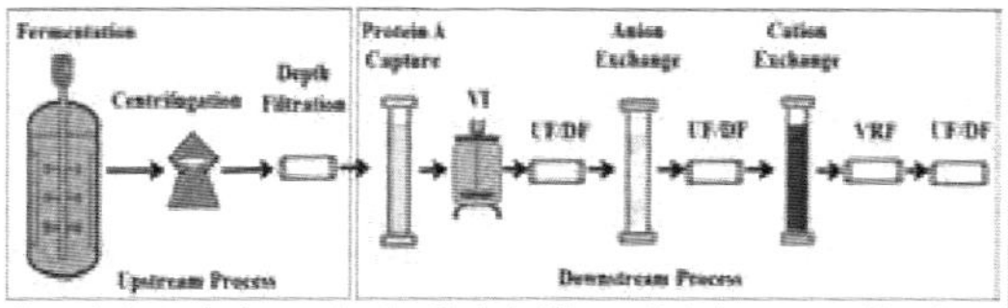

Figure 1. A standard monoclonal antibody (mAb) manufacturing process. VI = virus inactivation, UF/DF = ultrafiltration/diafiltration, VRF = virus retention filtration

Case Study 1: Decision tree analysis for rapid prediction of facility fit and debottlenecking

Aim

- Root causes of product loss
- Rules for the critical combinations of factors that lead to product loss
- Debottlenecking solutions

Methodology

Monte Carlo simulation data were analysed by a decision tree classification method, Classification and regression tree (CART).

Table 1. Summary of training dataset for CART classification

Classes partition	Description	No. of records
No	Batches with no unexpected mass loss at all	284
Light	Batches with unexpected mass loss less than 2kg	75
heavy	Batches with unexpected mass loss equal to or more than 2kg	41

Results

The pictorial CART tree provides a set of "if-then" rules with the threshold values for key parameters leading to unexpected product loss levels.

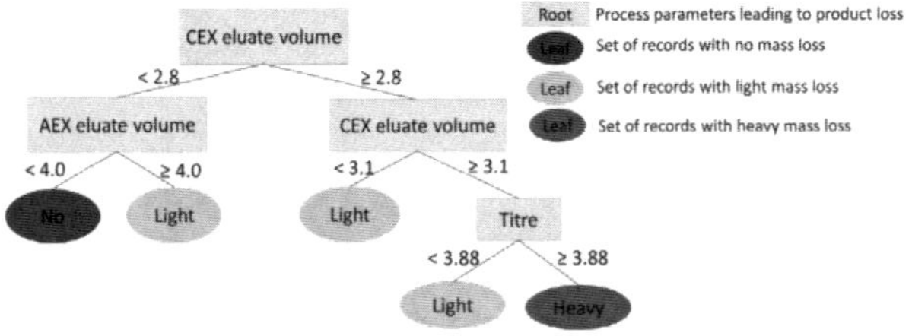

Figure 2. The optimal CART tree based on Monte Carlo simulation dataset. Numerical values are the threshold levels of the split points for split conditions

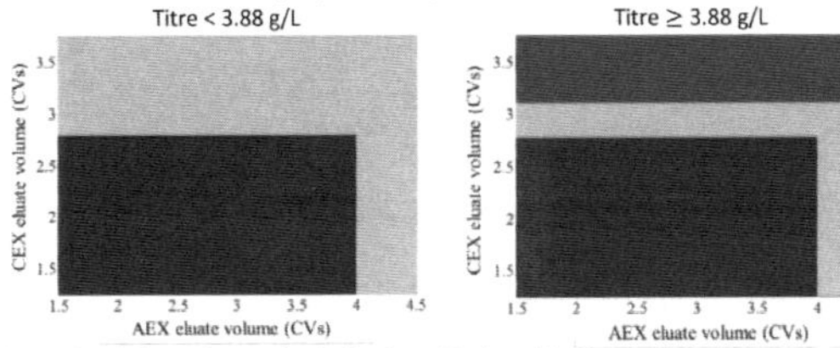

Figure 3. Windows of operation indicating critical combinations of AEX chromatography eluate volume and CEX eluate volumes under different titre that drive mass loss levels

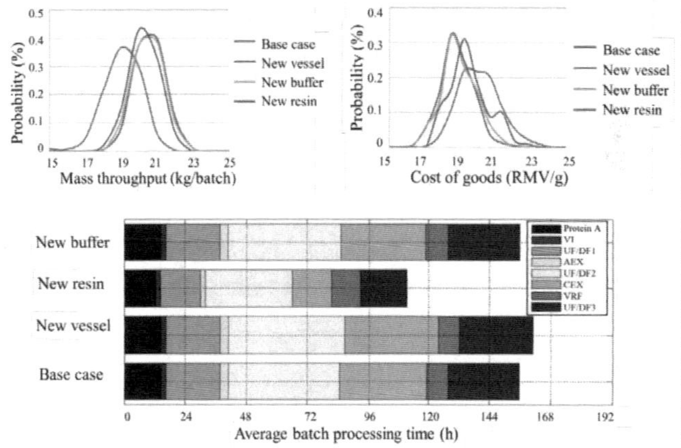

Figure 4. Comparison of three different solutions on mass throughput, COG and batch processing time

Case Study 2: Multi-objective optimisation for UF/DF facility design

Aim

- Ability of the existing UF/DF to deal with high-concentration products
- Optimal UF/DF design for both annual demand and cost

Methodology

- Multivariate regression method to explore the capability of existing facilities to produce high concentration formulations.
- Multi-objective Genetic Algorithm (GA) for optimal final UF/DF design

Results

1 Exploring the ability of given TFF systems for different product concentrations

2 Impact of different overconcentration ratios on mass loss due to UF/DF hold-up volume

3 Optimal UF/DF design of overconcentration ratio C_{oc} and number of cassettes N_c

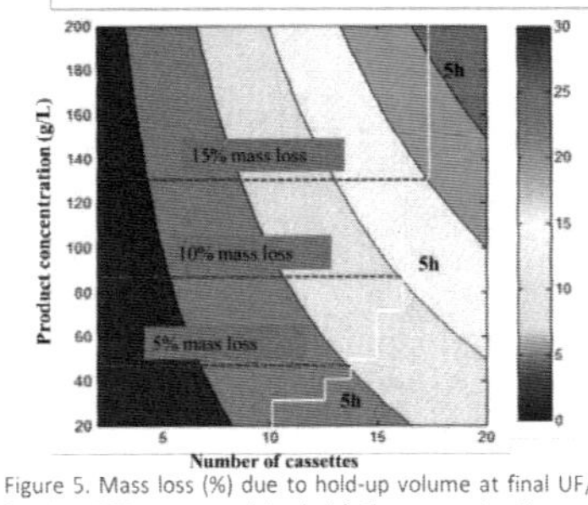

Figure 5. Mass loss (%) due to hold-up volume at final UF/DF step for given TFF system combined with 5hrs processing time constraint

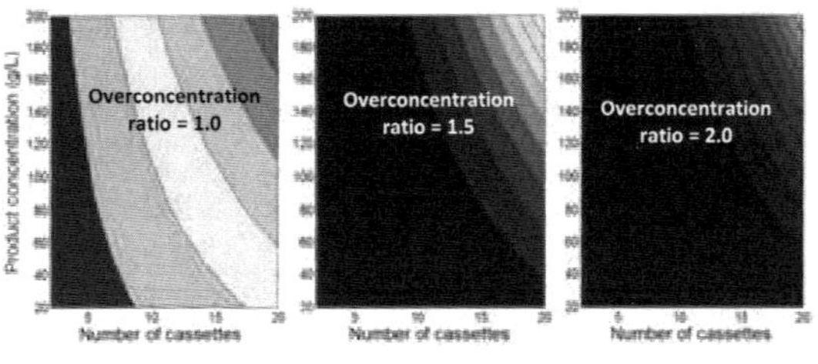

Figure 6. Mass loss (%) due to hold-up volume at final UF/DF step for given TFF system with different overconcentration ratios (1.0, 1.5 and 2.0)

Multi-objective GA: Object 1: maximise annual output
Object 2: minimise COG

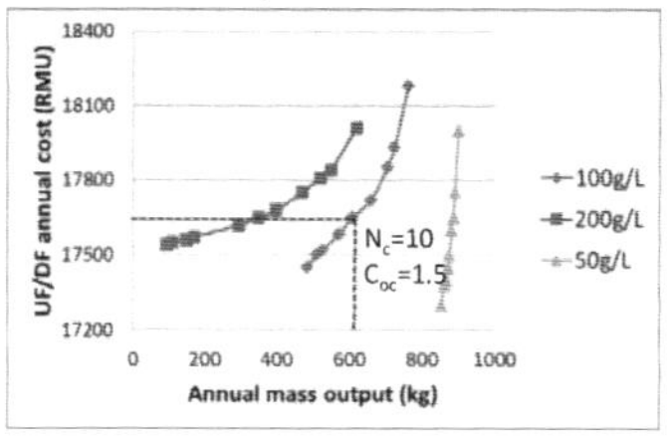

Figure 7. Pareto front chart of optimal combination of overconcentration ratio C_{oc} and number of cassettes N_c

Acknowledgements:
Funding from the UK Engineering & Physical Sciences Research Council (EPSRC) for the EPSRC Centre for Innovative Manufacturing in Emergent Macromolecular Therapies hosted at University College London with Imperial College London is gratefully acknowledged. Financial support from the consortium of industrial and governmental users is also acknowledged.

371193 Scale-up Strategies of High Shear Wet Granulation of a 50% Drug Load Compound: From 60L Bottom-Driven Granulator to 600L Top-Driven Granulator

Abstract:

We describe different strategies used for robust scale-up of a 50% drug load compound from a small scale bottom-driven granulator (60L Diosna) to a large scale top-driven granulator (600L Vector).

A 2^3 factorial design of experiments was conducted in bottom-driven 60L Diosna granulator, and a granule quality attribute design space was established. Initial studies conducted for transfering a 60L bottom-driven Diosna process to a 75L top-driven Vector granulator revealed that differences in the granulator bowl and impeller blade geometry required increasing the shear flow (impeller speed) in the top-driven granulator in order to match the granule/tablet attributes. Scale up of the process to a 600L top driven Vector granulator showed that the granulation end point was achieved with a longer wet massing time and with an impeller speed calculated using constant Froude number rule ($N_1\ D_1^{0.5}=N_2D_2^{0.5}$; N and D are impeller speed and blade diameter respectively). A longer granulation time for a larger vessel was suggested by Nakamura et. al.[1]who showed that in order to have dynamic similarity between the scales the granulation time should be adjusted such that the cumulative particle collision energy can be maintained at different scales. Alternatively, an increase in granulation fluid level was found to be a possible route for matching granule and tablet attributes across scales when impeller speed and granulation time are constrained from further increase.

[1]: H. Nakamura, H. Fujii, S. Watano, "Scale-up of high shear mixer-granulator based on discrete element analysis", *Powder Tech.* 236, 149-156 (2013).

372847 The Application of Novel Characterization Techniques to Understand Formulation-Process-Performance Interplay in Spray Dried Pharmaceutical Product Development

Abstract:

Spray drying has enabled the development of many poorly soluble compounds in the pipeline both for formulation of marketed products and as a delivery vehicle for safety studies. Non-uniformity/phase separation within the spray dried particles (SDP) could occur during the spray drying process, and may cause failure in product stability and performance. To fully characterize the SDPs to build understanding of formulation-process-performance interplay, however, remains an exceptionally difficult task. In this work, X-ray photoelectron spectroscopy (XPS) and high resolution focused ion beam scanning electron microscopy (FIB-SEM) were explored to close this gap.

XPS is a quantitative surface analysis technique with a penetration depth of 5-10 nm. During the formulation development of compound A and B, the surface concentration of SDP was investigated by XPS. In compound A, under extreme manufacturing conditions, the SDPs displayed "whiskers" by scanning electron microscopy, and significantly higher active pharmaceutical ingredient concentration on the surface by XPS which is consistent with surface crystallization. Further, SDPs of compound B prepared using different spray drying processes showed differences in surface composition, even in the absence of detectable crystallization, and the results were consistent with observed differences in dissolution performance. In addition, a single spray dried particle was cross-sectioned to obtain complete 3D structural and chemical imaging with FIB-SEM, which is a first in the pharmaceutical arena. Particles with size less than 25 um appear to be uniform, while the ones that are greater than 25 um consistently showed contrast in the backscattered images in their shells which is confirmed to correspond to phase separation by EDX.

Significant progress has been made in understanding the spray drying process through these new characterization tools, and this can lead to better QbD risk assessment and rational design of spray drying process. In the future, the expertise and knowledge will enable elaborate particle engineering, and drug product differentiating.

375715 Fluidized-Bed Impregnation of Active Pharmaceutical Ingredients Onto Porous Carriers

Abstract:

Impregnation is defined as the process of placing the chemical substances (including drugs) into the internal pores of a carrier. Impregnation of drugs onto porous carriers has been receiving increased attention in recent research activities, largely due to the drug dissolution profile change as a result of this process. Most of newly discovered drugs have poor solubility, so huge efforts have been applied to improve this problem. In impregnation, the solubility of drugs is improved by increasing the surface area of the drug. Furthermore, impregnation can improve the flow properties of drugs, which is very important in manufacturing of solid pharmaceuticals. Flow properties (i.e., cohesion, compressibility) affect mostly the content uniformity of the final dosage form. Assurance of content uniformity is very important especially for low dose drugs. The traditional method to improve flow properties is by the addition of lubricants and glidants. However, most of these additives have a negative impact on the dissolution of drugs because they are usually hydrophobic in nature. In impregnation, both dissolution and flow properties can be improved without the need of lubricants or glidants.

In this work, the impregnation of Acetaminophen (APAP) into a magnesium/aluminum silicate (Neusilin ^{R}) has been studied using a Mini-Glatt spray fluidized bed dryer/granulator. Four different weight percent's of APAP loading within the Neusilin: high (30%), medium (10%), low (1%) and ultra-low (0.1%), are examined. The experimental procedure includes 1) loading the fluidized bed with Neusilin until the top of the spray nozzle is reached, 2) warming up the Neusillin until a steady temperature is achieved, 3) spraying the APAP solution into the Neusilin, 4) drying the impregnated product. All the experimental conditions are kept constant except the APAP concentration. Ten Samples are taken from each experiment for characterization studies.

The sample content uniformity is measured using UV-spectroscopy at =247.5. Rotational shear testing is carried out in a FT4 powder rheometer to measure the sample flow properties. This test can be described briefly as the following: 1) A homogenous state is achieved by using rotating blades in a pre-conditioning step, 2) precompaction of the powder is effected using a vented piston, 3) the samples are sheared to obtain a yield point. This test has been carried out for both pure Neusilin and impregnated product to determine whether the flow properties of the impregnated product is affected by the presence of the drug or is only dependent on the pure carriers.

376882 Development of an E .coli Recombinant Protein Fermentation Process

Abstract:

Development of an *E .coli* Recombinant Protein Fermentation Process

Nitya Krishnan and Jane Gunson

In the late stage clinical development of an *Escherichia coli* (*E. coli*) recombinant protein fermentation process, changes to the host strain, production plasmid and glucose feeding strategy were implemented to increase product titer, maintain product quality and improve process consistency. The optimized fermentation process was developed using a small scale model where increased product titer and consistent product quality were achieved. Process understanding was expanded by the execution of univariate and multivariate studies. In addition, a small scale purification method was developed and used to isolate the recombinant product derived from the multivariate studies in order to evaluate product quality. Finally, the fermentation process was successfully transferred to manufacturing scale while maintaining the high product titer, product quality consistency and fermentation process performance.

Development of a Scale-down Plate Model for Raw Material Variability Analysis in Mammalian Cell Culture

Hedieh Rahimi, David Kolwyck, Gary Rogers, Jack Huang and Chetan Goudar (Cell Sciences & Technology, Amgen, Thousand Oaks, CA)

In recent years, cell culture processes in the biopharmaceutical industry have transitioned from complex to chemically-defined (CD) medium with an intent to enhance process robustness and consistency. Despite defined concentrations for all components in a CD medium, recent studies from our laboratory have shown that CD medium use does not ensure process consistency. Specifically, the quality and variability associated with CD medium components can significantly impact cell culture performance and critical quality attributes (CQAs) of the product. Consequently, rigorous selection and qualification of CD medium and reassurance of lot-to-lot consistency are essential for a robust cell culture process. While such studies are typically conducted in bioreactors, they are time and resource intensive which can limit their application. To overcome this limitation, we have developed a high throughput (HT) small-scale plate model to assess the impact of raw material variability on cell culture performance and product quality. Multiple medium powder lots which spanned several medium formulations and vendors were evaluated using the HT plate assay. While no significant differences in cell culture performance were seen, product quality was impacted. Comprehensive chemical analyses identified differences in trace levels of organic and inorganic impurities across medium types which were associated with product quality changes. Additionally, results from the HT plate studies were consistent with those from 2 L bioreactor studies suggesting the utility of this approach to screen variability associated with CD medium. .The simplicity of this new approach should enable rapid screening of cell culture medium resulting in more robust cell culture processes.

A Media Screening Engine for Live Viral Vaccine Process Development: Bridging High-Throughput DoE Screening and Upstream Process Development

Jason Rodriguez, Esther Lim, Chris Daniels, Kent Hamaker

Merck & Co., Inc.
770 Sumneytown Pike
West Point PA, 19486

Cell culture of complex biological molecules such as live viral vaccines (LVVs) requires robust media, which often contain high amounts of serum, to yield enough cell mass to support a strong infection and antigen yield for the final product. Media development can serve as a tool for process development of LVVs by allowing for optimization of critical components such as serum, which can burden LVV programs in terms of cost, raw material procurement, supply, lot-to-lot variability and regulatory concerns. Most importantly, the biggest impact of developing a robust culture media for any particular process lies in enabling higher product quality and/or quantity. To this end, we developed a screening workflow spanning multiple culture platforms to identify basal media and individual supplements that (1) increase cell growth and peak cell densities of a novel adherent cell line used for live viral vaccine (LVV) production and (2) increase viral spread and subsequent antigen yield during the infection process. Central to the screening platform is a high-throughput (HT) robotic screening system from HiRes Biosolutions, which enables screening of up to 2400 individual media formulations for cell growth and virus spread through HT immunofluorescence. Lead candidate media formulations were then tested in small-scale microcarrier culture to test cell growth and antigen yields to enable further process development in bench-scale bioreactors. In summary, our media development efforts identified alternative serum sources that supported LVV production and reduced our overall supply costs. In addition, we identified crucial supplements that increased cell growth and spread of LVV infection in cell culture that could potentially lower our serum demand in the future.

Design of a Large Scale Manufacturing Process and Site for Neural Stem Cells for Spinal Cord Injury Treatment.

Luqman Ahmad Mahir[1], Siti Nur Aishah Mohd Zain[1], Amulya Pervaje[1], Elaine Qian[1], Madeline Williams[1], Tiffany D Rau[2]

1. Vanderbilt University, Department of Chemical and Bimolecular Engineering, Nashville, TN.
2. Eli Lilly and Company,

Approximately 250,000 people in the United States live with chronic spinal cord injuries (SCI). Annually, there are between 12,000 and 20,000 new patients with SCI. Though there are many treatments currently available for this injury, total motor and neural function restoration is not typically achieved. As a result, a therapeutic able to aid in spinal cord restoration and remain compliant with current federal regulations is desired.

Recently, use of neural stem cells for treatment of this injury has been investigated because of the cell's multipotency and its ability to aid in spinal cord repair. Mesenchymal stem cells are able to differentiate into neural stem cells and can be separated from discarded umbilical cord tissue with relative ease. Discovering and obtaining proof of concept and then efficacy and safety for life improving therapeutics is only part of the path to bring products to the patients. The ability to make quantities large enough for clinical trials and then to meet commercial targets is a challenge especially in the area of stem cells. This work explores a proposed manufacturing process and facility design for a target market of 15,000 patients per year. With new treatments and ways to manufacture those treatments, it is believed that individuals with ailments such as spinal cord injuries will have a better quality of life.

Contact Information of Presenter:

Tiffany D Rau, PhD

Eli Lilly and Company

Center of Fermentation Excellence

tiffanyrau@aol.com and rau_tiffany_d@lilly.com

25 Timberly Drive Lake Charles, LA 70605

SIMULATION OF STIRRED TANK HYDRODYNAMICS USING MESH AND MESHLESS METHODS (356337)

Dragan D. Nikolic, Brian P. de Souza, Patrick J. Frawley
Synthesis and Solid State Pharmaceutical Centre, University of Limerick, Limerick, Ireland

Introduction

Engineering problems such as process design, scale-up and optimisation rely heavily on mathematical models of the process and a detailed description of a fluid flow in stirred vessels is a critical part. Hydrodynamics plays a major role in the dynamics of crystallisation phenomena in that it is responsible for the mixing, heat transfer and transport of particles throughout the crystalliser. The mixing in the vessel is always a very important aspect of design, because an adequate mixing is required for an efficient mass and heat transports and phase homogeneity. In the case of crystallisation, the phenomena such as nucleation, growth and in particular secondary nucleation are very sensitive to local hydrodynamics conditions and scaling may result in dynamic dissimilarity between cases and therefore result in large differences in final product quality.

The population balance (PB) framework has been adopted as the most fundamental approach for modelling particulate multiphase systems (1,2). The population balance modelling framework provides a deterministic description of the dynamic evolution of the crystal size distribution (CSD) by forming a balance to calculate the number of crystals in the crystalliser. In order to take into account local hydrodynamics conditions in a stirred reactor a common approach is to couple the population balance equations with turbulent computational fluid dynamics (CFD) modelling to correctly predict the key process performance indicators such as CSD, yield, recovery and reactor productivity (3,4). Although use of CFD for simulating particulate systems is becoming a standard approach, a number of issues still need to be addressed. One major problem is the computational expense of coupling a standard discretised population balance with a grid-based CFD code, as this approach requires a solution of an intractably large number of transport equations. Another, more feasible approach is to use a method of moments. However, even the methods of moments coupled with CFD do not provide a computationally efficient approach for scale-up and optimization studies due to a large difference in time resolution of up to ten orders of magnitude in some cases: a time frame of a typical crystallisation process is several hours compared to a time step necessary to achieve reasonable accuracy of only 10 microseconds. The other common problem encountered with the method of moments is that the CSD is lost and needs to be reconstructed from its distribution moments. Several reconstruction techniques were developed (5,6,7,8), however, no unified technique for reconstruction of a complete distribution from a finite number of moments is available in literature due to the fact that mathematically all the distribution moments up to infinity are required to achieve an accurate reconstruction (5). There is also a number of problems that needs to be addressed from the numerical perspective. So far the CFD models of stirred vessels were exclusively implemented in Eulerian reference frame using grid-based methods. Those methods suffer from many problems: convective transport should be simulated and can influence the integration time step and cause numerical problems, cumbersome numerical mapping is required to handle complex geometries, determination of free or moving boundaries and material interfaces is rather difficult and causes various numerical problems, the computational requirements poorly scale with the size of the system and the numerical methods required to solve large systems of equations are typically not capable to fully take advantage of the modern paralelization hardware and software systems. As a consequence, optimisation studies are computationally very expensive and, at the moment, not feasible in most cases.

In this work the above-mentioned issues were addressed using the Lagrangian frame meshfree Smoothed Particle Hydrodynamics (SPH) method for modelling of the fluid flow in stirred reactors. The SPH model was implemented and its potential benefits to modelling of crystallisation were discussed. The results from SPH simulations were compared to the available experimental data from the literature and to the simulation results using Eulerian frame grid-based CFD code available in the literature. Various aspects of the simulation results were assessed: capability to predict the fluid flow and a quality of prediction, the computational requirements, model reduction capabilities, existence of numerical problems and availability of important process performance indicators such as crystal size distribution. The developed SPH model was successfuly coupled with discretised population balance equations and benefits analyzed in terms of the quality of CSD prediction and computational requirements.

Meshfree methods

The key idea of the meshfree methods is to provide accurate and stable numerical solutions for systems of partial differential equations with a set of arbitrarily distributed nodes (or particles) without using any mesh that provides the connectivity of these nodes or particles. The meshfree methods found application to problems such as free surface flow, deformable boundary, moving interface, large deformation, mesh adaptivity and multi-scale resolution. A large number of meshfree methods exist and a comprehensive review of the history, development, theory and applications of the major meshfree methods is given in (9). All of them can be divided into three types: a) methods based on strong form formulations, b) methods based on weak form formulations, and c) particle methods (10). In this work the particle meshfree method Smoothed Particle Hydrodynamics is selected and applied to modelling of the fluid flow in stirred reactors.

SPH was originally invented for modelling astrophysical phenomena (11,12). Later, it was extended and applied to modelling of fluid mechanics and many other areas. SPH is a method for obtaining approximate numerical solutions of the equations of fluid dynamics by replacing the fluid with a set of particles and uses the integral representation method for approximation of field functions. The use of integral representation of field functions passes the differentiation operations on the field function to the smoothing (weight) function and has been proven stable for arbitrarily distributed nodes for many problems (10). SPH is a Lagrangian method in its nature, where the physical system is represented by particles moving in the Lagrangian frame according to the internal interactions and external forces, and thus evolving the system in time.

The method offers a number of attractive features. The first is that pure advection is treated exactly. The advection term disappears from the Navier-Stokes equations and velocity partial derivative becomes an ordinary time derivative allowing the momentum dominated flows to be easily handled. For example, if the particles are given a colour, and the velocity is specified, the transport of colour by the particle system is exact (10). The mass continuity equation is not necessary anymore since the mass balance is explicitly enforced using particles of a specified mass. Linear and angular momentum are explicitly conserved using the specially derived expressions for forces acting on particles (such as pressure, shear stress and surface tension). Population balance equations can now be applied to a smaller control volume (that is at a particle level) making the CSD available explicitly without reconstruction. Representation of multi-phase systems, free and moving surfaces is easy and not bound to the prescribed spatial grid like in grid-based methods. The fact that a fluid and crystals are represented by particles can be utilised to explicitly model phenomena relating to the solid phase such as secondary nucleation and breakage. Here, the local hydrodynamic conditions, particle size and velocity, and an angle of impact can be directly used to estimate the rate of secondary nucleation, with a direct linkage to experimental particle-wall collision measurements using the shadow imaging technique. The computational power required to integrate equations in time scales linearly with the size of the system (that is with the number of particles). Integration in time is typically performed using the explicit methods avoiding numerical problems common in grid-based methods. Grid-based numerical methods such as finite element or finite volume methods always require a solution of a large system of (non-)linear equations which is not relevant in SPH. Moreover, the use of weighted contributions rather than a linear system of equations allows for massive parallelisation, taking advantage of recent developments in CUDA GPU technology to significantly reduce simulation times. In a typical crystallisation model, the Navier-Stokes equations are coupled with a number of scalar transport equations such as temperature, concentration and the population balance equations (either moment equations or discretised number density function). In grid-based methods the additional transport equations lead to a significant increase in size of the numerical system and matrix storage while in SPH only to a marginal increase in memory and computational requirements. For instance, a hundred additional transport equations per cell resulting from discretisation of the number density function using the high-resolution van Leer flux limiter scheme with 100 bins causes only around 40% drop in performance. The same system implemented using the grid-based methods would be intractable. Large numerical systems typically solved on symmetric multiprocessor systems or computer clusters require some sort of communication and synchronization mechanisms. The only synchronisation technique used in SPH are primitive atomic operations used during calculation of forces between a pair of particles. Particle based SPH also offers much higher model reduction capabilities making lengthy optimization studies computationally feasible. Basically, the number of particles used to decribe a system can be decreased having the superposition effect on execution speed: use of larger particles decrease the total number of particles and larger particles are allowed to move a larger distance in space during every time step, both effects causing the simulation speed-up. Another advantage is that the resolution can be made to depend on position and time and refinement of particles is much easier to perform than the mesh refinement.

Obviously, there are some disadvantages of the SPH method. First, a very high computational power is required. This is solved using the modern general purpose graphics processing units (GPGPU) that allow a massive parallelisation. Second, typically a very large number of particles is required to model the system at the same level of details compared to grid-based methods. However, as it is discussed later in this work we can safely trade some details in hydrodynamics prediction to gain execution speed by reducing the number of particles (that is increasing the particle size) used to represent the fluid. Finally, fluids in SPH behave as slightly compressible and some numerical techniques are required to achieve incompressibility as it will be demonstrated in the section where an acceleration due to pressure forces is discussed.

Smoothed Particle Hydrodynamics theory

SPH method was developed for hydrodynamics problems in the form of partial-differential equations of field variables such as the velocity, density, energy etc. (10). In order to solve the above system of equations the problem structure must be first represented by a set of arbitrarily distributed particles. Then, the integral representation method is applied to obtain an approximation of field functions (kernel approximation). The kernel approximation is further approximated using particles by replacing the integration in the integral representation of the field functions and their derivatives with summations over the values at the neighboring particles in a local domain called the support domain.

The particle approximation is performed at every time step and the particle approximations are also applied to all terms related to field functions to produce a set of ODEs. Finally, the resulting ODEs are solved using an explicit integration algorithm to achieve a fast time stepping (10).

The essential formulation of SPH is that any field quantity can be determined through a weighted interpolation of surrounding field values (12). This weighting is characterized by the smoothing length, *h*, it decays with distance and it is given by the following integral equation:

$$A(\mathbf{r}) = \int_{\Omega} A(\mathbf{r}')W(\mathbf{r}-\mathbf{r}',h)d\mathbf{r}' \tag{1}$$

where *A(r)* is a field function at point *r* in space Ω, *r* is any point in Ω, and *W* is a smoothing kernel with *h* as its width. The numerical equivalent to Eq. 1 is obtained by approximating the integral interpolant by a summation interpolant:

$$A(\mathbf{r}) = \sum_j A_j \frac{m_j}{\rho_j} W(\mathbf{r}-\mathbf{r}_j,h) \tag{2}$$

where m_j is the mass of the particle *j*, r_j the position and A_j is the value of any quantity *A* at r_j. Basically, this equation states that any quantity can be represented by summation of quantities at nearby points multiplied by a weighting function, also known as a smothing kernel. A differentiable interpolant of a function can be constructed from its values using a differentiable kernel, and thus derivatives of an interpolant can be obtained by standard analytic differentiation. It is therefore a straightforward procedure to define the gradient and the Laplacian of a field function, under the assumption that the smoothing kernel is first and second order differentiable, respectively. The gradient of a field function is therefore given as:

$$\nabla A(\mathbf{r}) = \sum_j A_j \frac{m_j}{\rho_j} \nabla W(\mathbf{r}-\mathbf{r}_j,h) \tag{3}$$

and its Laplacian as:

$$\nabla^2 A(\mathbf{r}) = \sum_j A_j \frac{m_j}{\rho_j} \nabla^2 W(\mathbf{r}-\mathbf{r}_j,h) \tag{4}$$

Smoothing kernels

The use of different kernels in SPH is analogue to using different difference schemes in finite difference methods, thus the choice of smoothing kernel for a specific problem is of significant importance. The smoothing functions need to satisfy a number of conditions (10). The first is that the integral of the smoothing function over the support domain must be equal to one (*Unity condition*):

$$\int_{\Omega} W(\mathbf{r},h)d\mathbf{r} = 1 \tag{5}$$

The second condition is that the smoothing function should be equal to zero out of the support domain (*Compact support*):

$$W(\mathbf{r},h) = 0, ||\mathbf{r}|| > h \tag{6}$$

The third conditions is that the smoothing function should be non-negative in the support domain (*Positivity*):

$$W(\mathbf{r},h) \geqslant 0, ||\mathbf{r}|| \leqslant h \tag{7}$$

The fourth condition is that the smoothing function should be monotonically decreasing with the increase of the distance from the particle (*Decay*). The fifth condition is that the smoothing function should approach the Dirac delta function when its support domain approaches zero (*Delta function property*):

$$\lim_{h\to 0} W(\mathbf{r},h) = \delta(\mathbf{r}) \tag{8}$$

The sixth condition is that the smoothing function should be an even function (*Symmetric property*):

$$W(\mathbf{r}, h) = W(-\mathbf{r}, h) \tag{9}$$

The final property is that the smoothing function and its derivatives should be sufficiently smooth (continuous) to obtain good results (*Smoothness*).

Navier-Stokes equations

The basic physical quantities of an isothermal, viscous fluid are velocity u, mass-density ρ, and pressure p and they are considered as continuous fields in the fluid. The classical formulation of the motion for incompressible fluid flow over time t is governed by the Navier-Stokes equations coupled with the mass continuity equation:

$$\rho\left(\frac{\partial \mathbf{u}}{\partial t} + \mathbf{u} \cdot \nabla \mathbf{u}\right) = -\nabla p + \mu \nabla \cdot (\nabla \mathbf{u}) + \rho g + \mathbf{f}$$

$$\nabla \cdot \mathbf{u} = 0 \tag{10}$$

where μ is the viscosity of the fluid and f is the sum of internal and external force-densities acting on the fluid such as buoyancy, surface tension and collisions with rigid bodies. In the Lagrangian formulation, particles completely define a fluid which implies that the particles move with the fluid. Compared to the Eulerian view this means that any field quantity now depends on time only. The particles carry mass, position, and velocity, and will hold smoothed quantity approximations obtained from SPH. Thus, an acceleration for the Lagrangian fluid particle becomes an ordinary time derivative of its velocity and the advection term in Eq. 10 disappears. The basic Lagrangian formulation of the Navier-Stokes equations for an incompressible, isothermal fluid is therefore given by (10):

$$\rho \frac{d\mathbf{u}}{dt} = -\nabla p + \mu \nabla^2 \mathbf{u} + \rho g + \mathbf{f} \tag{11}$$

Regarding the mass-density, there are two approaches to evolve density in the SPH method (10). The first one is called the *summation density* which directly applies the SPH approximation to the density itself, which for any particle i is given by:

$$\rho(\mathbf{r}_i) = \sum_j \rho_j \frac{m_j}{\rho_j} W(\mathbf{r}_i - \mathbf{r}_j, h) = \sum_j m_j W(\mathbf{r}_i - \mathbf{r}_j, h) \tag{12}$$

and simply states that the density of a particle can be approximated by the weighted average of densities of particles in the support domain of that particle. Another approach is the *continuity density* which approximates the density according to the continuity equation using the concepts of SPH approximations with some additional transformations. The first approach is adopted in this work.

The pressure force at particle i, following the weakly-compressible approach to simulate incompressible fluids with SPH, yields:

$$\nabla p(\mathbf{r}_i) = -\rho_i \sum_{j \neq i} \left(\frac{p_i}{\rho_i^2} + \frac{p_j}{\rho_j^2}\right) m_j \nabla W(\mathbf{r}_i - \mathbf{r}_j, h) \tag{13}$$

The pressure gradient formula in Eq. 13 conserves the linear and angular momentum exactly as the forces acting between a pair of particles i and j are antisymmetric along the line of centers of the two particles. An equation of state is used to estimate the pressure from the density field:

$$p(\rho) = p_0 \left[\left(\frac{\rho}{\rho_0}\right)^{\gamma} - 1\right] \tag{14}$$

The reference density ρ_0 is set to the initial density of the fluid phase so that the pressure field is initially equal to zero. The stiffness of the equation of state can be adjusted with the two parameters p_0 and γ. For fluids it is common to set the exponent $\gamma = 7$ (13), and the pressure magnitude p_0 is given by:

$$p_0 = \frac{\rho_0 c^2}{\gamma} \tag{15}$$

The artificial sound speed c is chosen based on a scale analysis presented in (13) in order to limit the density variations to 1%. Tipically, it is set to at least one order of magnitude larger than the reference velocity to limit the fluid compressibility.

For incompressible fluids, the acceleration of particle i caused by shear forces is given by:

$$\mu\nabla^2\mathbf{u}(\mathbf{r_i}) = \frac{\mu}{\rho_i}\sum_{j\neq i}(\mathbf{u_j} - \mathbf{u_i})\, m_j \nabla^2 W(\mathbf{r}_i - \mathbf{r}_j, h) \tag{16}$$

Similarly to the pressure gradient formula given in Eq. 13 this formula also preserves the linear and angular momentum exactly. The Laplacian of the smoothing kernel in Eq. 16 should be constrained to be positive (10). Otherwise, the viscosity forces could increase the relative velocity and thereby introduce energy and instability into the system.

The formula for surface tension forces is adopted from (14) and takes into consideration molecular cohesion forces and surface area minimization forces. The cohesion forces are given by the following equation:

$$\mathbf{f}_{i\leftarrow j}^{cohesion} = -\sigma m_i m_j C(|\mathbf{r}_i - \mathbf{r}_j|)\frac{\mathbf{r}_i - \mathbf{r}_j}{|\mathbf{r}_i - \mathbf{r}_j|} \tag{17}$$

where i and j are neighboring fluid particles, m denotes mass and x denotes position of the respective particle, σ is the surface tension coefficient and C is a spline function specially created for the surface tension equation and defined as:

$$C(x) = \frac{32}{\pi h^9}\begin{cases}(h-x)^3 x^3 & 2x > h \wedge x \leqslant h \\ 2(h-x)^3 - \frac{h^6}{64} & x > 0 \wedge 2x \leqslant h \\ 0 & otherwise\end{cases} \tag{18}$$

The curvature forces are given by:

$$\mathbf{f}_{i\leftarrow j}^{curvature} = -\sigma m_i (\mathbf{n}_i - \mathbf{n}_j) \tag{19}$$

where n_i is the normal information obtained by applying the SPH approximation to the gradient of the smoothed color field as:

$$\mathbf{n}_i = h\sum_j \frac{m_j}{\rho_j}\nabla W(\mathbf{r}_i - \mathbf{r}_j, h) \tag{20}$$

Finally, the surface tension force can be written as:

$$\mathbf{f}_{i\leftarrow j}^{surf.tension} = K_{ij}\left(\mathbf{f}_{i\leftarrow j}^{cohesion} + \mathbf{f}_{i\leftarrow j}^{curvature}\right) \tag{21}$$

where K_{ij} is a symmetrized correction factor calculated by:

$$K_{ij} = \frac{2\rho_0}{\rho_i + \rho_j} \tag{22}$$

where ρ_0 is the rest density of the fluid, and ρ_i and ρ_j are the densities of the neighboring fluid particles. As we do not correct the fluid particle densities, a fluid particle with less than full neighborhood has $K_{ij}>1$, and a fluid particle with full neighborhood has $K_{ij}\approx 1$. Therefore, K_{ij} amplifies forces for the particles with neighborhood deficiency, while the forces remain the same for the particles with appropriate neighborhood.

SPH Implementation details

The SPH model of a stirred tank is implemented in c++ using Fluidix software (15). Fluidix is an NVidia CUDA-based parallel particle simulation platform which can be applied to practically any particle-based model, such as Molecular Dynamics, Dissipative Particle Dynamics and Smoothed Particle Hydrodynamics. 3D surface models can be imported from STL files to define system boundaries. Particles and triangle meshes are organized into efficient binary tree structures for highly optimized searching in both uniform and sparse distributions, allowing neighbour searching across particle sets as well as between particles and triangles of a 3D mesh. Due to the flexibility of c++ templates, particles can have any number of local variables, such as position, velocity, acceleration, density, temperature, pressure, and discretised number density function. A fully customized particle interactions can be executed in parallel with searches such as “each particle”, “each pair of nearby particles”, or “each particle colliding with a mesh”.

The simulation procedure consists of the initialization phase and the main SPH loop. First, triangular surface meshes are generated for stirred tank walls, baffles and a stirrer shaft with impellers. These meshes can be directly loaded into the Fluidix software. Then, uniformly distributed fluid particles of the specified size are generated inside the whole volume of the tank and the particles that are positioned inside the baffles and stirrer meshes are removed. An excess number of particles from the top of the reactor, up to the prescribed volume, is also removed. Now, so called “ghost” particles are created at mesh vertices. The “ghost” particles on the reactor surface are are used to simulate a heat transfer between the cooling fluid in the reactor jacket and the fluid in the reactor. The “ghost” particles on the impellers participate in impeller-fluid particle interactions. Basically, impeller particles exert pressure-gradient-like repulsive forces on fluid particles to produce the stirring effect according to the prescribed rpm of the stirrer. These forces are oriented in the direction of the normal to the impellers surface. The SPH main loop is implemented in a standard fashion. First, a density of all particles is estimated and internal forces evaluated such as the pressure gradient, shear forces, gravity and the surface tension. Stirrer mesh and its ghost particles are rotated and impeller-particles repulsive forces calculated. The system is now integrated in time using the second order Verlet scheme and particle-mesh collisions processed to update particle positions. This procedure is repeated until the defined time horizon is reached.

Case Study 1: Validation of Hydrodynamics

Application of SPH to crystallisation modelling first requires validation of hydrodynamics in a typical batch reactor vessel. In this paper, the hydrodynamics of a baffled stirred tank, driven by a Rushton turbine is considered (16). The vessel is cylindrical of diameter T=150mm, with four equispaced baffles of width 0.1T with a small clearence of 0.017T and a liquid height of H=T. A lid was positioned on top of the liquid to prevent air bubbles. The fluid was silicon oil and the rotational speed was 2672 rpm resulting in a tip speed of 7m/s and Reynolds number of 7300. The simulation results were compared to the experimental data obtained by a Laser Doppler Anemometry and simulation results using Reynolds Averaged Navier-Stokes and Large Eddy simulations from (16).

The SPH fluid particle sizes were chosen in the range of 0.5 to 2.5mm to study a dependency of the quality of fluid flow prediction on the particle size. As a smoothing function the quartic kernel from (11) was adopted with the smoothing length set to two particle diameters. Simulations were executed on a commodity NVidia GeForce 780 Ti graphics card with 2880 cores and 3GB of RAM and Dell Precision 7500 workstation running Debian GNU/Linux operating system.

Comparison of phase averaged axial profiles of radial and tangential velocity components at three radial locations is presented in Fig. 1 where *Exp* represents experimental data from (16), *RANS* represents simulation results using Reynolds Averaged Navier-Stokes method from (16) and *SPH* represents the simulation results from this work obtained for the lowest particle radius in this study (0.5mm). The kinetic energy of random velocity fluctuations and the dissipation rate of turbulent kinetic energy were not calculated in the SPH model.

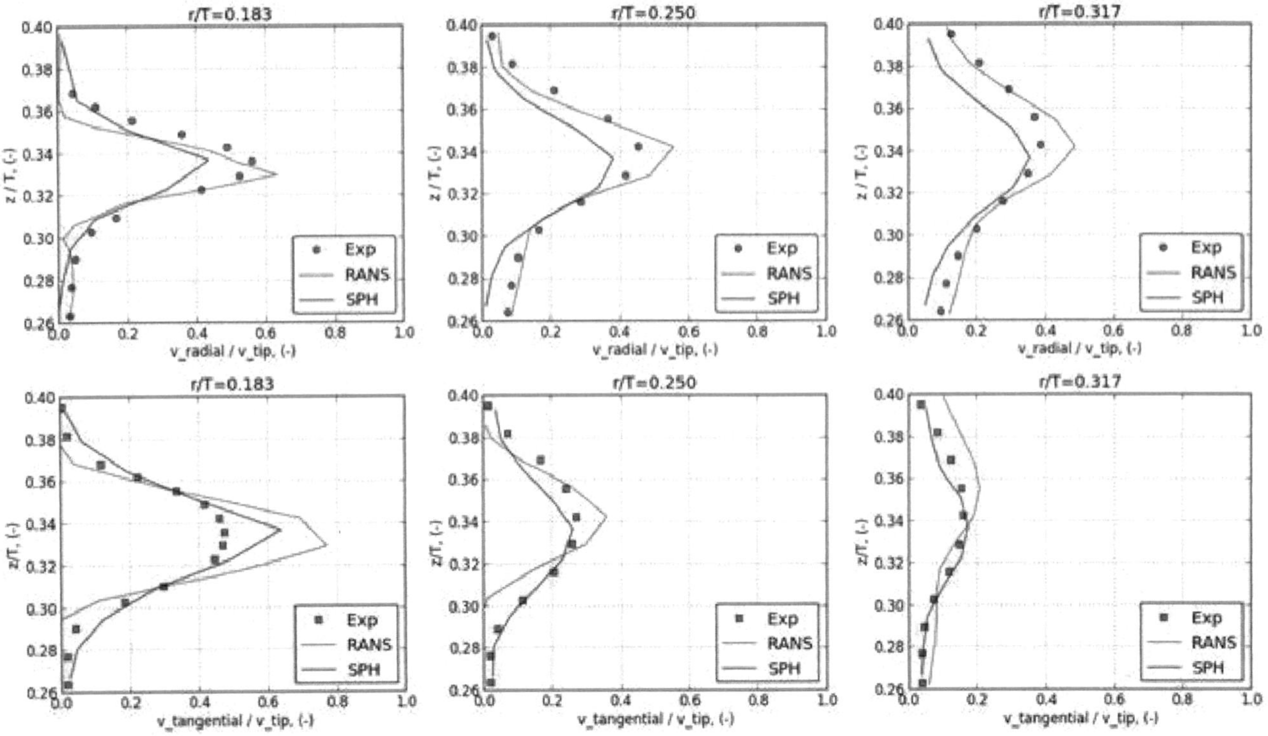

Figure 1. Phase averaged axial profiles of the radial (upper row) and tangential (bottom row) velocity components at three radial locations r/T=0.183 (left), r/T=0.250 (middle) and r/T=0.317 (right)

The number of particles in the stirred tank as a function of particle size and computational times needed to simulate one hour of real process as a function of particle count are shown in Fig. 2. Effect of particle radius on the quality of the mean flow prediction is depicted in Fig. 3 for particle sizes of 0.5, 1.5, 1.75 and 2.5mm. Each plot contains streamlines obtained by integration of a velocity vector field in an axial plane located 15^0 behind impellers. The velocity fields were averaged over 10 impeller revolutions.

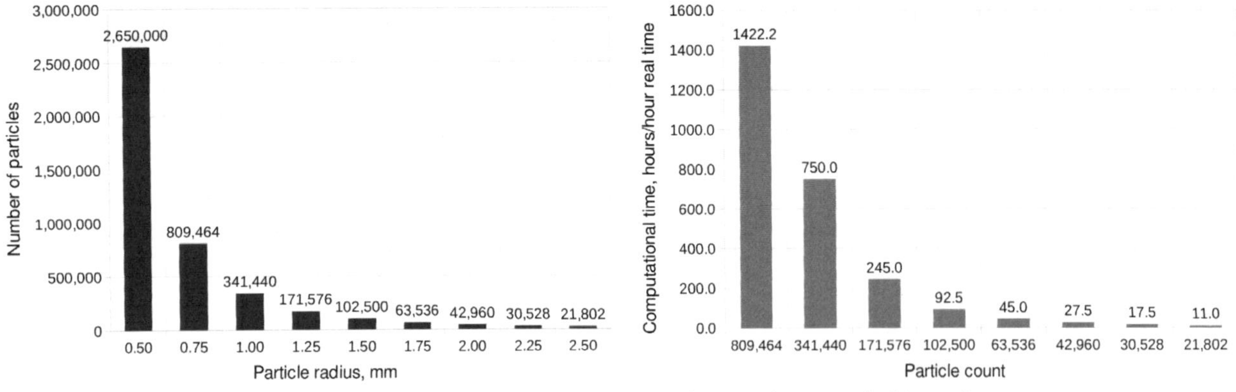

Figure 2. Number of particles as a function of particle size (left) and approximate computational time per hour of real time as a function of particle count (right)

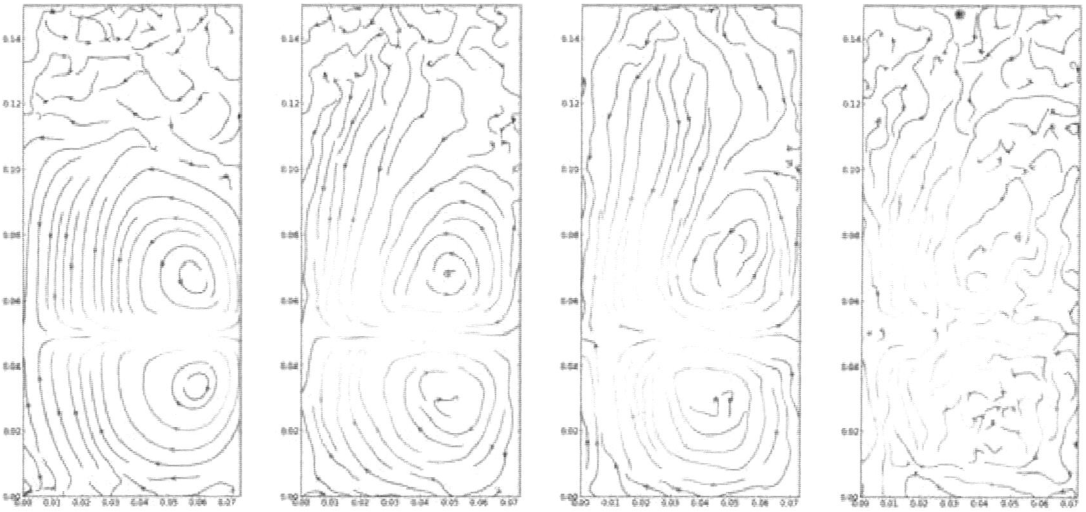

r_p=0.75mm r_p=1.50mm r_p=1.75mm r_p=2.50mm

Figure 3. Effect of particle radius on hydrodynamics prediction quality

The simulation results show that the SPH method is capable of predicting well the global mean flow, with marginally reduced accuracy when compared to the RANS simulations. The dominant flow features are well predicted. Two main circulation loops can be observed in Fig. 3: one below the impellers with downward flow near the tank wall, and one above the impellers with upward flow near the tank wall. The upper one does not reach the top part of the volume and that part of the volume remains porly mixed (approximately less than 20% of the volume). The streamlines in the unmixed zone appear as "squiggles". The reason for that is that in SPH particles always oscillate around their centers in random directions (i.e. like molecules do) even if the fluid is still. When subjected to an external force those small velocity fluctuations are cancelled out by a larger velocity amplitude caused by the external force. The upper circulation loop is more pronounced in RANS simulation compared to SPH and closer to experimentally obtained data. The radial velocity component is slightly underpredicted in the SPH and slightly overpredicted in the RANS simulation. Prediction of the tangential velocity component looks a bit better: it is slightly overpredicted at the tip location but shows a very good agreement with the experimental data at the farther points.

The SPH method requires huge a computational power to achieve the same level of prediction as grid-based methods as it can be seen from Fig. 2 and Fig. 3. However, it can be also observed that a good enough quality of the mean flow prediction can be obtained using larger particles. For instance, use of particles with radius of 1.5 or 1.75mm still produces a satisfatory mean flow prediction but dramatically reduces the required computational power for more than 30 times.

Case Study 2: Coupling CFD with Population Balance

CFD model developed using the SPH method can easily be extended with additional transport equations by applying a similar methodology like one previously described. Basically, the only difference is a set of additional function calls to apply particle approximations to all terms related to field functions in transport equations to produce an extended system of ODEs. Three new scalar transport equations were added: heat balance, mass balance (concentration of dissolved crystals) and a set equations resulting from discretisation of the number density function (Discretised Population

Balance method, DPB). In a preliminary study, the developed model was applied to growth-only cooling crystallisation of paracetamol from ethanol solution using the lab reactor and kinetics data from (17). The initial supersaturation was set to 1.5667 and an initial temperature 20^{0}C. The reactor is then cooled down to 0^{0}C through the reactor jacket. The population balance equation was discretised using the van Leer high resolution finite volume scheme with a flux limiter (18) using one hundred intervals. The performance of the developed CFD+DPB model and van Leer scheme was analyzed using the seed with sharp discontinuities in its CSD as shown in Fig. 4a. Since the process is growth-only the analytical solution for the ideal-mixing case is simply the initial CSD translated right by a distance $G{\cdot}t$ (growth rate multiplied by the time elapsed). The simulated CSDs after 600 seconds using the van Leer scheme for the case where the DPB was coupled with the ideal mixing model and with the SPH model are presented in Fig 4b.

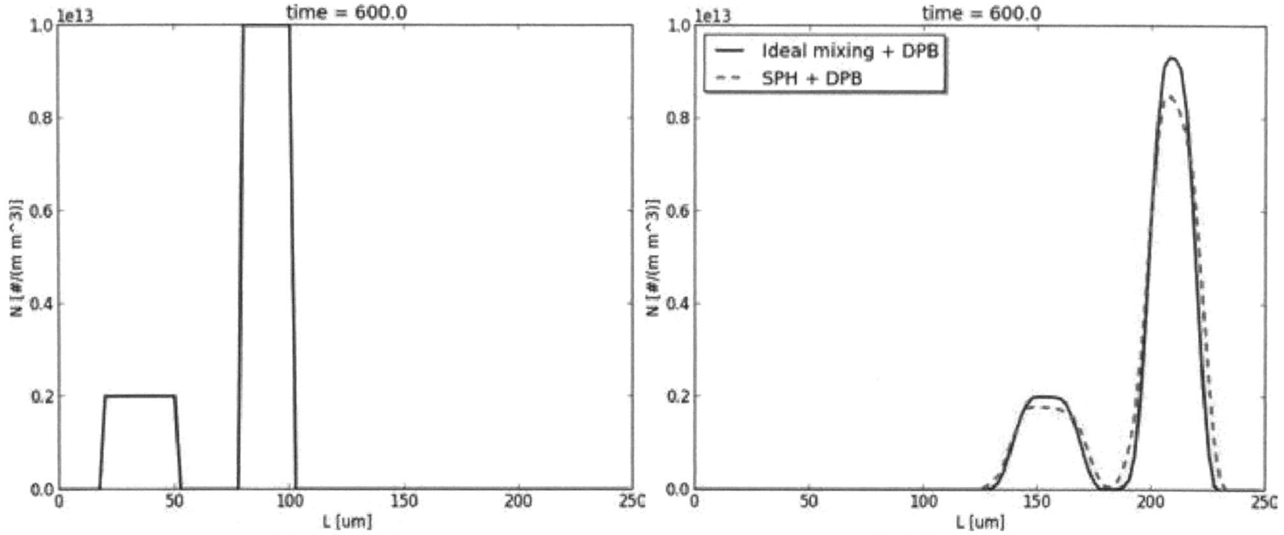

a) Inital CSD *b) CSD after 600s*

Figure 4. Simulation of the growth-only crystallisation using Discretised Population Balance method and van Leer high resolution finite volume scheme with flux limiter

The solid blue line in Fig. 4b shows the performance of van Leer discretization scheme in an ideal mixing case. The scheme successfully supressed oscillations in the solution and decreased the amount of the numerical diffusion present in all upwind schemes, although not completely. The dashed red line shows the CSD from the SPH model. Obviously, the nonidealities in the reactor cause that some fluid particles are well mixed while some other are located in poorly mixed zones (i.e. trapped in the zones around baffles or located in a poorly mixed top part of the volume) and contribute to the dispersion of the CSD.

A very important fact is that a simulation of coupled CFD and DPB is only around 40% slower compared to a pure hydrodynamics simulation. This is in a huge contrast to grid-based methods where one hundred additional equations would slow down simulation for few orders of magnitude.

Conclusions

The CFD model of a stirred tank has been implemented using the meshfree Smoothed Particle Hydrodynamics method and the simulation results have been compared to the experimental data and simulation results using the classical grid-based finite volume method available from the literature. The results clearly show that the grid-based methods produce a more detailed representation of the turbulent flow conditions in the reactor then the SPH method. The main reasons are the lack of reliable software codes that implement turbulent Navier-Stokes equations using the SPH method and a huge number of particles required to reproduce the same level of details. From the computational point, to achieve the comparable level of prediction the SPH method demands much higher computational power. Nevertheless, the meshfree SPH method is much more suitable for modelling of coupled hydrodynamics and crystallisation phenomena in stirred reactors for a number of reasons. It offers an exact model of a convective transport, extreme paralelization and model reduction capabilities, potential to directly utilize local hydrodynamics conditions for prediction of the secondary nucleation kinetics with a direct link to experimental particle-wall collision measurements using the shadow imaging technique, and the explicitly available CSD by solving a coupled system of Navier-Stokes and discretised population balance equations that is still almost intractable from the computational point of view using the classical grid-based CFD methods.

Acknowledgments: This research has been conducted as part of the Synthesis and Solid State Pharmaceutical Centre (SSPC) and funded by Science Foundation Ireland (SFI).

References

1. Ramkrishna D. (2000), "Population Balances: Theory and Applications to Particulate Systems in Engineering", *Academic Press*, New York.
2. Ramkrishna D., Mahoney A.W. (2002), "Population balance modeling, promise for the future", *Chem. Eng. Sci.* 57, 595–606.
3. Wei H.Y. (2010), "Computer-aided design and scale-up of crystallization processes: Integrating approaches and case studies", *Chem. Eng. Res. & Design*, 88, 1377.
4. Barrett M., O'Grady D., Casey E., Glennon B. (2011), The role of meso-mixing in anti-solvent crystallization processes, Chemical Engineering Science 66, 2523–2534.
5. John V., Angelov I., Öncül A.A., Thévenin D. (2007), "Towards the optimal reconstruction of a distribution from its moments", *Chem. Engi. Sci.* 62, 2890-2904.
6. De Souza L.G.M., Janiga G., John V., Thévenin D. (2010), "Reconstruction of a distribution from a finite number of moments with an adaptive spline-based algorithm", *Chem. Eng. Sci.* 65, 2741–2750.
7. Hutton K., Mitchell N., Frawley P.J. (2012), "Particle size distribution reconstruction: The moment surface method", *Powder Technology*, 222, 8–14.
8. Cogoni G., Frawley P.J. (2014), "Particle size distribution recontruction from a finite number of its moments using artificial neural networks", Joint Congress of ACTS-2014 and CGOM11, June 17-20, Nara, Japan.
9. Liu G.R. (2002), "Mesh Free Methods: moving beyond the finite element method", *CRC Press*, Boca Raton.
10. Liu G.R., Liu M.B (2003), "Smoothed Particle Hydrodynamics – a meshfree particle method", *World Scientific Publishing Co. Pte. Ltd.*
11. Lucy L.B. (1977), "A numerical approach to the testing of the fission hypothesis", *Astron. J.* 82 (12), 1013.
12. Gingold R., Monaghan J. (1977), "Smoothed particle hydrodynamics - theory and application to non-spherical stars", *Mon. Not. R. Astron. Soc.* 181, 37.

13. Morris J.P., Fox P.J., Zhu Y. (1997), "Modeling low reynolds number incompressible flows using SPH", *J. Comput. Phys.* 136 (1), 214-226.
14. Akinci N., Akinci G., Teschner M. (2013), "Versatile surface tension and adhesion for SPH fluids," *ACM Transactions on Graphics* 32, 6, 1–8.
15. MacDonald A. (2014), "Fluidix v1.0.3", OneZero Software (http://www.onezero.ca).
16. Hartmann H., Derksen J.J., Montavon C., Pearson J., Hamill I.S., van den Akker H.E.A. (2004), "Assessment of large eddy and RANS stirred tank simulations by means of LDA", *Chem. Eng. Sci.* 59, 2419–2432.
17. Mitchell N.A., Ó'Ciardhá C.T., Frawley P.J. (2011), "Estimation of the growth kinetics for the cooling crystallisation of paracetamol and ethanol solutions", *J. Crystal Growth* 328, 1, 39-49.
18. Qamar S., Elsner M.P., Angelov I.A., Warnecke G., Seidel-Morgenstern A. (2006), "A comparative study of high resolution schemes for solving population balances in crystallization", *Comp. Chem. Eng.* 30, 6–7, 1119-1131.

Understanding of pMDI-spacer interactions through CFD simulations

[1]Saurabh Sarkar, [2]Maziar Kakhi, [2]Prasad Peri, [1]Bodhisattwa Chaudhuri

1. *Department of Pharmaceutical Sciences, University of Connecticut,Storrs,CT-06269*
2. *US Food & Drug Administration, Silver Spring,MD-20993*

Despite being the primary therapeutic management for asthma for over 5 decades, pMDIs are frequently associated with high throat wall impaction and breath–actuation synchronization problems presenting significant dose delivery challenges, especially to pediatric and geriatric populations. Spacers or Valved Holding Chambers are used to address these problems by decreasing the spray momentum through increased flow residence time. However, there is considerable variability of performance with different pMDI –spacer combinations.. The purpose of the current study is to develop a validated multiphase multicomponent CFD model which can characterize flow and the resulting particle metrics (surface deposition, PSD etc.) when a suspension-based pMDI (Proventil HFA) is sprayed into a spacer. Such a model can help to shorten design-to-prototype lead times through increased process insight and control leading to more optimized products. The effect of different variables like volumetric rate of coaxially flowing air (28.3 or 11LPM) , spacer geometry (Aerochamber Plus or Optichamber Advantage) and delay in starting air flow after actuation (0 or 5 sec) are investigated. Flow behavior of the spray is simulated using the phenomenological RANS k-ω model with Lagrangian tracking of the dispersed phase (independent streams of propellant, cosolvent and drug) in a 2 way coupled mode.. The simulations were done with ANSYS Fluent 14.0.

The CFD model predictions compared well with experimental velocity and particle size data for actuation into unconfined surrounds. The results for inhaler-spacer systems show an increase of drug issuing from the spacers on increasing the volumetric flow rate as observed in experimental determinations. The recirculation observed near obstructions in axial path of the spray within the spacer is determined to be critical in enhancing spray retention within the spacer. Introduction of a delay is found to trap more drug in the spacer. The MMAD of the drug that escapes the spacer is smaller than what is injected for all cases indicating deposition of coarser particles within the spacer.

376761 Applying Poisson-Boltzmann Theory to Predict Excipient Concentration in Ultrafiltered Therapeutic Protein Formulations

Abstract:

The final bulk drug substance manufacturing step for most therapeutic monoclonal antibodies is a buffer exchange and concentration through an ultrafiltration membrane into a final formulation buffer. Properties such as pH and concentration of excipients in the resulting product solution are important, because they impact the stability of the frozen drug substance as well as the final product formulation composition.

Preferential rejection or retention of buffer components is commonly observed in this final production step; leading to a requirement to bias specific buffer components to higher or lower concentrations in order to achieve the targeted final composition. Here, we assess the applicability of a Poisson-Boltzmann model for prediction of the final product composition and pH as a function of protein charge, diafiltration buffer composition, and final product concentration.

374851 Fluid Bed Granulation: Towards a Comprehensive Process Model

Abstract:

Although fluidised bed granulation (FBG) is a widely-used unit operation, its practical application is often guided by empirical methods and the experience of the operating staff rather than by systematic and scientifically-based strategies. Employed in various branches of industry (for producing fertilisers, detergents, industrial and agricultural chemicals, food products, and also pharmaceuticals), granulation in general - and fluid bed granulation in particular - remained more of "an art than a science" [1].

As also demonstrated in other areas, the development of realistic mathematical models that are interlinked with high-precision in-line process measurements can yield powerful tools for a knowledge-based control of process and product quality for all particulate processes [2]. The complex interplay of many different variables and processes during fluid-bed granulation poses a significant challenge in developing such models. The associated effects may be grouped into one of three categories [3]: wetting and nucleation, consolidation and growth, as well as breakage and attrition. Every realistic model for a fluid bed granulator must incorporate these three effetcs. In addition, the model needs to account for all process-relevant parameters (e.g., air and liquid flow rates, binder properties, etc.) that influence the sub-processes as model input.

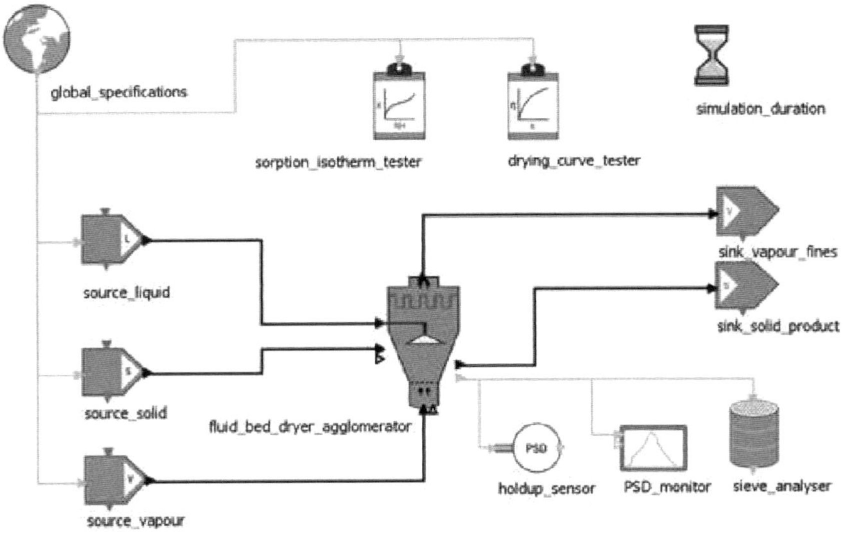

Figure 1: Schematic of the process modelling approach ("flowsheeting") using the custom fluid bed agglomerator unit.

In our work, we develop a model of the fluid bed granulator based on a basic model provided in the process simulation software gSOLIDS 3.1 (Process Systems Enterprise Ltd., London, UK). The basic model already contains agglomeration, drying, and elutriation of particles as separate phenomena. To make it suitable for batch processes, the extended model additionally takes into account the breakage of granulates, as well as the continuous introduction of spray. For breakage, different commonly used models are compared with respect to their applicability. The spray introduction is based on a new *user-defined* phenomenon for the wetting of particles. In addition, we also explored the possibility of introducing a liquid binder component as a part of an additional solid phase. We present results for the different approaches for wetting and for a selection of combinations of agglomeration and breakage kernels. The results are further compared to in-line measurements of particle size and moisture from industrial fluid bed processes of different scales.

References

[1] J. D. Litster, Powder Technol. **130**, 35 (2003).

[2] I. T. Cameron, F. Y. Wang, C. D. Immanuel, and F. Stepanek, Chem. Eng. Sci. **60**, 3723 (2005).

[3] S. M. Iveson, J. D. Litster, K. Hapgood, and B. J. Ennis, Powder Technol. **117**, 3 (2001).

The Simulation of Mixing and Free Surface Flows in Co-Rotating Twin-Screw Extruders

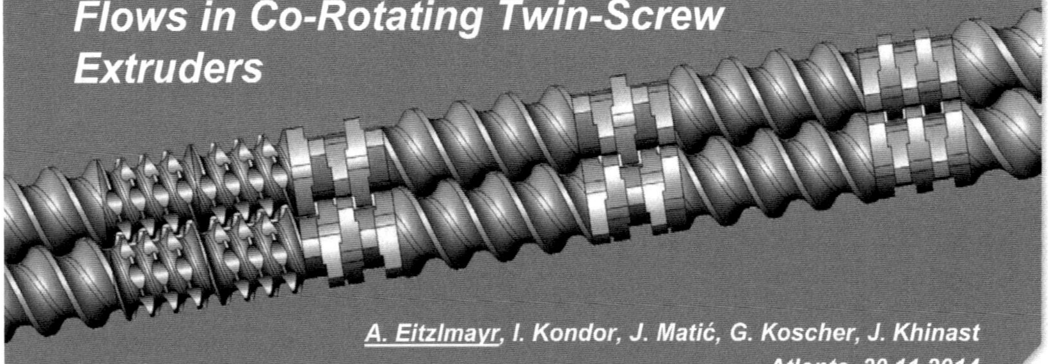

A. Eitzlmayr, I. Kondor, J. Matić, G. Koscher, J. Khinast
Atlanta, 20.11.2014

K1 Competence Center - Initiated by the Federal Ministry of Transport, Innovation and Technology (BMVIT) and the Federal Ministry of Economy, Family and Youth (BMWFJ). Funded by the Austrian Research Promotion Agency (FFG), Land Steiermark and the Styrian Business Promotion Agency (SFG)

Outline

- **Introduction**
  - → Simulation of Hot Melt Extrusion (HME)
  - → Used Approach
  - → Smoothed Particle Hydrodynamics (SPH)
- **Wall Boundary Conditions**
- **Results: Conveying Element**
  - → Comparison to CFD (Completely Filled)
  - → Partially Filled State
  - → Mixing Analysis
- **Results: Different Screw Elements**
- **Conclusions & Outlook**

Introduction

research center pharmaceutical engineering

Simulation of Hot Melt Extrusion

Challenges

- **Geometry:** complex, rotating.
- **Tight clearances** (resolution challenge).
- **Partially filled** (free surface flow).
- Complex **material behavior.**
- High viscosity → **dissipation** → nonuniform temperature.
- **Melting, Devaporization.**
- **But:** no turbulent flow.

Currently Feasible

- **Empirical Models (1D)**
- **First-principle Models (3D):**

  → **CFD:** viscous flow (completely filled).

  → **SPH:** viscous flow (also partially filled).

Introduction

research center pharmaceutical engineering

Simulation of Hot Melt Extrusion

Challenges

- **Geometry:** complex, rotating.
- **Tight clearances** (resolution challenge).
- **Partially filled** (free surface flow).
- Complex **material behavior.**
- High viscosity → **dissipation** → nonuniform temperature.
- **Melting, Devaporization.**
- **But:** no turbulent flow.

Currently Feasible

- **Empirical Models (1D)**
- **First-principle Models (3D):**

  → **CFD:** viscous flow (completely filled).

  → **SPH:** viscous flow (also partially filled).

Our Approach

research center pharmaceutical engineering

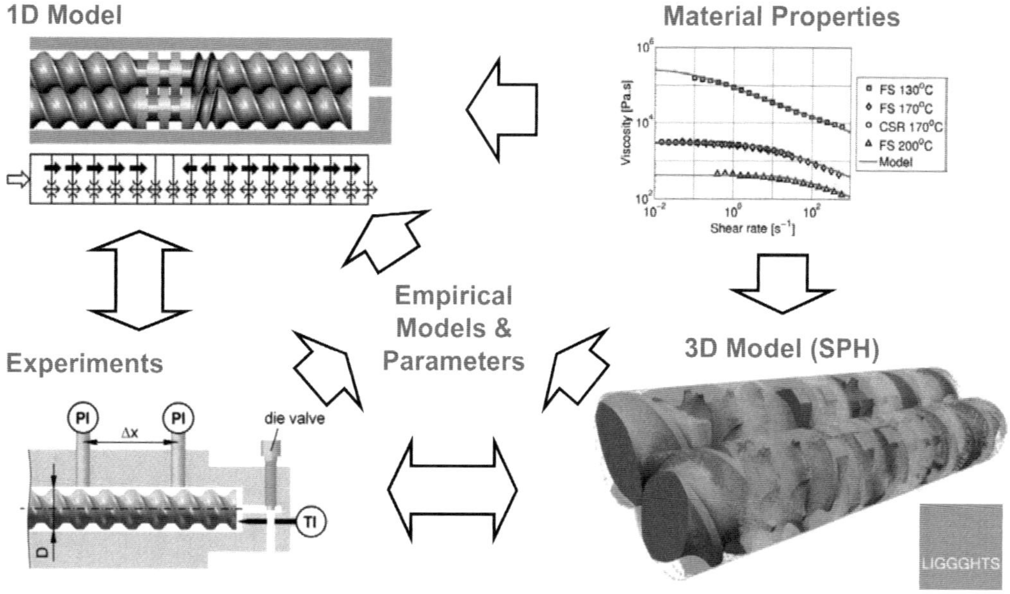

1D Modeling: Example Results

research center pharmaceutical engineering

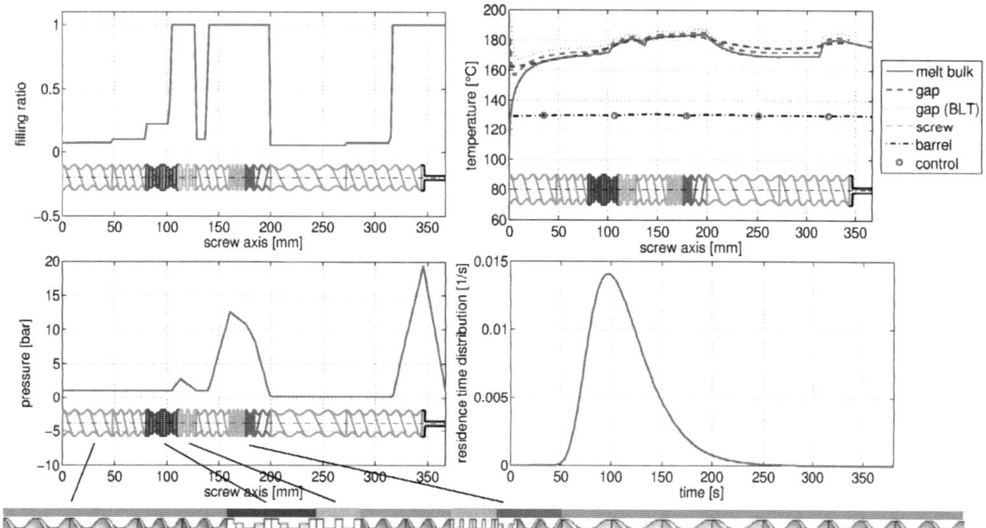

A. Eitzlmayr, G. Hörl, G. Koscher, G. Reynolds, Z. Huang, J. Booth, P. Shering, J. Khinast, AIChE J. 59 (2013) 11, 4440 – 4450.
A. Eitzlmayr, G. Koscher, G. Reynolds, Z. Huang, J. Booth, P. Shering, J. Khinast, Int. J. Pharm. 474 (2014) 157 – 176.

Introduction: SPH

research center pharmaceutical engineering

Smoothed Particle Hydrodynamics (SPH)

Lagrangian fluid elements ("particles") represent the flow

Benefits:

- **Mesh-free** (complex screw geometry)
- **Free surface** flows (partially filled screws)
- Inherent **scalar transport** (→ mixing)

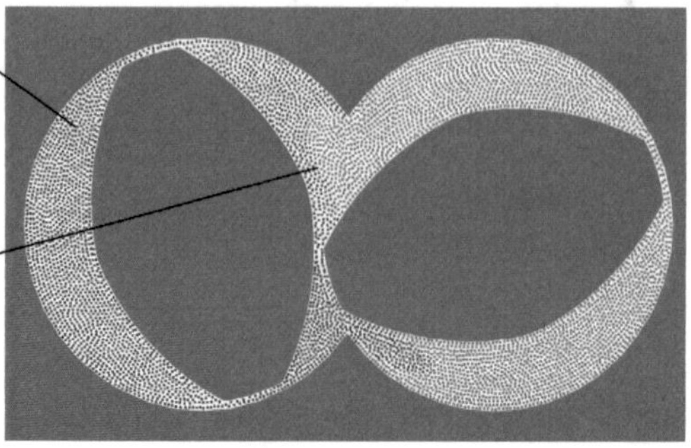

snapshot of a twin-screw cross-section

Used Software:

LIGGGHTS Open Source Particle Simulation Code (www.liggghts.com)

→ **Thanks to Dr. Christoph Kloss for cooperation**
DCS Computing GmbH, Linz, Austria

Introduction: SPH

research center pharmaceutical engineering

Smoothed Particle Hydrodynamics (SPH)

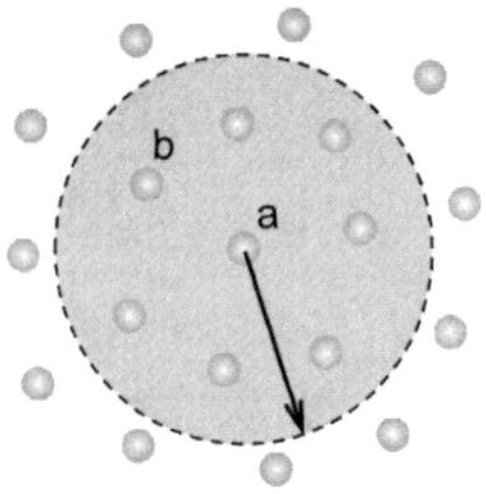

Summation interpolant*:

$$A_{S,a} = \sum_b A_b \cdot \frac{m_b}{\rho_b} \cdot W_{ab}$$

$$W_{ab} = W\left(\frac{|\vec{x}_a - \vec{x}_b|}{h}, h\right)$$

Kernel function* $W(q,h)$

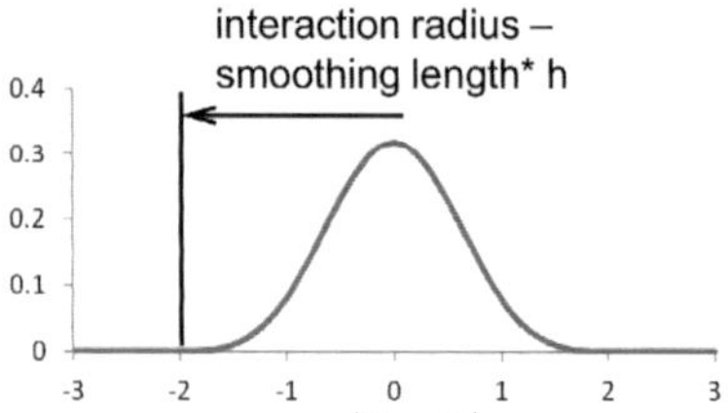

$$q = \frac{|\vec{x}_a - \vec{x}_b|}{h}$$

Continuity equation*: $$\frac{d\rho_a}{dt} = \sum_b m_b (\vec{v}_a - \vec{v}_b)\vec{\nabla}_a W_{ab}$$

Equation of state*: (weakly compressible) $$P_a = B\left[\left(\frac{\rho_a}{\rho_0}\right)^\gamma - 1\right]$$

Momentum equation*: $$\frac{d\vec{v}_a}{dt} = -\sum_b m_b \left(\frac{P_b}{\rho_b^2} + \frac{P_a}{\rho_a^2}\right)\vec{\nabla}_a W_{ab} + \sum_b \frac{m_b \cdot (\mu_a + \mu_b)}{\rho_a \cdot \rho_b}\left(\frac{1}{r_{ab}} \cdot \frac{\partial W_{ab}}{\partial r_a}\right)\vec{v}_{ab}$$

Morris model** (laminar flow)

* J.J. Monaghan, Smoothed particle hydrodynamics, Rep. Prog. Phys. 68 (2005) 1703 – 1759.
** J.P. Morris, P.J. Fox, Y. Zhu, Modeling Low Reynolds Number Incompressible Flows Using SPH, J. Comp. Phys. 136 (1997) 214 – 226.

Boundary Conditions

research center pharmaceutical engineering

Boundary Methods vs. Complex Geometry

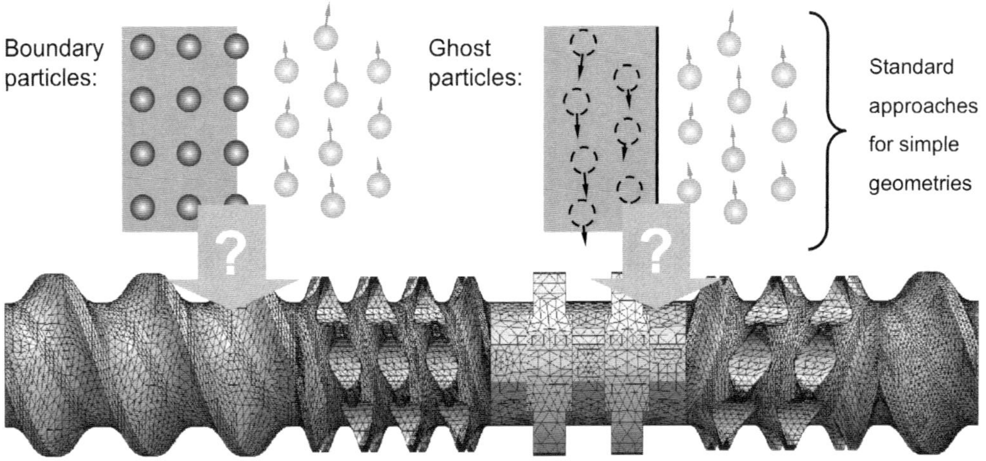

→ Novel "wall interaction" method* **applicable to complex geometries in the STL format.**

* A. Eitzlmayr, G. Koscher, J. Khinast, Computer Physics Communications 185 (2014) 2436 – 2448.

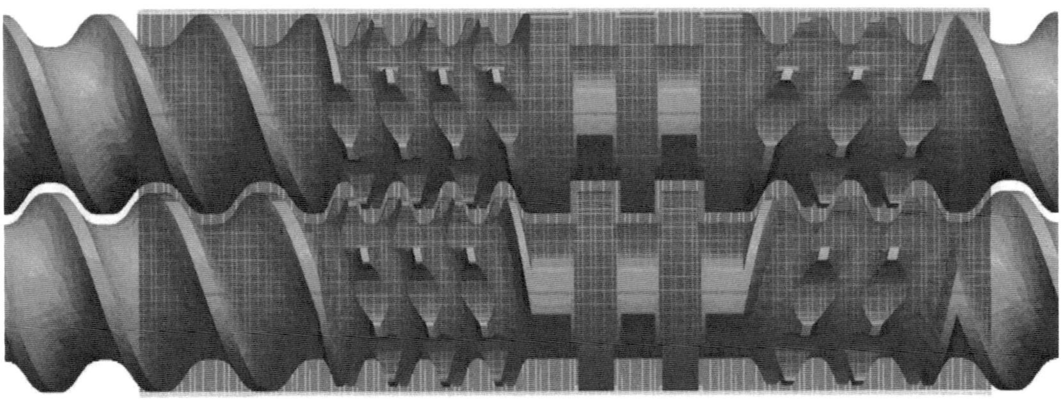

Validation Case

research center pharmaceutical engineering

Setup

Geometry:

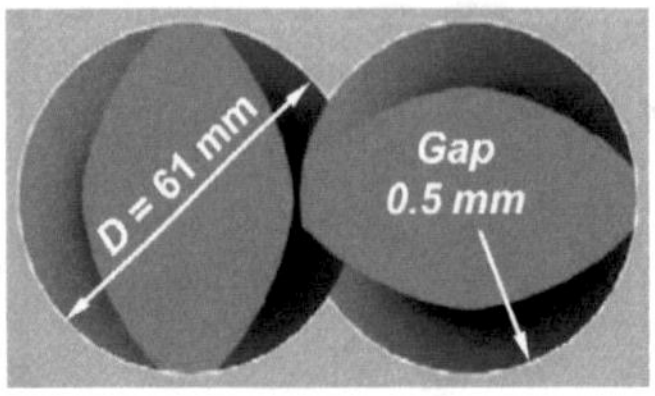

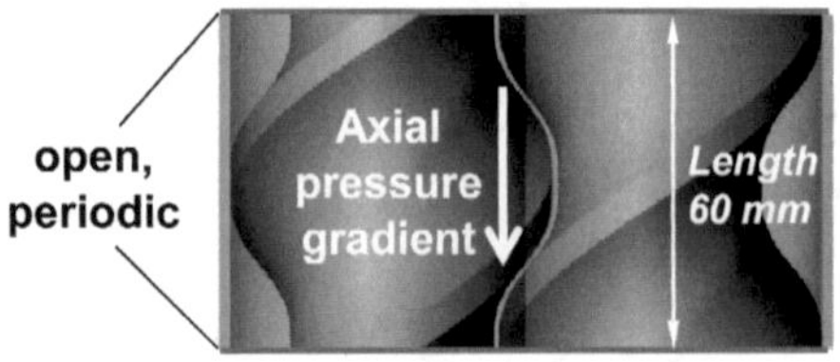

FVM: (Bierdel*)

Density ρ = 1000 kg/m³

Viscosity η = 1000 Pas (Newt.)

Speed n = 95 rpm

Re = 0.0059

$$\left(\mathrm{Re} = \frac{n \times D^2 \times \rho}{\eta} \right)$$

SPH:

Density ρ = 1000 kg/m³

Viscosity η = **1 Pas**

Speed n = 60 – 150 rpm

***Re* = 3.7** (creeping flow Re < 10)

Particle spacing: Δx = **0.5 mm** (ca. 1 Mio.)

Time step: Δt = **5 μs**

* Michael Bierdel. In: Kohlgrüber K, Wiedmann W. *Co-Rotating Twin-Screw Extruders.* Munich: Carl Hanser Verlag, 2008.

Validation Case

research center pharmaceutical engineering

Results

(A. Eitzlmayr, J. Khinast, Co-Rotating Twin-Screw Extruders: Detailed Analysis based on Smoothed Particle Hydrodynamics. Part 1: Hydrodynamics, submitted to Chemical Engineering Science)

Pressure characteristic

dim.-less pressure drop

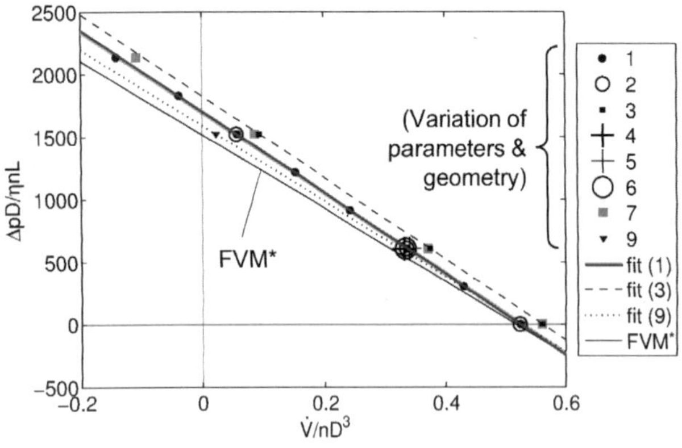

dim.-less flow rate

Axial velocity

(case: pressure drop = 0)

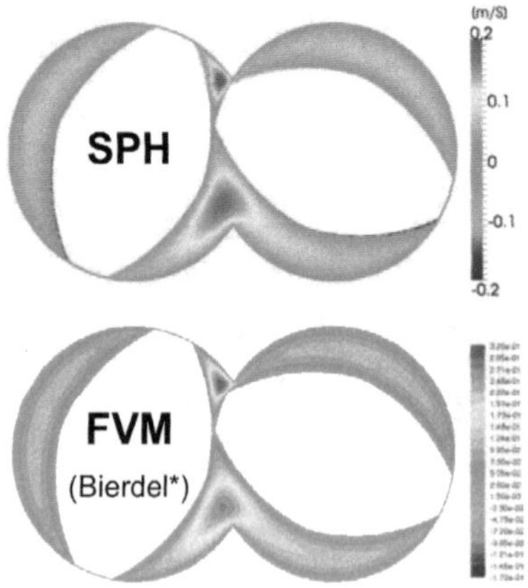

* Michael Bierdel. In: Kohlgrüber K, Wiedmann W. *Co-Rotating Twin-Screw Extruders.* Munich: Carl Hanser Verlag, 2008.

Validation Case

research
center
pharmaceutical
engineering

Results

(A. Eitzlmayr, J. Khinast, Co-Rotating Twin-Screw Extruders: Detailed Analysis based on Smoothed Particle Hydrodynamics. Part 1: Hydrodynamics, submitted to Chemical Engineering Science)

Pressure characteristic

dim.-less
pressure drop

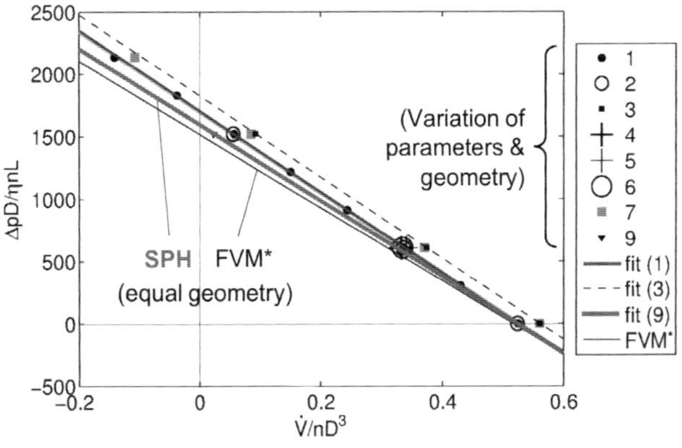

dim.-less flow rate

Axial velocity

(case: pressure drop = 0)

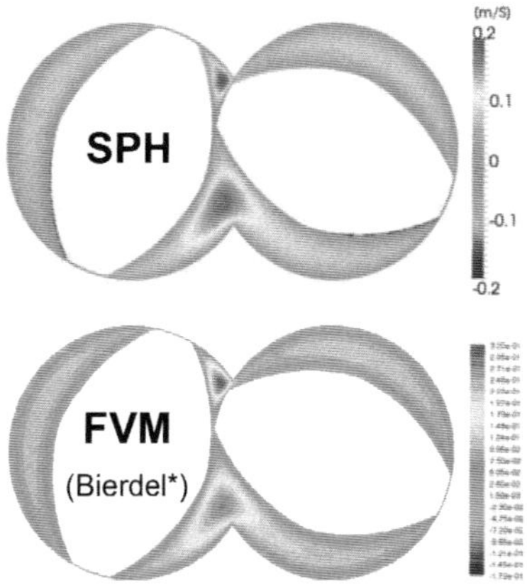

* Michael Bierdel. In: Kohlgrüber K, Wiedmann W. *Co-Rotating Twin-Screw Extruders*. Munich: Carl Hanser Verlag, 2008.

Validation Case

research
center
pharmaceutical
engineering

Results

(A. Eitzlmayr, J. Khinast, Co-Rotating Twin-Screw Extruders: Detailed Analysis based on Smoothed Particle Hydrodynamics. Part 1: Hydrodynamics, submitted to Chemical Engineering Science)

Power characteristic

dim.-less
driving power

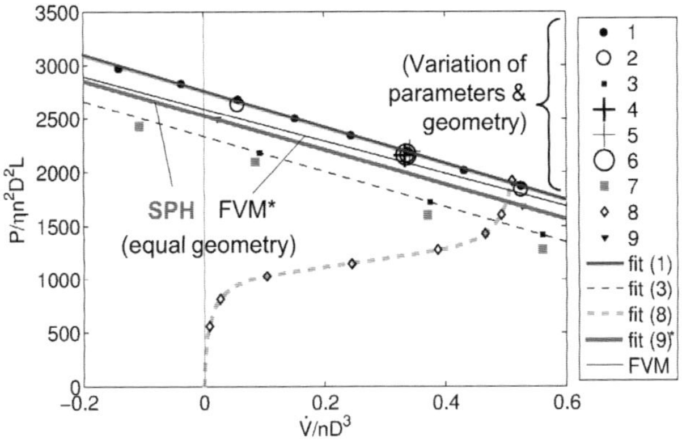

dim.-less flow rate

Axial velocity

(case: pressure drop = 0)

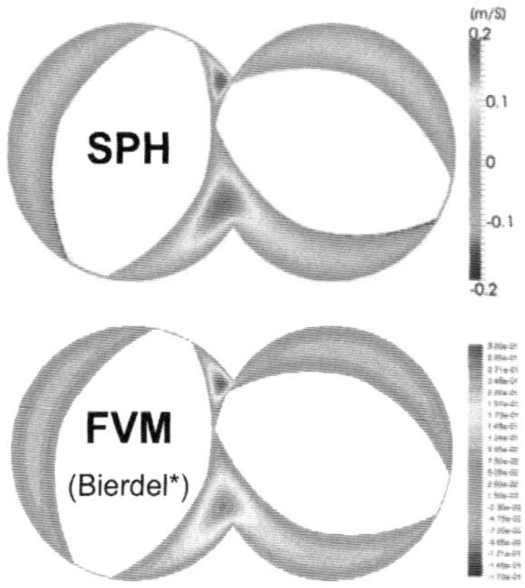

* Michael Bierdel. In: Kohlgrüber K, Wiedmann W. *Co-Rotating Twin-Screw Extruders*. Munich: Carl Hanser Verlag, 2008.

Partially Filled State

research center pharmaceutical engineering

Results

(A. Eitzlmayr, J. Khinast, Co-Rotating Twin-Screw Extruders: Detailed Analysis based on Smoothed Particle Hydrodynamics. Part 1: Hydrodynamics, submitted to Chemical Engineering Science)

Power characteristic

dim.-less
driving power

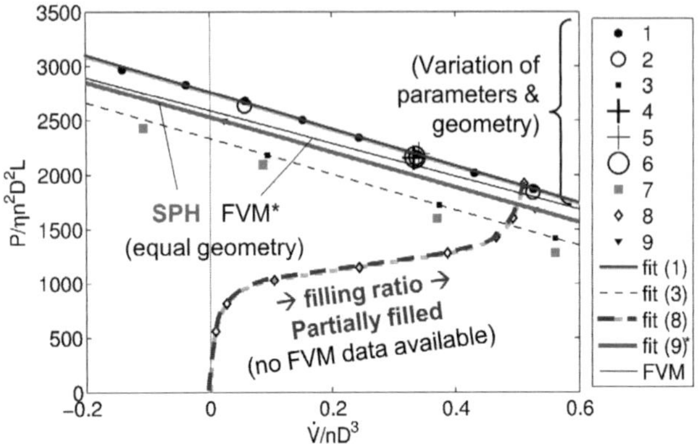

dim.-less flow rate

Axial velocity

(case: 50% filled)

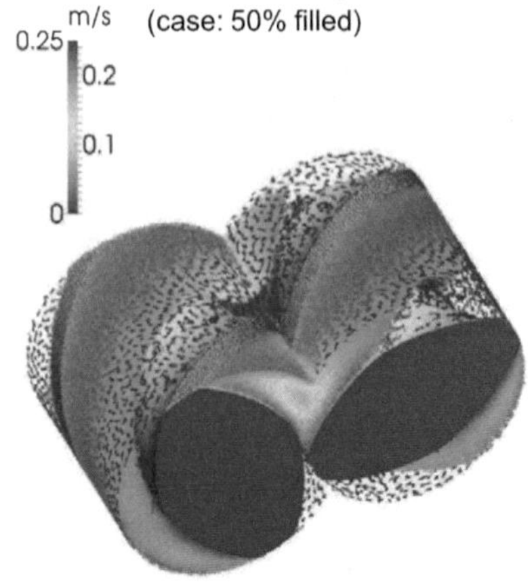

* Michael Bierdel. In: Kohlgrüber K, Wiedmann W. *Co-Rotating Twin-Screw Extruders*. Munich: Carl Hanser Verlag, 2008.

TU Graz UNI GRAZ JOANNEUM RESEARCH

Partially Filled State

research center pharmaceutical engineering

Results

(A. Eitzlmayr, J. Khinast, Co-Rotating Twin-Screw Extruders: Detailed Analysis based on Smoothed Particle Hydrodynamics. Part 1: Hydrodynamics, submitted to Chemical Engineering Science)

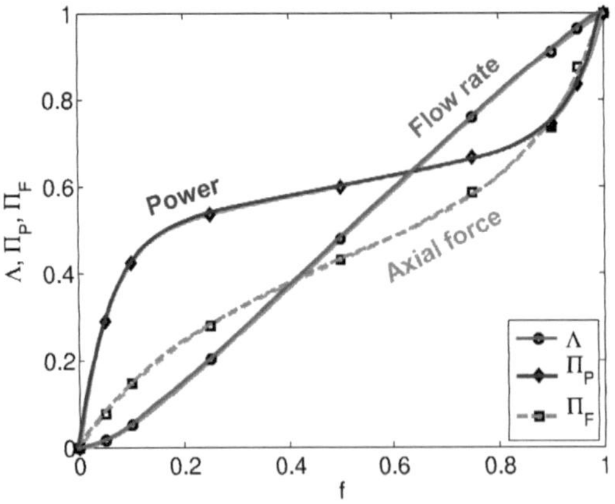

Filling ratio

Axial velocity

(case: 50% filled)

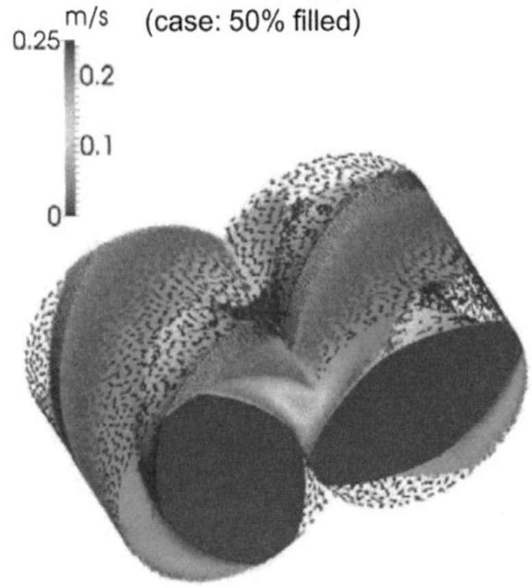

→ **Important input data for 1D modeling of extruders.**

TU Graz UNI GRAZ JOANNEUM RESEARCH

Mixing

research center pharmaceutical engineering

Evaluation

(A. Eitzlmayr, J. Khinast, Co-Rotating Twin-Screw Extruders: Detailed Analysis based on Smoothed Particle Hydrodynamics. Part 2: Mixing, submitted to Chemical Engineering Science)

Snapshots (example):

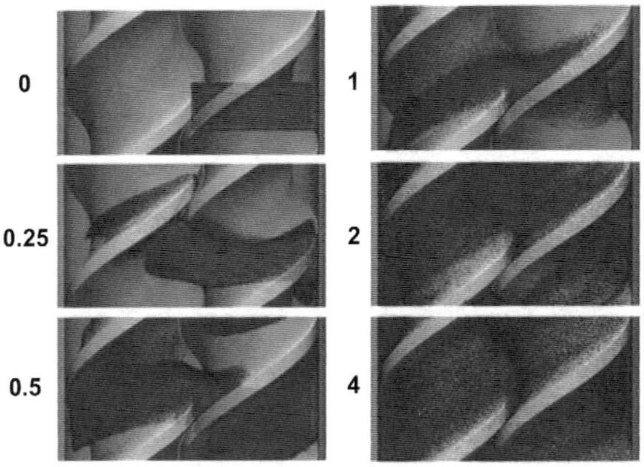

Intensity of Segregation: $$S = \frac{\sigma}{\sqrt{\frac{N_{tracer}}{N_{total}}\left(1 - \frac{N_{tracer}}{N_{total}}\right)}}$$

σ ... standard deviation of tracer content:

Overall:

Axial:

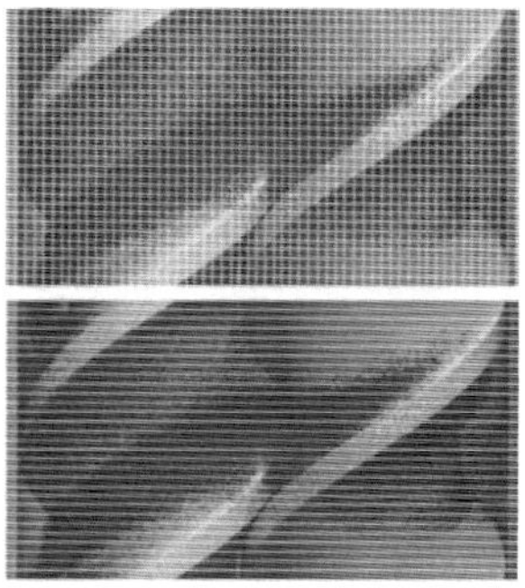

(spacing 2 mm)

Mixing

research center pharmaceutical engineering

Time Evolution

(A. Eitzlmayr, J. Khinast, Co-Rotating Twin-Screw Extruders: Detailed Analysis based on Smoothed Particle Hydrodynamics. Part 2: Mixing, submitted to Chemical Engineering Science)

Intensity of Segregation: $$S = \frac{\sigma}{\sqrt{\frac{N_{tracer}}{N_{total}}\left(1 - \frac{N_{tracer}}{N_{total}}\right)}}$$

Fitted curves: $S = S_0 e^{-kN}$

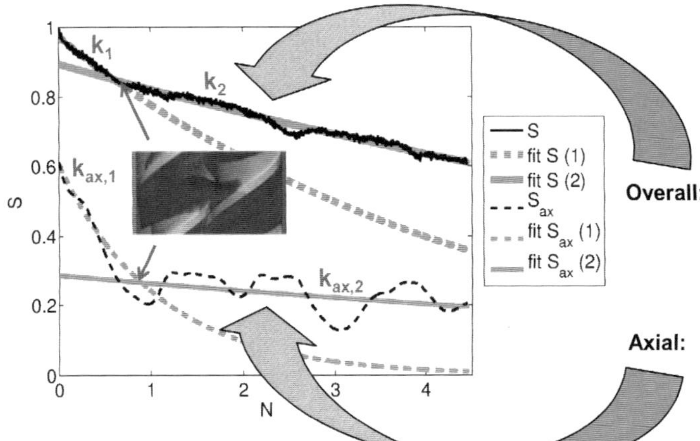

σ ... standard deviation of tracer content:

Overall:

Axial:

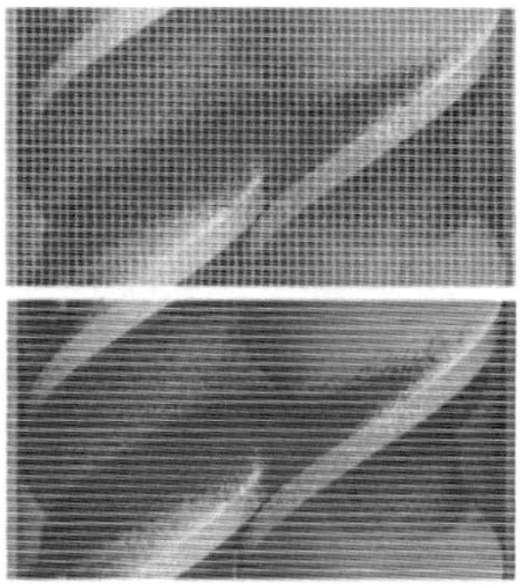

(spacing 2 mm)

Mixing

research center pharmaceutical engineering

Results

(A. Eitzlmayr, J. Khinast, Co-Rotating Twin-Screw Extruders: Detailed Analysis based on Smoothed Particle Hydrodynamics. Part 2: Mixing, submitted to Chemical Engineering Science)

Mixing rates vs. flow rate:

Mixing rate k is defined per revolution N:

$$S = S_0 e^{-kN} \quad \Rightarrow k = -\frac{1}{S}\frac{dS}{dN}$$

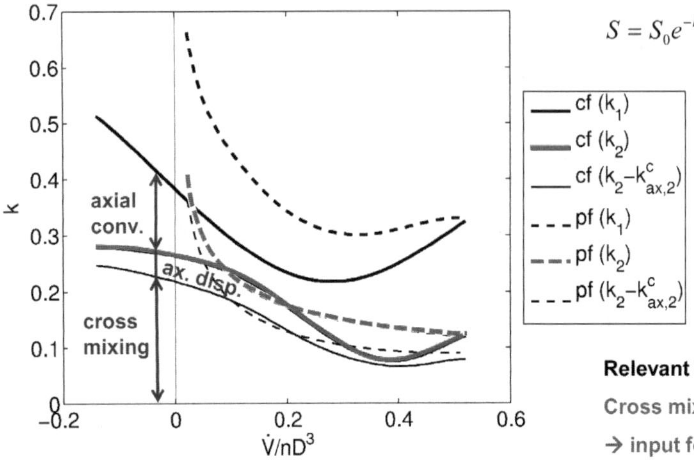

Relevant in reality:

Cross mixing + axial dispersion (k_2)

→ input for 1D modeling

The Simulation of Mixing and Free Surface Flows in Co-Rotating Twin-Screw Extruders
20.11. 2014, Atlanta, Page 19

Mixing

research center pharmaceutical engineering

Results

(A. Eitzlmayr, J. Khinast, Co-Rotating Twin-Screw Extruders: Detailed Analysis based on Smoothed Particle Hydrodynamics. Part 2: Mixing, submitted to Chemical Engineering Science)

Estimation of S for a given length L:

$S = S_0 e^{-k_2 N}$ (N with the average residence time)

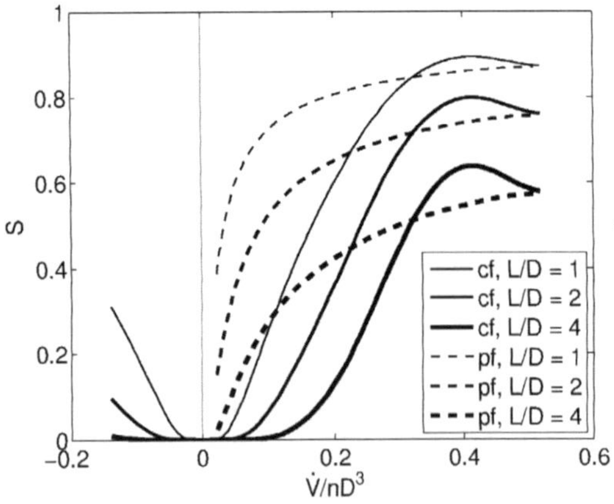

SMEC (energy input per kg)

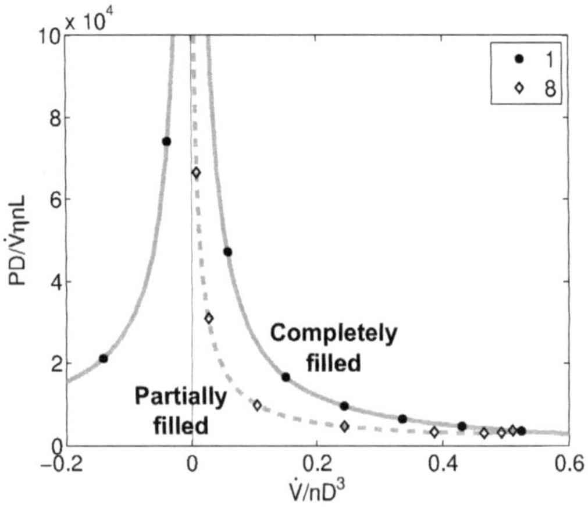

The Simulation of Mixing and Free Surface Flows in Co-Rotating Twin-Screw Extruders
20.11. 2014, Atlanta, Page 20

research center pharmaceutical engineering

Application to Different Screw Elements (D = 27 mm)

Conveying Elements

Pitch **30 mm** C30

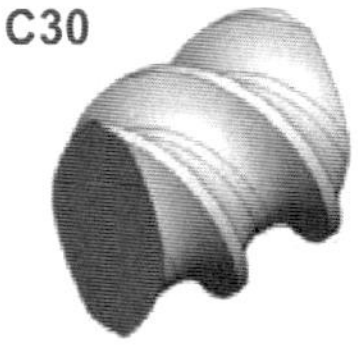

15 mm C15

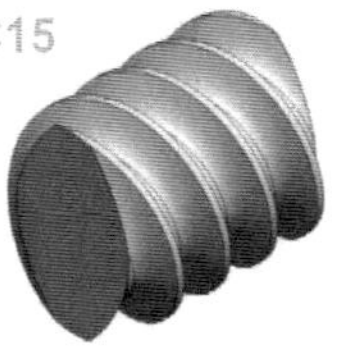

Mixing Element

M15

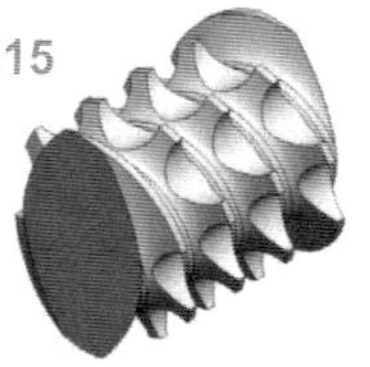

Kneading Elements

Stagger angle **30°** K30

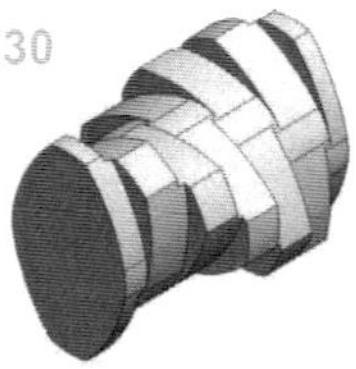

Stagger angle **90°** K90

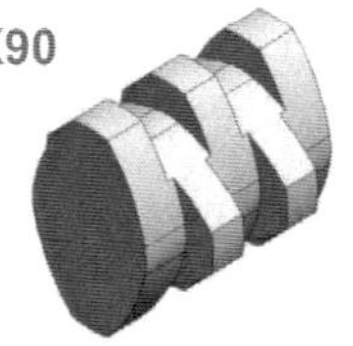

Screw Elements

research center pharmaceutical engineering

Pressure Characteristic (completely filled) **Flow Rate vs. Filling Ratio** (partially filled)

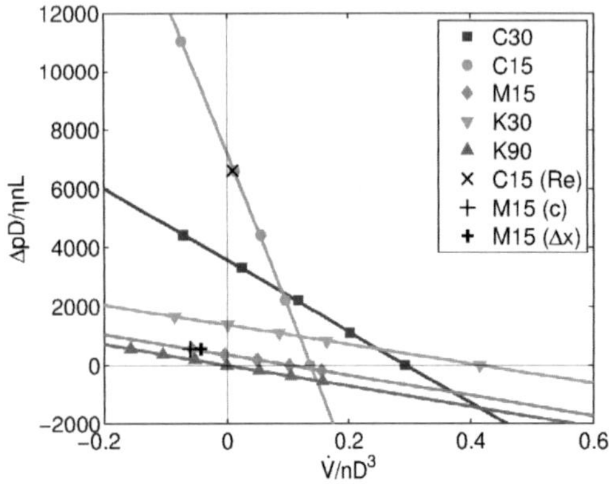

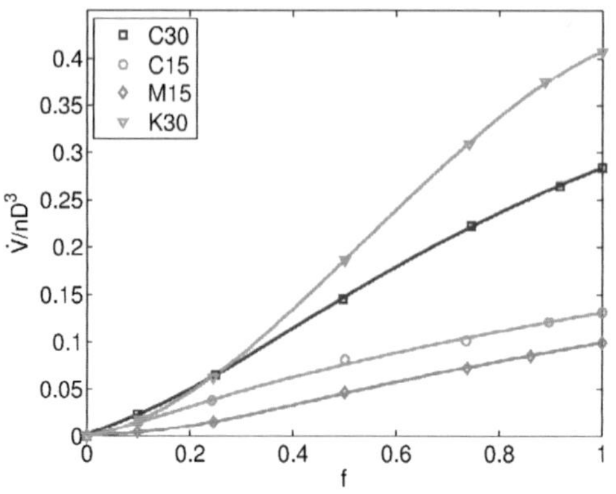

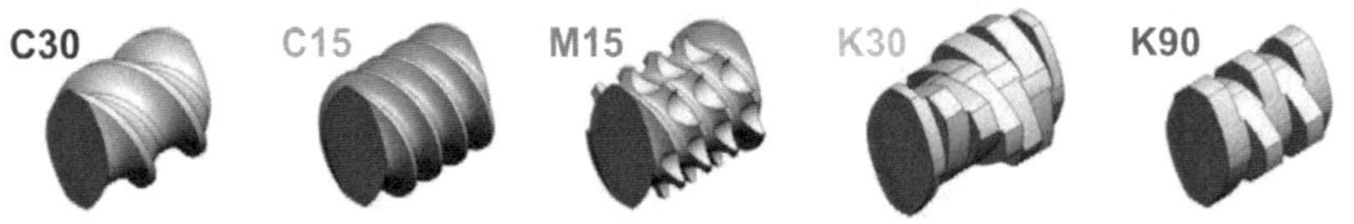

Screw Elements

research center pharmaceutical engineering

Power Characteristic **Overall Mixing**

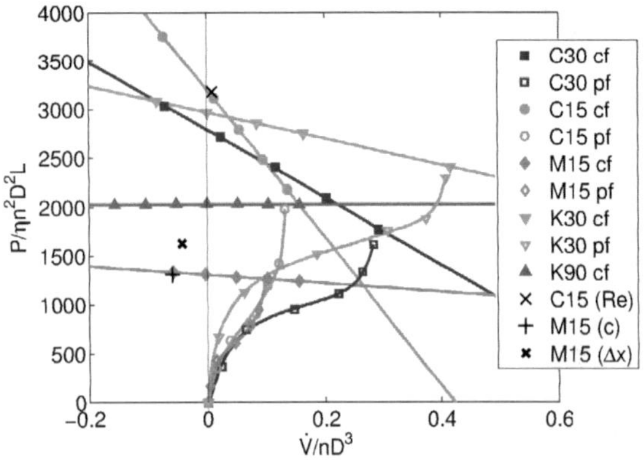

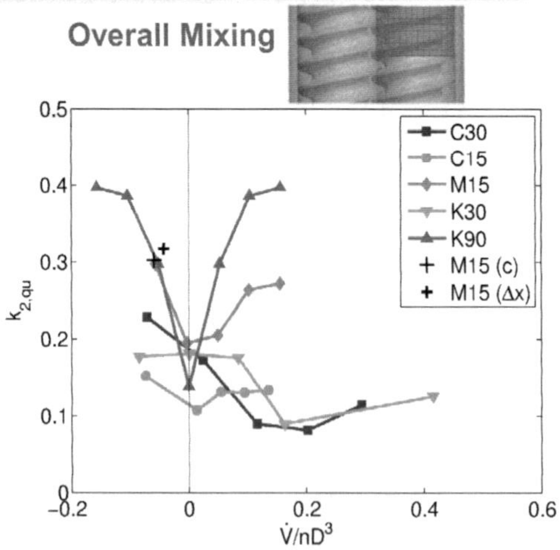

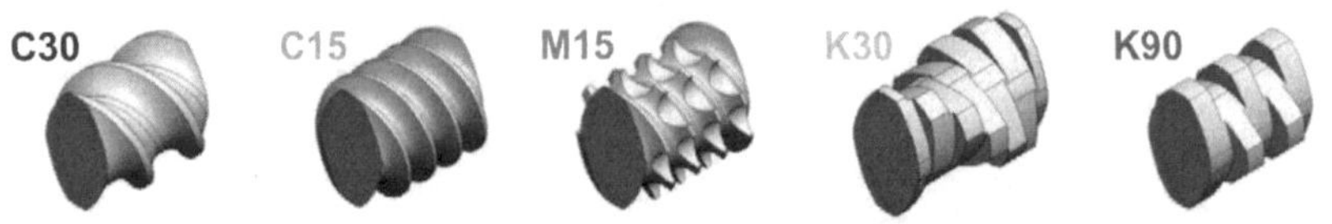

Conclusions & Outlook

research center pharmaceutical engineering

Conclusions

- Developed a **novel boundary method** for complex wall geometries in SPH.
- Achieved **good agreement with CFD** data (completely filled state).
- Beyond that, SPH allows the investigation of the **partially filled state**, and ...
- ... a detailed **analysis of mixing** based on tracer particles.
- Results can be used for the **parameterization of 1D models**.

Outlook

- Transfer to **GPU computing.**
- Extension of the SPH model to **non-Newtonian flow** including **thermal energy**.
- Study HME processes based on **1D modeling** (e.g., design, scale-up).

research center pharmaceutical engineering

Thank you!

Acknowledgements:
Leistritz Extrusionstechnik GmbH
DCS Computing GmbH

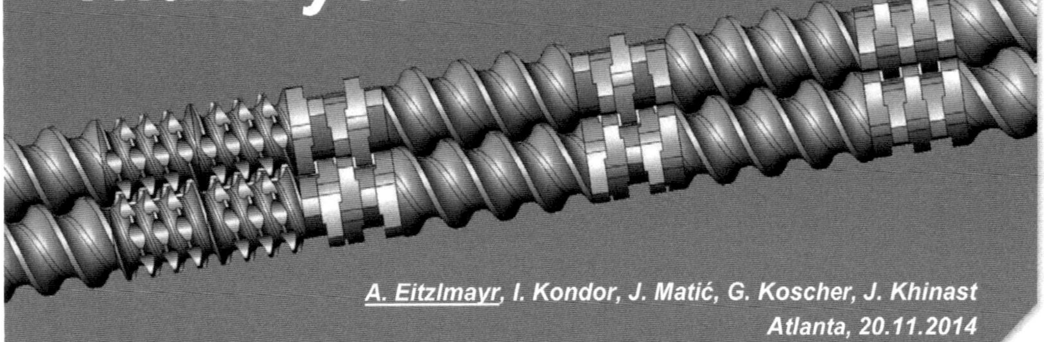

A. Eitzlmayr, I. Kondor, J. Matić, G. Koscher, J. Khinast
Atlanta, 20.11.2014

K1 Competence Center - Initiated by the Federal Ministry of Transport, Innovation and Technology (BMVIT), and the Federal Ministry of Economy, Family and Youth (BMWFJ). Funded by the Austrian Research Promotion Agency (FFG), Land Steiermark and the Styrian Business Promotion Agency (SFG).

385230 Case Study: Simulation-Based Process Optimization for a Monoclonal Antibody Cation Exchange Step

Abstract:

Case Study: Simulation-Based Process Optimization for a Monoclonal Antibody Cation Exchange Step.

The biopharmaceutical industry continues to explore options for improving process development efficiency to meet the needs of today's fast and competitive market. This presentation discusses the application of computer simulation to the optimization of a monoclonal antibody cation exchange (CEX) chromatography step. The chromatography simulation utilizes the General Rate Model for chromatography in combination with a Steric Mass Action binding isotherm to predict process performance over a range of operating conditions. Simulation was used to optimize the process throughput while maintaining acceptable product purity. The results suggest that application of computer simulation has the potential to significantly reduce the time and experimental resources required for process optimization.

PEGylation is a generic term for chemical conjugation of polyethylene glycol (or PEG) to biomacromolecules such as proteins. PEGylation is frequently used to improve the properties of therapeutic proteins. While this type of chemical reaction is easy to carry out, it typically results in the synthesis of complex mixtures of different PEGylated forms, i.e. mono-, di-, tri-, and higher-PEGylated protein. With many PEGylated therapeutic protein products the mono-PEGylated protein is usually the desired form. Formation of by-products in a PEGylation reaction represents material loss as well as the inconvenience of a challenging purification process that would follow. The problem of low selectivity of mono-PEGylation could be addressed through the use of specific reactions such as N-terminal PEGylation. However, even with such specific reactions, significant amounts of by-products are typically synthesized. Generally speaking, a trade-off between selectivity and yield of desired product is observed in most PEGylation reactions. In [1], we presented a systematic attempt at optimization of batch protein PEGylation. To this end, initially a first principles model (with the associated reaction network) was presented with kinetic rate constants determined using the experimental data. The dynamic model captures the effect of variation in initial condition as well as the pH at which the experiment is run. Having established the validity of the model beyond the fit data by additional experiments, the model is then utilized to determine the optimal pH, duration and initial recipe that maximize the formation of the mono-PEGylated product. The benefits of operating at the conditions suggested by the optimization problem are finally demonstrated though additional experiments.

In this contribution, we present several additional simulation and experimental results that validate both the structure and values of the identified parameters as well as explore additional means of operations. In particular, simulation results are generated to predict the productivity and yield for a fed-batch operation with parallel filtration of the reaction media. In addition, the rate kinetics are utilized in predicting operation using hollow fibre membranes for carrying out the reaction and separation. In summary, the experimental results are used to verify the structure of the proposed kinetic model, while the simulation results are used in an optimization framework to suggest optimal experimental operating conditions that are subsequently verified experimentally.

References

[1] X. Shang, B. Corbett, P. Mhaskar, and R. Gosh. Optimization of protein pegylation. AIChE Annual Meeting San Francisco, CA, 2013.

388174 Development of a Scalable Friedel-Crafts Acylation: A Mechanistic Approach

Abstract:

The commercial route of a late-stage asset under development was initially performed via a Friedel-Crafts reaction with chloroacetyl chloride followed by an alkylation. Due to the projected demand and manufacturing cost, the development of an efficient, scalable route was essential to the project. The first generation process was successful in laboratory scale reactions; however, steady erosion in yield with increasing batch size was observed in production scale campaigns. This scale dependence was found to result from polymerization of the input material. Mechanistic studies of the chemistry, aided by in-line FTIR, identified the unfavorable equilibrium between intermediates.

Efforts to understand and mitigate the risk of polymerization lead the team to an alternate approach. A second generation process was subsequently developed to replace the single lower yielding step with two robust high yielding reactions. This new process was found to provide a higher overall yield to the isolated product, as demonstrated at the vendor. This presentation will provide a case study on the challenges of the scale dependent first generation process and the mechanistic approach taken to identify and design the advantages of the second generation process.

Multiphase Transport Modeling for Vacuum Drying of Pharmaceutical Products

Aditya G. Dodda[a], Kostas Saranteas[b] and Michael A. Henson[a]

[a]Department of Chemical Engineering, University of Massachusetts, Amherst, MA 01003
[b]Process Chemistry and Engineering, Sunovion Pharmaceuticals, Marlborough, MA 01752

Abstract

Downstream processing of Active Pharmaceutical Ingredients (APIs) typically involves extraction, crystallization, filtration and drying. In addition to being the most energy intensive unit operation, drying is often a major manufacturing bottleneck due to its relatively long batch processing times. Furthermore, drying conditions must be carefully selected and maintained to avoid problems such as API degradation. As a result, prediction and optimization of drying dynamic are needed to ensure product quality and minimize operating time. In principle, drying dynamics can be accurately predicted by detailed models that capture the relevant physics. Traditional models used to understand drying of APIs are based on simple solvent diffusion and penetration mechanisms that fail to capture important heat and mass transport effects. More sophisticated transport models are invariably based on the assumption that all three phases (solid, liquid, vapor) are at the same temperature. We have found that these models can be parameterized to satisfactorily predict solvent concentration dynamics, but they lack the necessary detail to capture temperature dynamics within the drying cake.

In this presentation, we discuss the theoretical development and experimental validation of a new multiphase model for API vacuum drying. Drying experiments were performed over a range of temperatures and pressures using glass beads of different sizes and acetone as the solvent. An on-line mass spectrometer was used to measure vapor composition in the head space and thermocouples were used to measure temperature at different locations in the drying cake. We developed a two-phase drying model with a separate liquid phase and a combined solid–vapor phase to capture the relevant heat and mass transport dynamics. We show that the multiphase model qualitatively captures the complex drying dynamics revealed from our experiments. By using the measured vapor phase composition and cake temperature profiles to estimate unknown transport parameters, we show that the multiphase model is capable of producing quantitatively accurate predictions over the examined range of bead sizes, temperatures and pressures.

383740 Die Face Pelletizing of Sticky HME Formulations

Abstract:

Hot melt extrusion is a continuous process with growing importance for the pharmaceutical industry [1]. An extruder processes raw materials into homogeneous strands of molten material and offers, in that way, robust production opportunities for solid dispersions. Solid dispersions are required to face current drug development challenges such as poor water solubility or modified release of the active pharmaceutical ingredient (API).

The molten material is shaped in the downstream process into the product of the extrusion line. Often, pellets are needed as intermediates for tablet compaction, capsule filling or injection molding. Pellets can be produced by die face pelletizing or strand pelletizing. The pelletizing methods distinguish themselves in terms of material temperature where the cutting takes place. In strand pelletizing the emerging material is drawn as cylindrical strands from the extrusion die. It runs through a cooling section and is chilled near or below the softening point and cut by a knife roller. Thus, solid bodies are cut resulting in cylindrical shaped pellets. In contrast, die face pelletizing is executed with a rotating knife pressed on the extrusion die plate. Thus, the material is in a viscous state during cutting and can deform due to surface tension into spherical pellets [2], [3]. Spherical pellets lead to better flowability and enhance dosing accuracy in subsequent handling steps. In addition, the pellets are conveyed with a cooling air stream to the product container where they are separated from the air stream, e.g. with an impact separator. This arrangement results in a simple self-contained construction and offers high spatial flexibility. However, air cooled die face pelletizing was limited to few formulations because of stickiness [4], although the plastics industry uses a similar principle to process sticky materials for several decades. The plastics industry uses underwater pelletizing systems, where the melt stickiness is prevented by higher cooling intensities. However, water-cooled systems are not suitable in the pharmaceutical field due to water solubility of the API and high purity requirements.

The aim of the presented study was a comprehensive understanding of polymer melt stickiness and its avoidance during die face pelletizing. The influence of heat transfer on adhesion properties has been investigated and a hypothesis on melt stickiness proposed. The hypothesis relates heat transfer properties to the materials phase transition temperature. The proposed hypothesis is supported with probe tack investigations at defined heat transfer conditions. This realization triggered a novel die plate design, which thermally decouples the die face from the melt flow channel. Thus, the surface temperatures of the pelletizer are low as compared to underwater pelletizing so that stickiness is suppressed. The novel die plate's heat balance has been calculated and optimized in-silico and subsequently built and tested as a real prototype.

The prototype was incorporated into an extrusion line consisting of an 18 mm twin-screw extruder (Coperion ZSK 18) and a die face pelletizer (Automatik Plastics Machinery GmbH, Sphero®-THA). The setup has successfully been applied to manufacture pellets from different sticky pharmaceutical polymers (e.g. copovidone). The obtained pellets are almost spherical and in the size range of 1 mm. Thus, it has been proven that the new design enables processing of sticky materials so that a broader range of formulations can benefit from the advantages of die face pelletizing.

References:

[1] M. A. Repka, N. Langley, and J. DiNunzio, Eds., *Melt Extrusion - Materials, Technology and Drug Product Design*. New York, Heidelberg, Dordrecht, London: Springer, 2013.

[2] E. Roblegg, E. Jäger, A. Hodzic, G. Koscher, S. Mohr, A. Zimmer, and J. Khinast, "Development of sustained-release lipophilic calcium stearate pellets via hot melt extrusion.," *Eur. J. Pharm. Biopharm.*, vol. 79, no. 3, pp. 635–645, Nov. 2011.

[3] S. Bialleck, "Herstellung von Polysaccharidpellets mittels Schmelzextrusion," Rheinischen Friedrich-Wilhelms-Universität Bonn, 2011.

[4] D. Treffer, P. Wahl, D. Markl, G. Koscher, E. Roblegg, and J. Khinast, "Hot Melt Extrusion as a Continuous Pharmaceutical Manufacturing Process," in *Melt Extrusion: Equipment and Pharmaceutical Applications*, M. Repka, Ed. Springer Publishers, 2013.

372442 Continuous Purification of Artemisinin and Artesunate

Abstract:

Continuous purification of artemisinin and artesunate

Andreas Seidel Morgenstern,[1,2] Ju Weon Lee,[1] Zoltan Horvath,[1] Elena Horosanskaia,[1]

Heike Lorenz,[1] Kerry Gilmore,[3] Daniel Kopetzki,[3] D. Tyler McQuade,[3] Peter H. Seeberger[3,4]

1) Max Planck Institute for Dynamics of Complex Technical Systems, Sandtorstrasse 1, 39106 Magdeburg, Germany

2) Otto-von-Guericke-University, Chair for Chemical Process Technology, Universitaetsplatz 2, 39106 Magdeburg, Germany

3) Max-Planck-Institute of Colloids and Interfaces, Department of Biomolecular Systems, Am Mühlenberg 1, 14476 Potsdam, Germany

4) Freie Universität Berlin, Institute of Chemistry and Biochemistry, Arnimallee 22, 14195 Berlin

Germany

Artemisinin, contained in sweet wormwood, and its derivatives β-artemether, β-arteether and α-artesunate are currently the basis for the most effective ways to cure the malaria disease [1]. Continuous photocatalytical synthesis of artemisisin from the waste product dihydroartemisinic acid was recently reported to be capable to produce in a laboratory flow reactor up to 3.5 kg/L/day with approx. 70 wt. % yield [2, 3]. The possibility of further synthesizing the mentioned derivatives in a continuous way could be demonstrated just recently for artesunate [4].

For a model target compound it was shown, that the outlet of a continuously operated flow reactor could be successfully fed directly into a simulated moving bed (SMB) chromatography unit in order to carry out both processes continuously [5, 6].

We will present results for various scenarios of the continuous isolation of artemisinin and artesunate from the effluent of continuously operated reactors. One of the investigated options is based on applying a 3-zone open-loop SMB process with an additional regeneration zone. The potential of combinations of chromatography and crystallization processes [7, 8] will be also discussed.

References

[1] World Malaria Report, World Health Organization, Geneva, 2011.

[2] F. Lévesque, P.H. Seeberger,
Angew. Chem. Int. Ed., 2012, 51, 1706 – 1709.

[3] D. Kopetzki, F. Lévesque, P.H. Seeberger,
Chem. Eur. J., 2013, 19, 5450 – 5456.

[4] K. Gilmore, D. Kopetzki, J.W. Lee, Z. Horvath, D. T. McQuade, A. Seidel-Morgenstern,
P.H. Seeberger, manuscript submitted

[5] A.G. O'Brien, Z. Horváth, F. Lévesque, J.W. Lee, A. Seidel-Morgenstern, P.H. Seeberger,
Angew. Chem. Int. Ed., 2012, 51, 7028 – 7030.

[6] J. W. Lee, Z. Horváth, A.G. O'Brien, P.H. Seeberger, A. Seidel-Morgenstern,
Chem. Eng. J., 2014, in print

[7] E. Horosanskaia, A. Seidel-Morgenstern, H. Lorenz,
Themochimica Acta, 2014, 578, 74 - 81.

[8] H. Lorenz, A. Seidel-Morgenstern,
Angew. Chem. Int. Ed., 2014, 53, 1218 – 1250.

382048 Tablet Coating Scale-up Based on DEM Simulations: Going from Lab- to Industry-Scale

Abstract:

Introduction

Pan coating is a commonly used unit operation to apply a functional layer around tablet cores. The tablet cores are placed in a rotating drum, and a coating liquid is sprayed onto the moving tablets. The coating fulfils different functions; a special application is active coating, where the coating layer itself contains an active pharmaceutical ingredient (API). In this case, the variation of coating mass between tablets has to be consistently small.

To guarantee a consistently high uniformity, a large amount of process understanding is needed. To this end, a number of experimental investigations were done. In recent years, a strong trend emerged to complement this with numerical simulations[1,2].

Due to lengthy simulation times, these investigations often are restricted to small-scale cases. This limits the statements that can be done for the industry (full) scale process, and generally means that no or few large-scale data points are available for scale-up considerations.

In this work, the application of an active coating was investigated from lab scale to industry scale using Discrete Element Method (DEM) simulations (both commercial software and in-house code). The influence of process parameters on the quality attributes is studied for each scale. Further, it is studied how far scale-up rules known in literature[3] are applicable, and new guidelines are considered.

Material & Method

Tablet coating process

The investigated tablet is a Gastrointestinal therapeutic systems (Adalat® GITS, Bayer Pharma AG, Germany) coated with an aqueous suspension containing candesartan cilexetil as API. Different sizes of pan coaters (BFC, L.B. Bohle, Germany) were used. The process parameters were set according to experimental runs and available formulation data. As the variation of this API between tablets has to be small, a high inter-tablet coating mass uniformity is vital. The central aim therefore was to improve the inter-tablet coating uniformity.

Discrete Element Method (DEM) simulations

For the simulations, a commercial software (EDEM 2.5.1, DEM Solutions, UK) and an in-house developed code (XPS, Research Center Pharmaceutical Engineering GmbH, Austria) were combined to cover the scale range. The XPS software exploits the massively parallel structures of modern graphic cards to simulate number of particles that is sufficiently large to cover even the industrial scale with reasonable computation times. The material properties came from measurements[4]. Different methods for the modeling of a spray in DEM simulations were implemented[5].

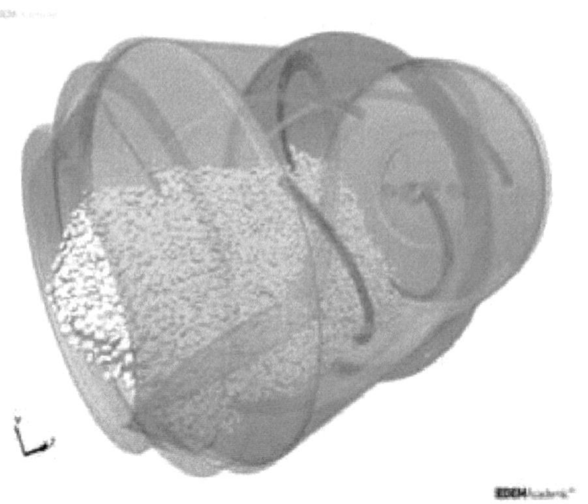

Figure 1: Snapshot of the Discrete Element Method simulation of the lab-scale coating drum. Conclusion

In this work, two implementations of Discrete Element Method simulations were combined to describe a tablet coating process from lab scale to pilot scale. The parameters needed from simulation were taken from measurements where possible. For the pilot scale, comparison of the simulation result to available measurement data was done. On one hand, valuable insights into the process mechanics were gathered on each separate scale. On the other hand, covering the whole scale range allowed for a critical treatment of existing scale-up rules and the investigation towards new guidelines.

References

1. Ketterhagen, W. R., am Ende, M. T. & Hancock, B. C. Process modeling in the pharmaceutical industry using the discrete element method. J. Pharm. Sci. 98, 442–70 (2009).

2. Toschkoff, G. & Khinast, J. G. Mathematical modeling of the coating process. Int J Pharm (2013).

3. Mueller, R. & Kleinebudde, P. Prediction of tablet velocity in pan coaters for scale-up. Powder Technol. 173, 51–58 (2007).

4. Just, S. et al. Experimental analysis of tablet properties for discrete element modeling of an active coating process. AAPS PharmSciTech 14, 402–11 (2013).

5. Toschkoff, G. et al. Spray models for discrete element simulations of particle coating processes. Chem Eng Sci 101, 603–614 (2013).

Scalability and Impact of Freeze/Thaw Conditions on the Drug Substance Quality of Monoclonal Antibody Formulations

William Rayfield[1], Heera Khan[1], Sunitha Kandula[1], Matt Flamm[2], Nihal Tugcu[1]
[1]Process Development and Engineering, BioProcess R&D, Merck Research Laboratories, Kenilworth, NJ.
[2]Informatics Information Technology, Merck Research Laboratories Information Technology, West Point, PA.

Abstract: Biopharmaceutical production often involves an extended storage of bulk protein prior to drug product fill. A majority of biologics undergo numerous degradative pathways at 2-8°C over long time periods, so it is essential to ensure that the product quality is maintained during the storage period. In addition, when shipping biologics in the liquids phase, the shear and agitation stress that can occur during transport can potentially be detrimental to product stability[1]. Thus, it is important to store the bulk drug substance at subzero temperatures to retard the degradation kinetics and extend the shelf life. The freezing of bulk protein at subzero temperatures can be performed in an uncontrolled or controlled manner. Uncontrolled freezing/thawing of bulk protein at high concentration (>25 g/l) in conventional large storage containers (>2L) can lead to cryoconcentration which excludes solute molecules, leading to zones of higher protein and excipient concentrations that can reach nine times higher than the initial protein concentration. Cryoconcentration can result in changes in the formulation environment via the Donnan effect which can cause a negative impact on product quality (high molecular weight species, analytical ion-exchange chromatography profile, etc). This paper will discuss the impact of freeze and thaw on monoclonal antibody drug substance quality at multiple scales over various timeframes using different storage container types. Commonly used storage temperature ranges (-20°C, -40°C, -70°C) were evaluated. The study investigated the modeling of both controlled and uncontrolled freezing for scale-up from lab scale to production scale. The information obtained in this study is critical for designing freezing/thawing conditions for a manufacturing process.

---

[1] S. Singh, P. Kolhe, W. Wang, S. Nema. Large Scale Freezing of Biologics. BioProcess International, Vol 7, No. 10, Nov 2009, pp. 34-42.

386882 Scale-up of Agitated Drying: Effect of Shear Stress and Hydrostatic Pressure on API Powder Properties

Abstract:

Scale-up of agitated drying processes to minimize particle size changes in Active Pharmaceutical Ingredients can be challenging. Particle agglomeration or attrition problems due to agitated drying are often discovered upon the initial scale-up from the lab to the plant. Traditional laboratory drying equipment has not successfully reproduced the degree of agglomeration or attrition observed at scale. This discrepancy may be attributed to the ability of particulate solids, such as crystalline API's, to transfer stresses from the normal direction into the shearing direction. As batch size increases during scale-up, the compressive and shearing forces experienced by the API increase. This work characterizes the propensity for particle attrition to occur at different stages of the drying process by employing a laboratory agitated dryer that has been modified to measure impeller torque and reproduce the range of compressive and shearing forces observed during scale-up. A workflow for assessing risk of API attrition at manufacturing scale is described.

Mass Transfer Effects in a Nitro Reduction

Steve Richter and Brian Kotecki

The reduction of nitro groups with hydrogen over a heterogeneous catalyst is a useful and common step in pharmaceutical synthesis. This multistep hydrogen reduction is highly exothermic and proceeds through nitroso and hydroxylamine intermediates, which can be unstable and pose a potential process safety risk. During scale up of a process step, an investigation was performed into the effect of mass transfer of hydrogen into the liquid phase and the effect on the reaction under operating regions dominated by reaction kinetics or mass transfer conditions. The intensity of mixing affected the accumulation of these unstable intermediates and is dependent on the reactor geometry and operating parameters. A strategy to address process safety concerns and strategies for examining mass transfer and kinetic effects for scale-up are outlined.

This presentation was sponsored by AbbVie. AbbVie contributed to the design, research, and interpretation of data, writing, reviewing, and approving the publication. Steve Richter and Brian Kotecki are AbbVie employees.

368151 Kinetics Model for Designing Grignard Reactions in Batch or Flow Operations

Abstract:

Grignard reactions are one of the oldest, yet quite complicated to understand and design for, at different scales of operation. The heterogeneity of the solid-liquid reaction means that there is an interplay of mass transfer and reaction kinetics rates. In designing a Grignard reaction for scales ranging from laboratory to manufacturing, using either batch or continuous operation, a key component of process design development is to obtain an understanding of kinetics that are relatively free of mass-transfer. A fit-for-use reaction kinetics model was developed to parameterize the reaction and mass transfer rates of the Grignard reaction across different scales from 250mL to 500 gallon and for different grades of magnesium. The model includes three parameters, one ($k_{SL}a$) representing the mass-transfer component and two (k and E_a) representing the Arrhenius dependent rate constant. The reaction kinetics model was then applied to design the magnesium recharge frequency by taking operational variability into account. Furthermore, the model is coupled with equipment heat transfer characteristic to understand the process dynamic of the CSTR in the presence of process disturbances. Based on the kinetics model and energy balance, reactor temperature was determined to be suitable to detect process failure to ensure process safety at manufacturing. The resulting process understanding enabled the team to define the final operation strategy to ensure a safe and robust process to be scaled up to commercial scale production. This exercise can be extended as a platform for other similar Grignard reactions in the future.

Mixing modeling applied to a mass transfer limited heterogeneous reaction with lithium hydroxide to reduce cycle time variation.

Christopher Morrison and
Qunying Dai

Overview

- Understanding the Problem
- Potential Solutions
- Mixing
  - Effects
  - Modeling
  - Calculating Suspension of Solids
  - Understanding Issues Observed from Lab to Plant
- Conclusion

Problem(s)

Getting to know your reaction

- Plant scale – variable reaction time
- Lab scale reactions consistent … but …
- Round bottom flask vs. jacketed laboratory reactors differences
- Safety concerns with ~120°C exotherm with LiOH

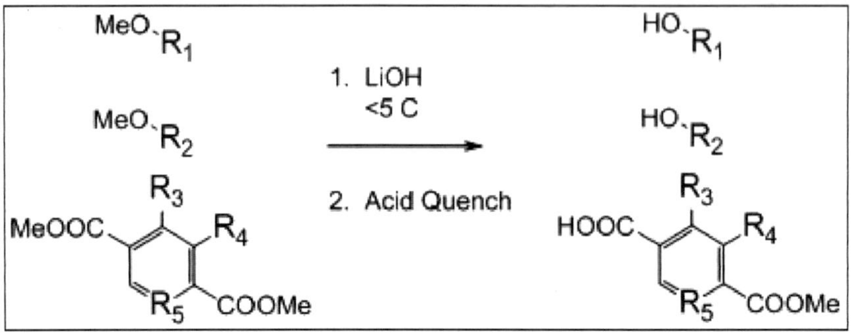

Potential Solutions

Pulling out all the tricks

- Solubility Modification to Avoid a Mass Transfer Limited System
  - Changing to a solvent that lithium hydroxide has a higher solubility
  - Changing bases to avoid the solubility issue
  - Increasing the process temperature to increase lithium hydroxide solubility

- Mass Transfer Limited Options
  - Milling lithium hydroxide to achieve smaller particle size and increased surface area
    - In process wet milling
    - Dry milling
  - Mixing Analysis

Mixing Effects

Recognizing that the reaction is mass transfer limited

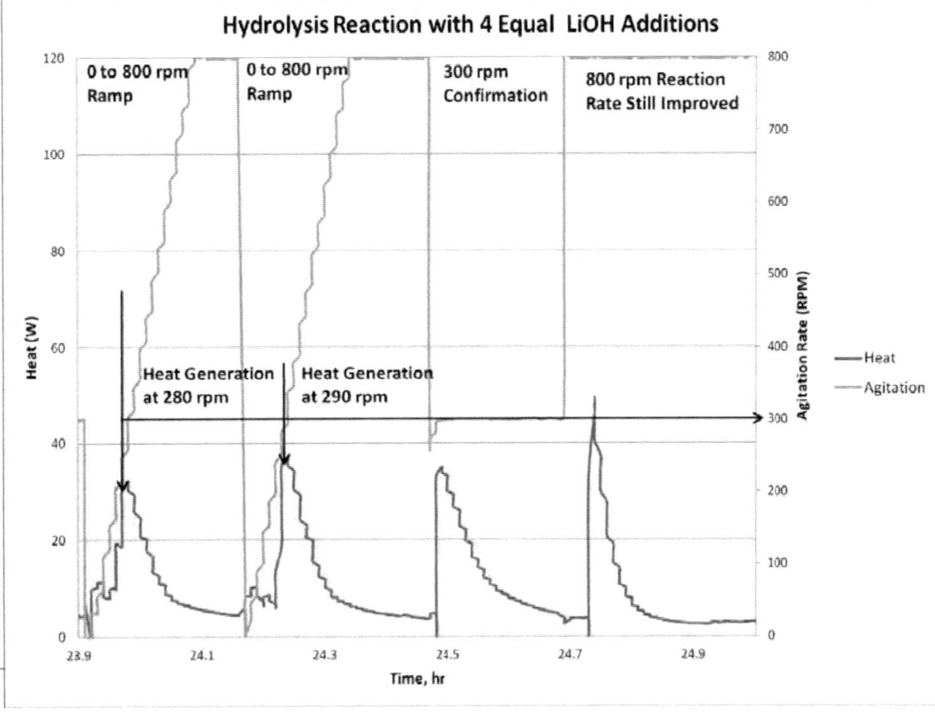

Mixing Modeling Scale-up Parameters

- Power per Mass or Volume (W/kg or W/L)
  - Most widely used scale up parameter for homogeneous rxns and many heterogeneous rxns
  - Energy imparted into the system divided up by the unit mass or volume
- Tip Speed (m/s)
  - Used to either generate or avoid attrition of particles
  - Estimate shear stresses on the system during scale up can be especially important for non Newtonian fluids
- $\theta_{95\%}$ (s)
  - Primarily used for blending of heterogeneous mixtures
  - Used to estimate time of how long until a blend takes to reach 95% homogeneity
- Zwietering Just Suspended Solids, Njs (rpm)
  - Used for heterogeneous rxns to ensure materials are freely moving
  - Agitation rate at which all solids will be moving, but there is still concentration gradients of solids in the slurry

Mixing Modeling to the Rescue

Mixing modeling using the FMP_Design_Guide_v4

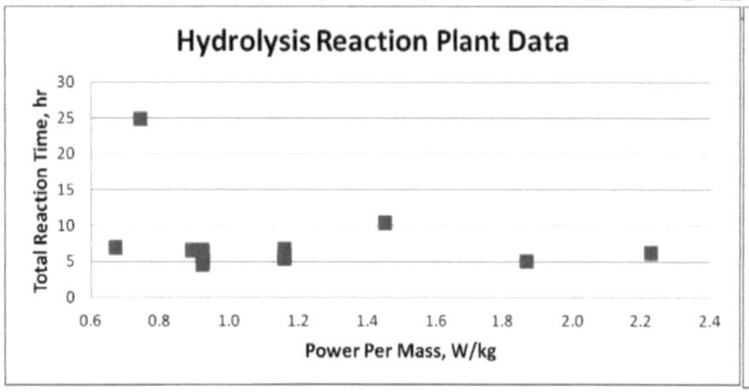

Unable to explain or predict reaction time.

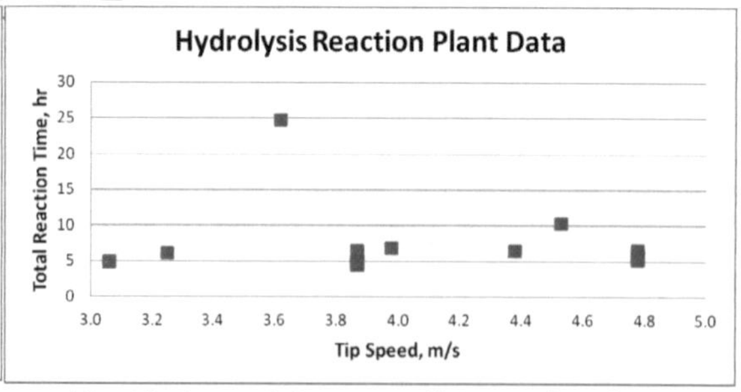

Unable to explain or predict reaction time.

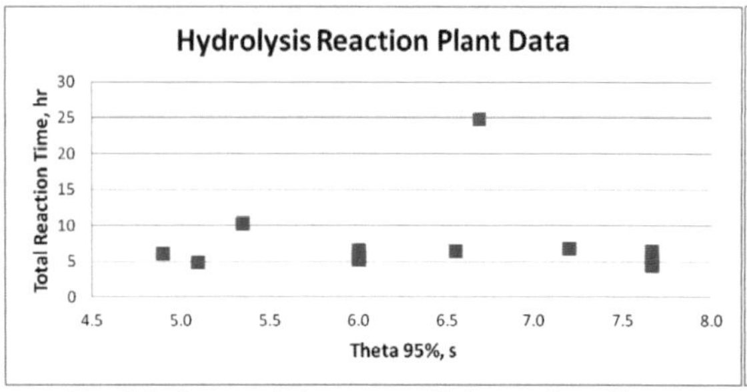

Unable to explain or predict reaction time.

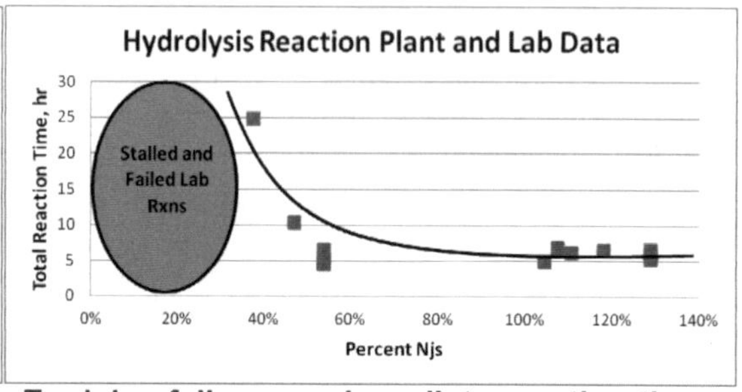

Explains failures and predicts reaction time.

Solid Suspensions and How to Calculate Njs

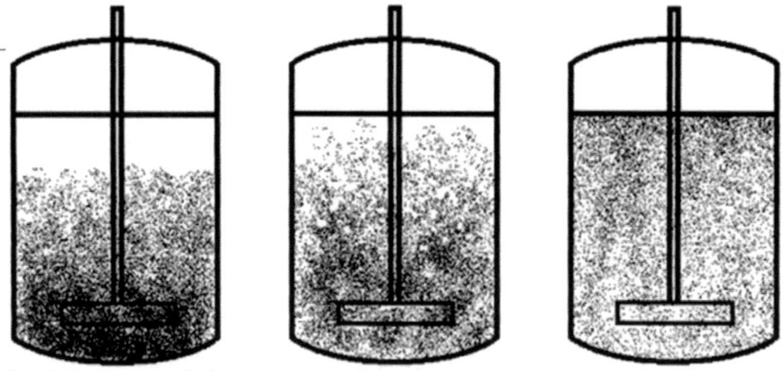

Partially Suspended:
Concentration gradient
Some solids are moving

Just Suspended:
Concentration gradient
All solids are moving

Fully Suspended:
Uniform concentration
All solids are moving

$$N_{js} = S\nu^{0.1}\left[\frac{g(\rho_s - \rho_l)}{\rho_l}\right]^{0.45}\frac{X^{0.13}\,d_p^{0.2}}{D^{0.85}}$$

Material / System Related

Where:

S: Zwietering Njs constant (function of impeller and tank geometry)
D: impeller diameter (m)

Equipment Related

g: gravitational constant (m²/sec)

Location Related

ν: kinematic viscosity (m²/sec)
ρ_s: solid density (kg/m³)
ρ_l: liquid density (kg/m³)
X: solids loading ((kg solids/kg liquid) x 100)
d_p: particle diameter (m)

Challenges to be Aware of During Scale Up

Understanding the equipment effects on the process

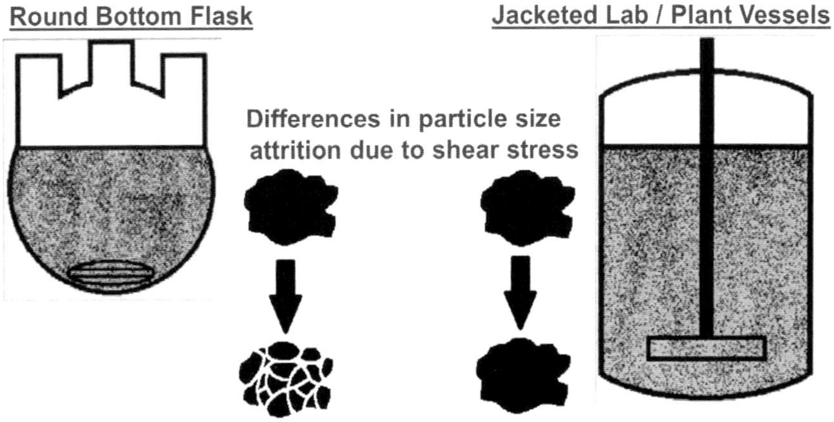

Effects of surface area to volume ratio on cooling rate and addition rate limitations

Typical Heat Loss*		
Vessel	Heat Loss IW/kg/K)	Time for 1°C Loss at 80°C
2500 L rxr	0.054	21 min
100 mL beaker	3.68	17 sec

*Source: Table 11.9 from Pharmaceutical Industry R&D to Manufacturing, 2011.

Conclusion

- By first intent mass transfer limited reactions are to be avoided if at all possible, due to the various restrictions from the chemistry this was not avoidable.

- Zwietering Njs is able to predict success or failure of the process and has been demonstrated from several gram lab scale to ~60 kg plant scale.

Potential Additional Work:

- Grenville, Mak & Brown Njs may be a less conservative correlation than Zwietering Njs and could be evaluated if a more accurate prediction is required. Zwietering Njs was selected due to its widely accepted use and available data on nearly all reactor configurations and impellers.

Acknowledgements

Thank you for your work on the chemistry, support, and our many discussions on this issue:

- Qunying Dai

Thank you for the many discussions around this work and peer reviews:

- Huan Wang
- Kathy Guillen
- Rahn Mckeown
- Harvey Zimmerman
- Chris Werner
- GSK Particle Science and Engineering group

Questions?

385758 A Drop-on-Demand Manufacturing System for the Production of Melt-Based Pharmaceutical Dosage Forms

Abstract:

In recent years, the FDA has encouraged pharmaceutical companies to place renewed emphasis on the development of more innovative, efficient manufacturing methods with the use of process analytical technologies and the QbD approach. As part of this renewed emphasis on improvement of manufacturing, the pharmaceutical industry has also begun the selective transition from traditional batch processing to continuous manufacturing [1].

As a part of the NSF Engineering Research Center for Structured Organic Particulate Systems, a mini-manufacturing process for drug formation has been under development. The process utilizes the drop-on-demand (DoD) printing technology for predictable and highly controllable deposition of active pharmaceutical ingredients (API) onto an edible substrate, such as a polymeric film or placebo tablet, using a semi-continuous operation suitable for low volume production of personalized dosage forms [2].

In this paper, we present a process narrative of the pilot facility and discuss automation and supervisory control of the DoD manufacturing system along with the analysis of the drug forms created which consist of polymer and drug constituents. Implementing a supervisory control system for the manufacturing process, including automation and closed-loop control, is essential for producing individual dosage forms with precise control of dosage amount, morphology and formulation composition. The system consists of a positive displacement pump, xy-staging, imaging system and temperature controllers. The drop volume is monitored using the imaging system, to ensure consistent drop size and thus dosage amount. The xy-staging allows creating precise drop positioning while printing.

Using this process, different drug formulations including solvent-polymer-API and polymer-API systems, i.e. co-melt systems, can be produced. Polymers are added to the printing material to help control drug morphology, material properties and the formulation composition [3]. Multiple temperature controllers are used to control the temperature of the process elements including material reservoir, pump, tubing and nozzle. Temperature control on the process allows maintaining printable rheological and material properties for the production of melt-based dosage forms. Temperature control on the substrate is crucial, since crystallization temperature has an effect on product quality, influencing the bioavailability of the drug [4]. After drop deposition, changes in drug morphology can be monitored using a spectroscopic technique, such as Raman or NIR, and corrected by temperature control action on the substrate. Using the proposed temperature control strategy on the substrate, one can tailor the crystallization behavior to compensate for these variations and achieve consistent drug morphology. Finally, the drug forms created using co-melt systems will be analyzed.

1. K. V. Gernaey, A. E. Cervera-Padrell, and J. M. Woodley, "A perspective on PSE in pharmaceutical process development and innovation," *Computers & Chemical Engineering*, vol. 42, pp. 15–29, Jul. 2012.
2. L. Hirshfield, A. Giridhar, L. S. Taylor, M. T. Harris, G. V. Reklaitis, "Dropwise Additive Manufacturing of Pharmaceutical Products for Solvent-based Dosage Forms," *Journal of Pharmaceutical Sciences*, vol. 103, issue 2, pages 496-506, February 2014.
3. N. S. Trasi and L. S. Taylor, "Effect of Additives on Crystal Growth and Nucleation of Amorphous Flutamide," *Crystal Growth & Design*, vol. 12, no. 6, pp. 3221–3230, Jun. 2012.
4. E. Icten, Z. K. Nagy, G.V. Reklaitis, "Supervisory Control of a Drop-on-Demand Mini-manufacturing System for the Production of Pharmaceuticals," *Proc. of 24th Eur. Symp. Comput. Aided Process Eng. ESCAPE 24*, vol. 24, June 2014.

Sensor Network Design and Value of Information Analysis for Continuous Pharmaceutical Manufacturing

Traditionally the pharmaceutical industry has employed batch manufacturing but with the advancement of Process Analytical Technology (PAT) by the FDA, there has been an impetus to move towards continuous manufacturing. This paradigm shift to continuous processing needs real time online measurement of Critical Quality Attributes (CQAs) and hence development of various sensing schemes for their measurement. The sensing schemes are new in the pharmaceutical industry for online measurement and are on a learning curve both for the industry and the FDA. One way to build confidence in on-line measurement of CQA's is to use multiple sensors for monitoring the same CQA. Since, product quality is the main issue in pharmaceutical manufacturing, one strategy is to place as many redundant sensors as possible for each measurement. However, cost considerations implies that strategies needs to be balanced between benefit and cost. At present, there has been no clear cut approach on sensor network design in the pharma manufacturing setting, which includes sensor placement and number of sensors to be used for the measurement of any given quality attribute.

In this work, an approach to sensor network design for continuous dry granulation line is studied. The existing literature on sensor networks has been in the gas and fluid processing areas of chemical and petro-chemical industry and has not considered the constraints posed by solid processing. The complexity of the physics underlying the processing and operation of the continuous granulation line adds to the problem as it makes it difficult to relate various quality parameters across locations at which a specific property is measured (nodes).

This study includes consideration of sensor placement for the measurement of various quality attributes on the manufacturing line and the benefit associated with redundant sensors. Data reconciliation and gross error detection are carried out to mitigate the bias associated with faulty sensors and correctly evaluate quality parameters. A value of information analysis is presented to calculate the additional benefit associated with multiple sensors on a given node of the sensor network. The methodology developed is general, includes identical and non-identical sensors on the network, and calculates the value for all sensor combination across the node. A data-driven model is derived to relate various quality parameter across multiple nodes. The analysis is expanded to include values from multiple nodes to provide the system wide value gained by having redundant sensors at different nodes. The results from the study are used to determine the optimal number of sensors on a node and can be used to carry out the economic analysis and determine the quality measurement and controller performance.

References:

1) M. Bagajewicz, *Value of Information in linear systems*, AIChE Journal, Vol. 52(2), 2006, 638-650

2)M. Bhushan and R. Rengaswamy, *Design of Sensor Network Based on the Signed Directed Graph of the Process for Efficient Fault Diagnosis*, Ind. Eng. Chem. Res., Vol. 39 (4), 2000, 999-1019

390112 Modeling of Residence Time Distribution in Continuous Solid Oral Dose Pharmaceutical Manufacturing Processes

Abstract:

Modeling of Residence Time Distribution in Continuous Solid Oral Dose Pharmaceutical Manufacturing Processes

M. Sebastian Escotet-Espinoza, Amanda Rogers, Fernando J. Muzzio and Marianthi Ierapetritou

Department of Chemical and Biochemical Engineering, Rutgers University, Piscataway, NJ

The pharmaceutical regulatory environment has recognized the need for a systematic approach to process development through the implementation of engineering modeling and optimization techniques in order to manufacture safe and effective products with their quality *built* into the process [1]. This philosophy of Quality-by-Design (QbD) has been applied in other industries to reliably create products with the desired quality attributes. Quality is assured through the development of reliable and robust processes, designed based on the knowledge of process principles. The transition from batch to continuous pharmaceutical operations can facilitate the development of processes within the QbD framework. Continuous processing allows real-time control, mitigates the need for scale-up studies and can help to reduce product variability [2]. In continuous manufacturing processes the major challenges are the coordination of equipment input-output properties, and the formulation of predictive process control strategies. To achieve equipment coordination and predictive capabilities, the relationships between critical quality attributes (CQAs), material properties (CMPs) and process parameters (CPPs) need to be correlated sequentially between the multiple unit operations in product manufacturing. Mathematical models characterizing individual unit operation performance as a function of CPPs (e.g., screw speed, residence time) and CMPs (e.g., particle size distribution) have been developed. In addition, multi-unit processes can be characterized with respect to the temporal-space location of particles in the system [3].

The residence or space-time is a critical equipment and process parameter that characterizes the period of time a particle stays in one or more unit operations in a continuous system. The residence time distribution (RTD) is defined as the probability of particles exiting the equipment at a given time period based on the flow patterns inside of the unit [4]. Mathematical models for the RTD have previously been developed for chemical engineering applications in order to characterize the influence of operating conditions, material properties and unit geometry on the degree of non-ideal behavior (i.e., back-mixing). Pharmaceutical unit operations for which the RTD has been experimentally studied include feeding, blending, and granulation processes [5]. In previous work, the RTD of convective mixers was extensively studied and correlated to lumped and discrete parameters such as the Peclet number and effective particle diffusivity [6].

In this work, we focus on developing the mathematical modeling tools for RTDs in continuous solids-handling unit operations as a predictive tool for process outcomes and characterization of unit performance. Our research focuses on modeling the residence time distribution in feeding (i.e., feeders) and blending (i.e., convective mixers) operations as a function of operating conditions. The overall goal of the research is to model the RTD using historic and recently developed RTD models. Data from both pulse and step change experiments for the convective blender have been reduced to fundamental unit operation parameters (i.e., Peclet number) using the Fokker-Planck Equation (i.e., PFR with axial mixing). The effective diffusivity of the system has been studied with respect to CMPs of acetaminophen-excipient blends in order to establish an initial method for correlating material properties to unit operation. Critical unit operation parameters (e.g., mass hold-up, blade speed) and blend properties (e.g., composition, particle size distribution) can then correlated to in order to develop new RTD methods. The relative standard deviation (RSD) of the convective mixer with respect to the pulses experiments has been studied in conjunction to the RTD in order to establish a predictive control strategy for the unit operation. RTD can be used to characterize the propagation of transient disturbances across sequential unit operations in flowsheet simulations. The RTD can provide traceability along the continuous line and delineate continuous products based on the feed strategy. Manufactured products created using different batches of starting material can be traced in order to comply with process validation. This will facilitate the prediction and rejection of out-of-spec (OOS) products by tracing disturbances along the process. Overall, RTD models can add significant value to the transition from batch to continuous operations by providing reliable predictive tool for behavior of unit operations based on their residence time.

References

1. USFDA, *Pharmaceutical cGMPs for the 21st century: a risk-based approach*, U.S.D.o.H.a.H. Services, Editor. 2002: FDA, Rockville, MD.

2. Boukouvala, F., et al., *Computer-Aided Flowsheet Simulation of a Pharmaceutical Tablet Manufacturing Process Incorporating Wet Granulation.* Journal of Pharmaceutical Innovation, 2013. **8**(1): p. 11-27.

3. Gernaey, K.V. and R. Gani, *A model-based systems approach to pharmaceutical product-process design and analysis.* Chemical Engineering Science, 2010. **65**(21): p. 5757-5769.

4. Danckwerts, P.V., *Continuous flow systems: Distribution of residence times.* Chemical Engineering Science, 1953. **2**(1): p. 1-13.

5. Gao, Y., F.J. Muzzio, and M.G. Ierapetritou, *A review of the Residence Time Distribution (RTD) applications in solid unit operations.* Powder Technology, 2012. **228**(0): p. 416-423.

6. Gao, Y., et al., *Characterizing continuous powder mixing using residence time distribution.* Chemical Engineering Science, 2011. **66**(3): p. 417-425.

Determination of Design Space for Oral Pharmaceutical Drugs

Kalliopi Chatzizaharia and **Dimitris Hatziavramidis**

School of Chemical Engineering

National Technical University of Athens, Greece

14AIChE Annual Meeting, Atlanta, GA

Design Space

Quality-by-Design initiative by pharmaceutical regulatory agencies of the European Union, Japan and USA enables implementation of changes in formulation and manufacturing processes to manufacture a pharmaceutical product within a multidimensional Design Space, proposed by the manufacturer and having been demonstrated to assure improved process capability and **product quality**, without the need for further **regulatory approval**

Product Quality → Critical Quality Attributes: $\mathbf{y} = (y_1, \ldots, y_n)^T$ →

Critical Material Attributes (of drug components) &

Critical Process Parameters: $\mathbf{x} = (x_1, \ldots, x_n)^T$

Design Space

$DS = \{\tilde{\mathbf{x}} \in X \mid E[\, y_j \mid \mathbf{x}] = \hat{y}_{j|\mathbf{x}} \in L_j \ \forall j = 1,..,m \wedge L_1 \cup \ldots \cup L_m = L\}$

L: set of **specifications** for y_j

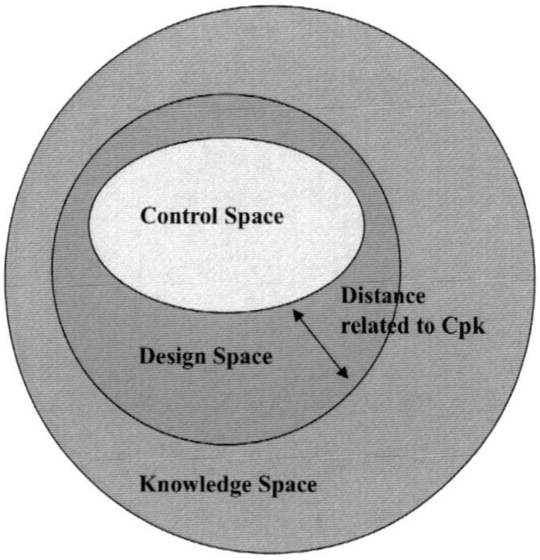

Oral Drugs

Drug Quality: efficacy, safety, reliability

- **Efficacy**: **drug stability, bioavailability**
  membrane **permeation, absorption, dissolution rates** →
  therapeutic polymorph (crystallization conditions)
  concentration of **A**(ctive) **P**(harmaceutical) **I**(ngredient) in **blood,** c_{db}
  concentration of excipients (**disintegrants**) in **G**(astro) **I**(ntestinal) **T**(ract), c_{eg}

- **Safety**: $c_{dL} < c_{db} < c_{dU}$; $c_{db} < c_{dL} \Rightarrow$ **no** therapeutic effect
  $c_{dL} < c_{db} \Rightarrow$ **toxic** effect

- **Reliability**: it does what it is supposed to do, according to its label
  powder **blend** (API+excipients) **composition** remains the **same** throughout processes → **powder flowability, tablet strength** & **friability** → **concentration** of excipients
  (glidants, lubricants, binders, etc.), **grain size**

Bioavailability & Bioequivalence

Bioavailability (BA): "the rate and extent to which the API is absorbed from a drug product and becomes available at the site of action"

BA depends on: API release from oral form, balance among dissolution, elimination, metabolism and absorption rates, solubility and membrane permeability

Bioequivalence (BE): property of drugs which have similar plasma concentration profiles

Biopharmaceutical Classification System: identifies drug classes for which in vivo BE measurements are waived in favor of in vitro-dissolution measurements

Whenever a waiver can be granted, two drugs are assumed to be **bioequivalent** if their **dissolution profiles** are similar

EMEA & FDA: "Any methods to prove similarity of dissolution profiles are accepted as long as they are justified"

Design Space Determination Methods

- **Ignoring** data **uncertainty**: **Response Surface (RS)**
  Assumptions: Errors statistically independent & normally distributed
  - **Overlapping responses**
  - **Optimized responses**
- **Accounting** for data **uncertainty & correlation structure**: **Bayesian Posterior Predictive Approach (BPPA)**
  Assumptions: 1. Prior distribution honestly reflects genuine information
  2. Any uncertain quantity is referred to as a random variable

$$p(\boldsymbol{\theta} \mid \mathbf{y}) = \frac{p(\mathbf{y} \mid \boldsymbol{\theta})\, p(\boldsymbol{\theta})}{p(\mathbf{y})} = \frac{p(\mathbf{y} \mid \boldsymbol{\theta})\, p(\boldsymbol{\theta})}{\int_{\boldsymbol{\theta}} p(\mathbf{y} \mid \boldsymbol{\theta})\, p(\boldsymbol{\theta})\, d\boldsymbol{\theta}}$$

$p(\boldsymbol{\theta})$: probability of the parameters $\boldsymbol{\theta}$ from **prior times**

$p(\mathbf{y} \mid \boldsymbol{\theta}) = L(\boldsymbol{\theta}, \mathrm{y})$: **likelihood** of $\boldsymbol{\theta}$ parameters for fixed data $\mathbf{y}$

$p(\boldsymbol{\theta} \mid \mathrm{y})$: **posterior probability**

Design Space Determination Methods

Desirability function ($d_{\mu i}$)

$$d_{\mu i} = \begin{cases} [(\hat{y}_{\mu i} - y_{\mu i}^{min})/(T_{\mu i} - y_{\mu i}^{min})]^a, & y_{\mu i}^{min} \le \hat{y}_{\mu i} \le T_{\mu i} \\ [(\hat{y}_{\mu i} - y_{\mu i}^{max})/(T_{\mu i} - y_{\mu i}^{max})]^b, & T_{\mu i}^{min} \le \hat{y}_{\mu i} \le y_{\mu i}^{max} \\ 0, \ \hat{y}_{\mu i} < y_{\mu i}^{min} \text{ or } \hat{y}_{\mu i} > y_{\mu i}^{max} \end{cases}$$

$\hat{y}_{\mu i}$: estimated mean response; $y_{\mu i}^{min}$, $y_{\mu i}^{max}$, $T_{\mu i}$: minimum, maximum desired limits and target for $\hat{y}_{\mu i}$ respectively; and a, b are input parameters that determine the shape of the reliability function

Composite desirability (D): geometric mean of p individual $d_{\mu i}$

$D = (d_{\mu 1}\, d_{\mu 2} \ldots d_{\mu p})^{1/p}$ $\qquad$ $0 \le d_{\mu i}$ and $D \le 1$

Three-stage tableting process

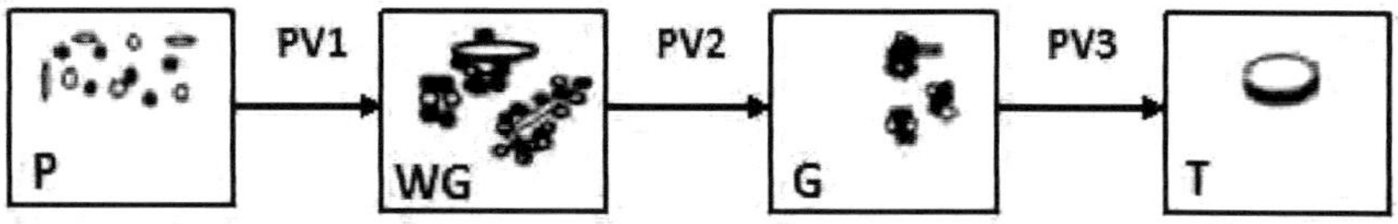

P(owder) W(et) G(ranulation) **Drying** G(ranule) T(ableting)

Westerhuis, J.A. et al. (1997) *Multivariate modelling of the tablet manufacturing process with wet granulation for tablet optimization and in-process control.* Int. J. Pharm. 156(1), 109–117

Box-Behnken design: corners and midpoints of edges of a cube

Design factors - CPPs & CMAs: water amount for wet granulation - *Water* (ml), granulation time - *Time* (min), moisture of granules *Moisture* (%), compression force - *CompF* (kN); excipients: *HPC* (%) & *MCC* (%)

Response variables - CQAs: tablets crushing strength (*CS*), disintegration time (*DT*) and ejection force (*EF*)

Regression Model

logCS= $0.215*MCC+0.219*Moisture+0.102*CompF+0.006*Water+2.834*(Water)^2-0.001(CompF)^2-0.008*Moisture*CompF-4.691e\text{-}04*MCC*Water-8.959$

logDT= $0.190*MCC+0.095*HPC-1.280*Moisture+0.074*CompF+0.004*Water+0.184*(Moisture)^2+3.070e\text{-}05*(Water)^2-4.442e\text{-}04*MCC*Water-5.475$

EF shows no significant terms

CQAs	Data Range	Target [Specifications]
CS (N)	4.00 - 75.00	66.07 [30.90 - 75.86]
DT (s)	2.00 - 421.00	288.40 [69.18 - 416.87]
EF (N)	41.00 - 358.00	274.00 [218.55 - 329.45]

Three-stage tableting - Determination of Design Space

Overlapping Responses

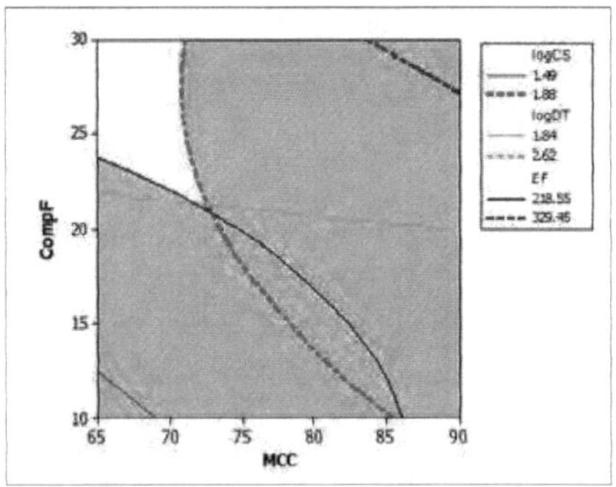

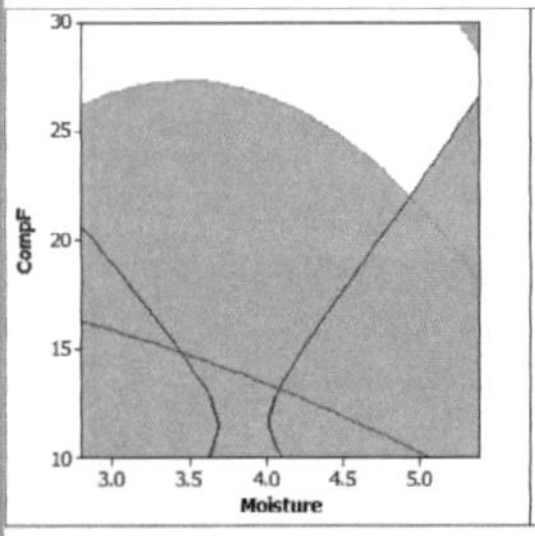

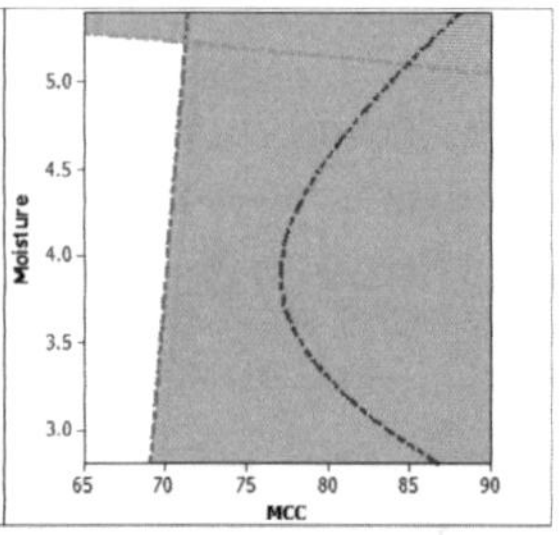

Optimized Responses

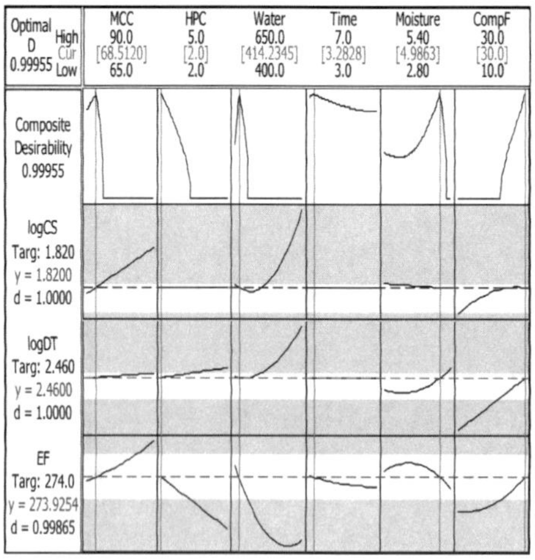

Design Space determination by Optimized Responses method

CPPs- CMAs values	Response Surface Analysis	Bayesian Method
MCC -optimum	68.51	-
HPC -optimum	2.00	-
Water -optimum	414.24	-
Time -optimum	3.28	-
Moisture -optimum	4.98	-
CompF -optimum	10.00	-
MCC_l	77.94	75.00
HPC_l	5.00	5.00
$Water_l$	650.00	400.00
$Time_l$	3.00	5.00
$Moisture_l$	2.80	3.10
$CompF_l$	15.15	30.00
MCC_u	69.28	65.00
HPC_u	2.00	3.00
$Water_u$	400.00	450.00
$Time_u$	3.00	3.00
$Moisture_u$	5.20	4.20
$CompF_u$	30.00	30.00

Desirability of Methods for Design Space determination

Specifications	CS (N)	DT (s)	EF (N)	Method's Composite Desirability
Target	66.07	316.23	274.00	
yl	30.90	69.18	218.55	
yu	75.86	416.87	329.45	
Response Surface Design Analysis	CS	DT	EF	0.85
Target	66.07	288.40	273.93	
yl	41.69	69.18	218.55	
yu	75.86	416.87	293.19	
Bayesian Method	CS	DT	EF	0.74
yl	57.54	234.42	269.15	
yu	72.44	416.87	281.84	

In vitro Dissolution of an oral drug

Experimental data on **in vitro dissolution** of an **oral drug** by ELPEN (Greek Pharmaceutical Company)

Tablet mass = 312 mg

Formulation: API and 3 excipients: main x_1 and x_2; x_3: 2-5% w/w

Comparison of dissolution profiles

Dissolution kinetics models: zero and first order kinetics, Hixson–Crowell, Weibull, Higuchi, Baker–Lonsdale, Korsmeyer–Peppas and Hopfenberg

Model-independent comparison of dissolution profiles

- **Ratio tests**: ratios of parameters (e.g., %drug dissolved, AUC, mean dissolution time) from release assays of the reference and test products at the same time
- **Pair-wise procedures**: Rescigno index (ξ_i), difference factor (f_1) and **similarity factor** (f_2)

Comparison of Dissolution profiles

Dissolution efficiency (*DE*) ratio of area under the dissolution curve up to a testing time point to the area of the rectangle that describes 100% dissolution up to the same time point

$$DE = 100 * \frac{\int_0^t d * dt}{d_{100} * t}$$

d: function of % drug dissolved at time t

Dissolution Area Difference (*DAD*) factor

$$DAD = \left|1 - \left(\frac{\int_0^t d_{exp} * dt}{\int_0^t d_{ref} * dt}\right)\right|$$

d_{exp} : dissolution function of each run; d_{ref} : dissolution function of reference drug dissolved throughout time t

- ***min DAD*** ⇒ best similarity of the dissolution profiles
- ***Calculation method***: multiple segment trapezoidal rule (steep changes in dissolution-vs-time curves)

Comparison of Dissolution profiles

Similarity factor (f_2)

$$f_2 = 50 * \log\left\{\left[1 + (1/N)\sum_{i=1}^{N}(\overline{x_{ti}} - \overline{x_{ri}})^2\right]^{-1/2} * 100\right\}$$

N: number of time points; $\overline{x_{ti}}$: mean % drug dissolved of test product
$\overline{x_{ri}}$: mean % drug dissolved of reference drug

- $50\% \leq f_2 \leq 100\%$: similar dissolution profiles
- Drug dissolved within 15 min > 85% ⇒ similar dissolution profiles without further mathematical evaluation

Mixture Design for experimental data

Mixture Design of a generic oral tablet formulation

- **Design factors**: mixture components proportions
- For a 3-component mixture: $0 \leq x_i \leq 1$ and $\sum x_i = 1$, i = 1, 2, 3
- **Simplex Centroid design points**: vertex points ($x_i = 1$), centroid (x_i s equal) and axial points ($x_i = 0$)

Two replicates of the center point (process variability and curvature evaluation)

Response variables - CQAs: tablet weight, hardness and dissolution profile

Reference - Generic drug Dissolution profiles Comparison

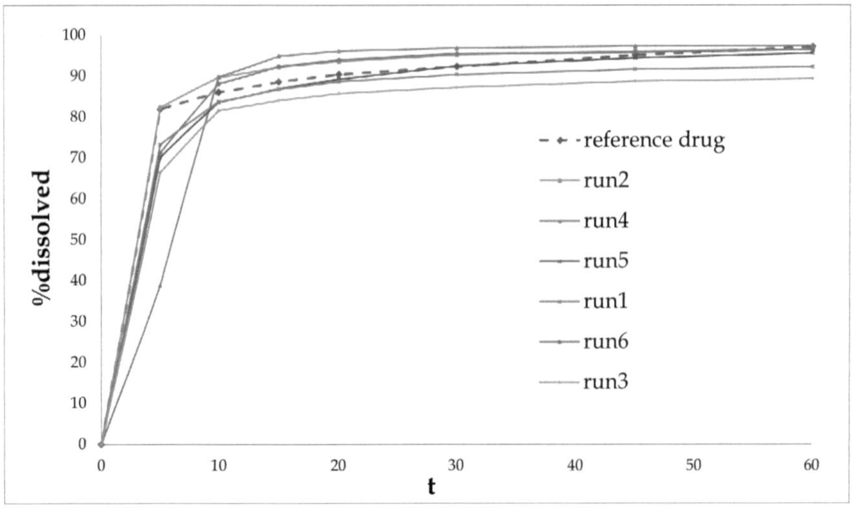

- **Reference dissolution profile**: first order kinetics equation

  $d(t)=a + b\, e^{kt}$

  d: drug percent dissolved at time *t*; a=91.05, b=-91.29, k=-0.57, R^2= 0.987

Reference - Generic drug Dissolution profiles Comparison

Similarity factor (f_2) and **Dissolution Area Difference** (*DAD*)

Experimental Runs	DAD	f2_15 (%)	f2_60 (%)
run 1	0.391	66	69
run 2	0.043	78	79
run 3	0.185	54	57
run 4	0.236	33	40
run 5	0.083	60	67
run 6	0.049	62	67

Mixture Regression Model

weight = $314.29x_1+309.06x_2+315.64x_3-2.81x_1x_2$

hardness = $17x_1+10.73x_2+359.87x_3+588.22x_1x_2$

$f2\_15 = -296.78x_1-195.01x_2+753.42x_3+1076.85x_1x_2$

$f2\_60 = -208.94x_1-110.04x_2+646.15x_3+755.41x_1x_2$

CQAs	Data Range	Target [Specifications]
weight (mg)	310.15 - 312.26	312 [304.20-319.80]
hardness (N)	154.400 - 169.40	170 [165.75-174.25]
f2_15	33 – 78	100 [50-100]
f2_60	40 – 79	100 [50-100]

Design Space determination

Overlapping responses

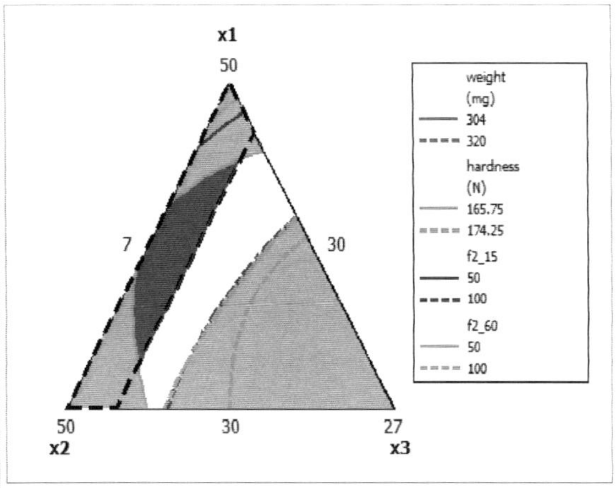

Optimized responses

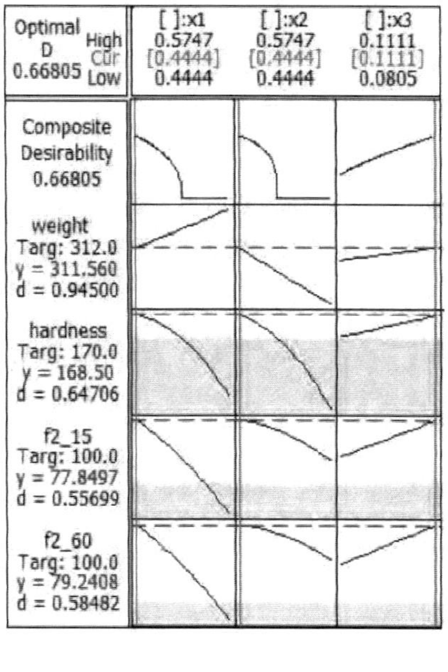

Design Space determination

Component values	Mixture Design Analysis	Bayesian Method
x1-optimum	38.66	-
x2-optimum	38.66	-
x3-optimum	9.67	-
x1-l	43.33	50
x2-l	36.67	30
x3-l	7.00	7
x1-u	37.67	35
x2-u	39.67	45
x3-u	9.66	7

Composite Desirability for Design Space determination methods

Specifications	weight (mg)	hardness (N)	f_{2_15} (%)	f_{2_60} (%)	Method's Composite Desirability
Target	312.00	170.00	100.00	100.00	
yl	304.00	165.75	50.00	50.00	
yu	320.00	174.25	100.00	100.00	
Mixture Design Analysis	weight	hardness	f_{2_15}	f_{2_60}	0.64
Target	311.56	168.50	77.85	79.24	
yl	311.60	165.42	56.66	60.12	
yu	311.50	168.35	78.89	80.29	
Bayesian Method	weight	hardness	f_{2_15}	f_{2_60}	0.57
yl	311.90	168.30	50.44	63.11	
yu	312.30	169.20	77.71	79.00	

Conclusions

- ☑ **DS determination**: multi-response optimization and overlapping responses
- ☑ **Multi-response optimization**: DS bounds in neighborhood of optimum conditions
- ☑ **Overlapping responses**: DS bounds from the global lower and upper specification limits
- ☑ **Method effectiveness measure**: Composite Desirability
- ☑ **Best performance**: **Experimental Design analysis -** proposed when the data are complete and there is little uncertainty
- ☑ **Bayesian approach**: proposed if data involve high uncertainty and correlation structures
- ☑ **Bioequivalence** introduced in **Experimental Design** and **DS determination** as a **CQA** with f_2 values

Acknowledgement

I would like to acknowledge financial support in the form of a scholarship from the National Technical University of Athens

Thank you for your attention!

369877 Integrated Dynamic Real Time Optimization and Advanced Feedback Control of Continuous Tablet Manufacturing Process

Abstract:

It is desired for pharmaceutical companies to satisfy the frequently changing globalized market demands, to optimize resources and inventory and to maximize the revenue such that the higher amount of capital and resources invested can be recovered with sufficient profit margin within a limited reduced effective patent life. Higher regulatory constraints and relatively inefficient QBT (Quality by Testing) based open-loop batch manufacturing also contributes towards suboptimal manufacturing. Therefore, a novel manufacturing technique based on continuous processing, integrated with dynamic real time optimization and an efficient advanced control system through which the pharmaceutical manufacturing can be optimized is required.

Recently pharmaceutical manufacturing efforts are focused on continuous manufacturing integrated with inline/online monitoring tools (PAT). In such a system, feedback control and real time optimization can be utilized to achieve better performance and quality production with minimum cost. Advanced control systems utilizing Model Predictive Control (MPC) and classical control structures (PID) have been shown to exhibit many advantages compared to MPC or PID alone [1-2]. In such a hybrid control scheme, MPC associated with complex process dynamics, is placed at a supervisory level and provides the set points to the regulatory level PID controllers.

In this work, Dynamic Real Time Optimization (DRTO) has been integrated with a hybrid control system for a continuous tablet manufacturing process for optimal automatic feedback control and QbD-based efficient continuous manufacturing. In the proposed optimal manufacturing approach, the integrated DRTO provides the optimal operational set points for the plant control system in real time optimizing an objective function. The DRTO takes into consideration the capital fixed cost, the operating cost, the market fluctuations, the product inventory/deficiency, the product quality assurance strategy, the regulatory constraints, and the product rejections. The control system then makes sure to deliver the required product with desired quality with minimum resources and time. The overall profitability depends on the combined performance of DRTO and control system. Therefore, a robust optimization strategy and an efficient control system have been integrated for maximum return of revenues. DRTO integrated with hybrid control strategy ensures the maximum possible profit irrespective of the market demand fluctuations. A systematic generic framework including the methods and tools through which the cost and profit can be estimated and the DRTO can be integrated with hybrid control system has been developed. A DRTO and control system integration algorithm, ontological data base, model library, and supporting tools (system identification toolbox, simulation tool, optimization tool, MPC toolbox) are the integral part of the framework. The algorithm provides the stepwise guidelines for DRTO-hybrid control system integration. The database provides the cost analysis data needed to perform optimization and has been build based on extensive literature and industrial survey while the model library consisting of the process flowsheet model is used to simulate the process in closed-loop scenario. The process models have been developed in simulation tool utilizing input-output step response data and system identification toolbox with which the classical control system and MPC (designed using MPC toolbox) has been integrated. On top of control architecture, the DRTO is then integrated using optimization toolbox for real time optimization and control. Finally, the real time optimization and control framework is integrated with the continuous tablet manufacturing pilot plant through real time monitoring tools, control hardware and software and OPC communication protocol. The application of the integrated DRTO and hybrid control framework has been demonstrated through a flowsheet simulation of continuous tablet manufacturing process.

The objective of this presentation is two-fold: first to highlight the integration of DRTO with advanced hybrid control architecture, and second to demonstrate the performance of the integrated system.

[1]. Singh, R., Sahay, A., Muzzio, F., Ierapetritou, M., Ramachandran, R. (2014). Computers & Chemical Engineering, DOI: 10.1016/j.compchemeng.2014.02.029.

[2]. Singh, R., Ierapetritou, M., Ramachandran, R. (2013). European Journal of Pharmaceutics and Biopharmaceutics, 85(3), Part B, 1164-1182.

365441 Influence of Controlled Fluid Shear on Nucleation Kinetics in Glycine Aqueous Solutions

Abstract:

Understanding crystal nucleation from solution is important for rationally designing and controlling manufacturing processes that involve crystallization. Both batch and continuous industrial crystallization processes typically involve fluid flow where supersaturated solutions and/or suspensions of growing crystals are pumped or agitated. It is well known that fluid shear can induce secondary nucleation and it has been experimentally demonstrated that primary nucleation can be affected by fluid shear as well (1,2). In order to gain further experimental evidence to facilitate improved understanding of effects of fluid shear on primary nucleation, we investigated nucleation in aqueous glycine solutions under well-defined flow conditions at a constant temperature.

Supersaturated glycine solutions were prepared by nanofiltering and slow cooling, and they were then subjected to controlled fluid shear in Couette cells under isothermal conditions. Average shear rates between $25s^{-1}$ and $1000s^{-1}$ were studied. Crystallization in the Couette cell was monitored using a custom built instrument that allowed transmission and light scattering measurements to be taken simultaneously in situ; the onset of nucleation was associated with a rapid increase in scattering and decrease in transmission, so induction times were readily obtained from these measurements. The role of surface was also studied by changing the diameter of the inner cylinder of the Couette cell; this allowed surfaces of $2.5cm^2/cm^3$ solution, 5 cm^2/cm^3 solution and 10 cm^2/cm^3 solution to be investigated. Due to the stochastic nature of nucleation, experiments were repeated multiple times and a statistical analysis was performed to show that the number of repetitions was sufficient for accurate trends to be deduced.

Nucleation induction times in sheared solutions were found to be considerably lower than those in unsheared solutions, while a great care was taken to avoid seeding either in solution or at solid-liquid interfaces. This suggested that fluid shear had a profound influence on primary nucleation glycine from aqueous solutions. In sheared solutions, a large number of small crystals formed in close succession, causing solutions to rapidly go turbid, so it is proposed that shear enhanced the formation of primary nuclei which then initiated extensive secondary nucleation (3). Increasing the average shear rate was found to reduce the induction time in a power law relationship with the exponent of -0.98 ± 0.1. This indicated that the total strain applied had an important effect on nucleation kinetics. Increasing the surface area was also found to notably decrease the induction times, with a power law relationship with the exponent of -0.88 ± 0.2, which indicated that the surface played an important role in nucleation kinetics as well. In situ dynamic light scattering measurements showed that colloidal scale liquid-like clusters (nanodroplets) observed in mesostructured amino acid aqueous solutions (4) had their mean size increasing upon continued shearing of solutions. This is consistent with these mesoscale clusters (in the bulk liquid phase and/or at the solid-liquid interface) playing a direct role in the primary nucleation pathway in this system as proposed recently (1), where larger clusters correlated with shorter nucleation induction times.

(1) Jawor-Baczynska, Sefcik and Moore, Crystal Growth & Design, 13, 470 (2013)

(2) Liu and Rasmuson, Crystal Growth & Design, 13, 4385 (2013)

(3) Kadam, Kramer and ter Horst, Crystal Growth & Design, 11, 1271 (2011)

(4) Jawor-Baczynska, Moore, Lee, McCormick and Sefcik, Faraday Discussions, 167, 425 (2013)

377543 Analysis of Crystallization Kinetics Based on CSD Estimates from PAT Tools

Abstract:

Nucleation and growth are the two dominating phenomena in crystallization processes. In our study, a partial crystal size distribution (CSD) was estimated for the size range from 50 μm to 500 μm by combining in situ measurements from both FBRM and ATR-FTIR. Experimental results show that the evolution of the CSD was better tracked by this framework, compared to a qualitative interpretation using chord length distribution (CLD).

To explore the process and the corresponding kinetics, seeded and unseeded batch cooling crystallization experiments were designed to start with different initial conditions and proceed through different cooling rates. The general procedure is: (1) a concentrated ethanolic solution of paracetamol is cooled to supersaturated conditions, followed by a constant-temperature plateau; (2) either seeds are added or primary nucleation occurs on the constant-temperature plateau; (3) after desupersaturation of the solution is complete, the solution is further cooled to the final temperature. The estimated CSDs from CLDs were consistent with our expectations of the design of the crystallization runs, such as initial concentrations and cooling rates. These CSDs were also verified by sieving.

A full population-balance model was solved in order to fit the experimental data and estimate kinetic parameters. A numerical method (conservation/solution element scheme) was applied to the population balance to simulate the experiments efficiently and was used in parameter estimation. Classical nucleation theory was used for primary nucleation, secondary nucleation was related to relative supersaturation and mass of crystals, and the growth rate was assumed to be size-independent.

Using the estimated parameters, the simulated runs for unseeded crystallizations suggest that the final CSD was strongly influenced by the relative importance of primary versus secondary nucleation when they occur simultaneously in the clear supersaturated solution; if secondary nucleation is dominant at that moment, then most of the crystal mass is concentrated in small size ranges. An optimal two-step cooling profile was designed to produce a final size distribution with more large crystals and it was validated experimentally.

Heterogeneous Nucleation of APIs on Engineered Biocompatible Polymer Surfaces

Li Tan, Rachel M. Davis, Prof. Allan S. Myerson, Prof. Bernhardt L. Trout

Date: May 12th, 2014

Continuous manufacturing of pharmaceuticals have significant advantages over batch processing in terms of economics, reliability, safety, and sustainability. Directly crystallize active pharmaceutical ingredients (APIs) on polymer excipient surfaces can achieve impurity separation and drug-excipient composite formation in a single step, which eliminates many solid handling steps traditionally needed in downstream processing. In order to develop such a process, we must understand the kinetics of heterogeneous nucleation of small molecule drugs on biocompatible polymer surfaces, and study how heterogeneous nucleation affects polymorph formation. The chemical interactions between APIs and polymer substrates plus the effect of surface geometries are the focuses of the present investigation.

We engineered surfaces by first manufacturing a rigid silicon mold using interference lithography to create nanometer-sized and mono-dispersed pillars of varying angles on a flat surface. Negatives of the mold were transferred to polyvinyl alcohol (PVA) films via nano-imprint lithography to create films with nano-pores. The films were submerged in solution as substrates for cooling crystallization of two API compounds, aspirin and indomethacin. A high throughput system was used to collect heterogeneous nucleation kinetics data on the patterned films at low supersaturation settings. Experimental results show that the induction time of aspirin on these surfaces were reduced by more than an order of magnitude with favorable surface chemistry, with further reduction achieved by matching surface geometry. Matching angles on the substrate surface and angles between dominant faces resulted in the greatest reduction in induction time. X-ray diffraction studies revealed favorable interactions between dominant habit faces of aspirin to PVA via hydrogen bonding. Crystallization experiments with indomethacin as the model compound show that in addition to enhancing nucleation rates, polymer chemistry and surface geometry also alters the distribution of polymorphs obtained in the experiments. Correlation between final polymorph distribution on film surface and initial nucleation events were established using Raman spectroscopy. Computational analyses were also used to corroborate experimental results and hypothesis presented.

Impact of surface chemistry on crystallization of acetaminophen

Using polymeric surfaces is a novel way to control crystallization however, the mechanism is still ambiguous. In the present study, chitosan (CS) films that were doped with different molecules were employed to investigate the impact of surface chemistry on the heterogeneous crystallization behavior of acetaminophen (APAP).Crystal growth rate and crystallization kinetic measurements of APAP on various substrates were performed. Fourier transform infrared spectroscopy was used to probe intermolecular interactions between APAP and CS films. The results were interpreted with the aid of molecular simulations. It was found that substrates with more hydrogen bonding groups that could interact with APAP promoted its crystallization.

Development of a robust API crystallization in a multi-component solvent mixture: Using high throughput automation as an enabling technology to develop comprehensive solubility maps

Authors: **Ben Cohen**[1], Michelle Mahoney[1], Brenda Remy[1], Jun Qiu[1], Chris Sfouggatakis[1], Zhongping Shi[1], Chenchi Wang[2]

[1] Chemical Development, Bristol-Myers Squibb Co.,
[2]Drug Product Science & Technology, Bristol-Myers Squibb Co.,

Session: 2B07 Crystallization of Pharmaceutical and Biological Molecules

Knowledge of solubility is essential to process design of API crystallizations. Detailed solubility enables a basis for defining robust and optimized process conditions, such as dissolution, seeding and isolation conditions. Determination of solubility maps is often performed by manual methods. However, as the number of components grows, and hence the number of possible process combinations grows, obtaining a comprehensive solubility map becomes more time-intensive. An example of this complexity was observed in the final crystallization of a BMS API currently in development. API solubility was found to be affected by numerous factors in the crystallization: the relative amounts of the three solvents, the amount of base, and the amount of reaction byproduct, as well as temperature.

To streamline the construction of the solubility map, a high-throughput workflow was developed to measure API solubility using automation equipment. The new workflow generated accurate solubility, with units of mass fraction through the use of an internal standard, over a wide temperature range. The data set enabled construction of a reliable solubility map which led to improved understanding of the crystallization. It helped establish process ranges around important operations such as dissolution, seeding, anti-solvent addition and isolation, as well as acceptance criteria for the upstream reaction and workup. The detailed solubility knowledge provided a basis for defining optimized process conditions within a robust design space.

The Role of Excipient in Maintaining Process Control and Productivity of Direct Compression Tabletting from Laboratory- to Pilot-Scale

True L. Rogers[1] (TLRogers@Dow.com), Kathryn O. Hewlett[1] (KOHewlett@Dow.com), Tejas Gunjikar[2] (TGunjikar@Dow.com), Karen M. Balwinski[1] (KMBalwinski@Dow.com), Louis F. Ruocco[3] (Ruocco@Dow.com), Jörg Theuerkauf[1] (JTheuerkauf@Dow.com)

[1] The Dow Chemical Company, Midland, MI USA

[2] The Dow Chemical Company, Mumbai, India

[3] The Dow Chemical Company, Philadelphia, PA USA

Purpose: Modified-release (MR) matrix tablets are typically produced using hydroxypropyl methylcellulose (HPMC) as the rate-modifying polymer. Several ingredients in the matrix tablet formulation, particularly the active pharmaceutical ingredient (API) and HPMC, may be irregularly shaped and prone to poor flow and segregation. Although direct compression would be the most streamlined and cost-efficient manufacturing process, it is often not a viable option due to flow and segregation issues especially at larger scale. In this study, HPMC matrix tablets were produced via direct compression on a rotary press at laboratory-scale. Two of the formulations were then selected for pilot-scale tabletting trials. The purpose of this talk is to compare process control, productivity, and achievability of desired end-product attributes at both laboratory- and pilot-scale.

Methods: The matrix tablet model formulation selected for scale-up contained 49.4% propranolol HCl, 40.1% HPMC, 9.3% pre-gelatinized starch, 0.7% colloidal silica, and 0.5% magnesium stearate. The targeted tablet weight was 162 mg, so the target API dose per tablet was 80 mg. Either the standard CR grade of HPMC, METHOCEL™ CR, or a new direct compression grade of HPMC, METHOCEL™ DC2, was used in the formulation to make tablets via direct compression. All ingredients except magnesium stearate were blended. Magnesium stearate was then briefly blended into the powder mixture, and the powder blend was subsequently compressed to matrix tablets (8.0-mm diameter). Tablets were collected every 5 or 10 minutes during each tabletting trial, depending upon scale. Pilot-scale direct compression represented a 33X increase over laboratory-scale.

Results: Process control was gauged by tablet-to-tablet reproducibility in attributes such as individual tablet weight, API content per tablet, and individual tablet breaking force. Comparisons of these attributes by scale and by the HPMC morphology grade utilized will be shown. Although tablets containing METHOCEL DC2 were slightly lower in breaking force, the press could be operated at a 1.6X greater speed at pilot-scale, achieving the targeted tablet weight with higher process control than tablets containing METHOCEL CR. As a result, a 200,000 tablet-per-hour production rate was achieved with DC2 tablets weighing 164.4 ± 1.1 mg. A 125,000 tablet-per-hour production rate was achieved with CR tablets weighing 163.6 ± 1.6 mg. Tablets exhibited satisfactory unit-dose uniformity within 90-110% of the labeled claim of 80 mg API per tablet. The expected level of modified-release performance was achieved (f2 = 78) with 75-80% API being released from the matrix tablet within 12 h.

™ Trademark of The Dow Chemical Company ("Dow") or an affiliated company of Dow

Figure 1. Process run charts for individual tablet weight measurements collected throughout the pilot-scale tabletting trials for a) DC2 matrix tablets produced at 200,000 tablets/h and b) CR matrix tablets produced at 125,000 tablets/h.

a) b)

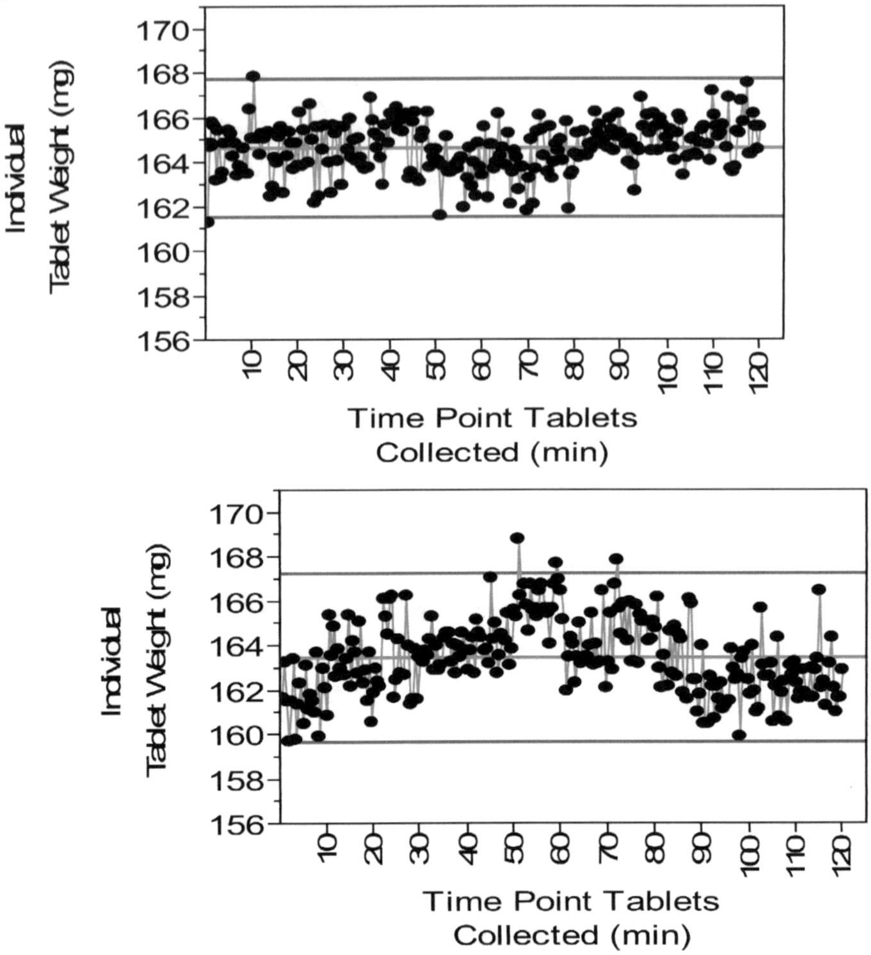

Conclusions: A key learning from this study is that matrix tablets produced by direct compression required the use of METHOCEL DC2 in order to achieve the desired end-product attributes and greater process control. Direct compression is well known as the most streamlined of the tabletting processes but often is not a feasible option due to flow or segregation issues. Streamlined manufacturing, greater production capacity, and satisfactory process control rendered direct compression a viable matrix tablet manufacturing option as long as the more flowable METHOCEL DC2 excipient was utilized.

Acknowledgements: The authors gratefully acknowledge Vaishali Dhavse and her team at Rubicon Research (Mumbai, India) for the scale-up tabletting studies.

392529 Scale-up of Wet-Granulation Process: Case Study of a Risk Assessment Based Technology Transfer to Manufacturing Site

Abstract:

Scale-up and transfer to commercial manufacturing site of a high drug load (50%) wet granulation process was successfully implemented by identifying and defining a proven and acceptable range (PAR) for the critical process parameters. Critical quality attributes of the final products were correlated to in-process material attributes during development through implementation of well designed DOEs. Process analytical tools (PAT) and conventional analytical tests were used to characterize these intermediates. The presentation will detail a holistic approach of in-process attribute and parametric controls combined with appropriate finished product tests to successfully scale-up and transfer, process and product to commercial manufacturing site.

369535 Plant-Wide Hybrid MPC-PID Control of a Continuous Direct Compaction Tablet Manufacturing Pilot-Plant for QbD Based Pharmaceutical Manufacturing

Abstract:

It is highly desired to implement a closed-loop feedback control system into a pharmaceutical manufacturing plant to enable the transition toward a more desirable Quality by Design (QbD), rather than Quality by Testing (QbT) based manufacturing of the next generation of pharmaceutical products. This approach utilizes an optimal utilization of resources, while satisfying the strict regulatory constraints, flexible market demands, operational complexities, and economic limitations.

In different industrial sectors such as chemical and petrochemical a closed-loop control process operations has proven to be a very effective operational strategy over the last 30 years [1]. Hoverer, in comparison to classical control schemes (PID), Model Predictive Control (MPC) is more complex to implement and is computationally more expensive. Therefore, a hybrid strategy that takes advantage of both MPC and PID is an optimal alternative for pharmaceutical manufacturing. The control hardware, software and sensor integration and thereby implementation of an advanced control system into the actual pharmaceutical pilot-plant for efficient closed-loop operation is still a challenging task because of different level of complexities (e.g. solid handling) involved [2].

In this work, a plant-wide advanced hybrid MPC-PID control system has been implemented into direct compaction continuous tablet manufacturing pilot-plant. The critical control variables includes, drug concentration, powder level in an instrumented hopper, tablet weight and hardness. The control loops have been closed using industry standard control platform (e.g. DeltaV), PAT data management tool (synTQ), OPC communication protocol and fieldbus devises. For drug concentration control, PAT system is used to read the NIR spectral data at blender outlet and communicate it to the MVA (Multi-Variable Analysis) model performing PCA/PLS to provide the API concentration.

A chemometric tool is used for real time NIR prediction and a PAT data management tool is used to manage the process data and to send the signal to the control platform using OPC communication protocol. The instrumented hopper level (placed in between blender and tablet press) is controlled by manipulating the speed of a rotary valve. The hopper level is monitored by webcam. The webcam takes the image of powder level in real time and passes it to the MATLAB image analysis toolbox where the actual powder level has been calculated. The measured level is then sent to control platform via MATLAB OPC toolbox as the input to MPC block.

In the tablet press, the tablet weight and hardness are controlled through a cascade control arrangement using two master loops and one slave loop. Master loops are used to control the weight and hardness and provide the set point for the slave controller, which controls the main compression force by manipulating the fill depth. They share a common slave controller, meaning that only one master controller is activated at a time. The tablet weight is measured and controlled more frequently than hardness. Check master (Fette) has been used for tablet weight and hardness measurement in real time.

The main challenges in this implementation are the integration of multiple hardware/software platforms including Process Control System, NIR analyzer, Multi-Variable Analysis (MVA) software, Process Modeling Software, and Bus Technology. The performance of hybrid MPC-PID scheme has been compared with cascade PID and cascade PID with Smith predictor schemes. Results demonstrate enhanced performance of critical quality attributes (CQAs) under hybrid MPC-PID scheme compared to cascade PID/Smith predictor schemes, thus illustrating the potential of hybrid MPC-PID control system in improving pharmaceutical manufacturing operations.

The objective of this presentation is two-fold. First to highlight the plant-wide implementation of hybrid MPC-PID control system into a continuous tablet manufacturing pilot-plant, and second to demonstrate the performance of the implemented control system.

[1]. Singh, R., Ierapetritou, M., Ramachandran, R. (2013). European Journal of Pharmaceutics and Biopharmaceutics, 85, 1164–1182. 1.

[2]. Singh, R., Abhishek Sahay, A., Muzzio, F., Ierapetritou, M., Ramachandran, R. (2014). Computers & Chemical Engineering, DOI: 10.1016/j.compchemeng.2014.02.029.

Introduction

A pilot-plant dosator nozzle capsule filling machine (MG2 Labby – output up to 3,000 cph) and a large manufacturing scale capsule filling machine (MG2 Planeta – output up to 100,000 cph) were used to develop a scale-up strategy for filling low fill-weight inhalation products into capsules. The QbD-based strategy consists of identifying the critical process parameters (CPP) and material attributes (CMA) on quality attributes of products using the same screening Design of Experiment (DOE) for both machines. The DOE was used to identify the CMAs and CPPs for fill weight and weight variability (RSD). In addition, the Design Space based on these CMAs and CPPs was used to predict fill weight for other powders.

Materials and Methods

The DOE (D-Optimal with Design statistics G-Efficiency) includes four process parameters of the capsule filling machine at three levels: 1- Diameter of the nozzle (dia), 2- Size of dosing chamber (cha), 3- Depth of powder bed in the bowl (layer), and 4- Speed of capsule filling (speed). The DOE also considers two interactions between process parameters: 1- Size of dosing chamber and powder layer depth; and 2- Diameter of the nozzle and size of dosing chamber.

The materials used for the DOE with the Planeta include: Respitose SV003 (sieved Lactose; x50=73.4), Respitose ML001 (milled Lactose; x50=52.4) and Lactohale 100 (sieved Lactose; x50=155.2). The model was validated using Inhalac 250 (sieved Lactose; x50=60μm). The materials used in the Labby include those used for Planeta and also Inhalac 230 (sieved Lactose; x50=112.3), Lactohale (blend of Lactoses; x50=66.5), Respitose SV010 (coarse sieved Lactose; x50=127.6) and Respitose ML006 (fine milled Lactose; x50=18.2). The model developed in Labby was also validated using Inhalac 250. In addition to particle size and shape, a large number of material attributes are assessed: bulk (BD) and tapped density (TD); compressibility; 3 flow indexes (Carr index, Basic flowability energy, and ffc); cohesion, wall friction, and air permeability.

Results

The results show that the same CPPs and CMAs, with almost the same weight coefficients, affect fill weight for Labby (Figure 1) (Llusá, Faulhammer, 2013) and for Planeta (Figure 2). The linear models capture 98% and 97% of the variability in fill weight for Labby and Planeta respectively and they can be used to accurately predict the fill weight (confidence interval = 95%) of a powder that is not in the development set. However, the model developed in Labby loses some of its prediction ability for fill weight when used for Planeta (confidence interval = 90%).

Regarding fill weight variability (RSD), the same CPPs (even if with different weight coefficients) were found to affect RSD for both machines. However, a different CMA was identified for each machine. The latter invalidates the use of the model developed in Labby to predict the RSD for Planeta. The linear model developed in Labby has also a poor performance predicting RSD for capsules filled with Labby using powders that are not part of the powder set used in the model development. Moreover, quadratic models did not improve the fit and predictive performance of the model. The CPPs and CMAs can explain 50% of the variability in the set for Planeta (Figure 3) and 59% in the set for Labby (Figure 4). The RSD values obtained in the execution of the DOE are quite different for the two machines and Planeta yielded consitently lower RSD than Labby for the same process parameters (the process control settings are also turned off). The powders, environmental conditions and operating procedure (SOP) in the execution of the DOE for Labby and Planeta were the same. Therefore, the differences in RSD for these two machines may be attributed to construction details that are germane to each machine. For example: 1- the gap between tip of the nozzle and bottom of the rotary container. This parameter, which affects the powder precompression in the nozzles, was modified for the Labby to match the design of the Planeta machines. Also, as a result of the different dimensions and geometries of the machines, dynamic process parameters are different. For example: 1- the vertical dosator speed inside the bed of powder (Augsburger, 2011), 2- rotation speed of the bowl.

Conclusions

The process understanding and Design Space developed with the Labby (pilot plant scale) can be safely used to set the process parameters to obtain a specific fill weight in the industrial scale machine. Regarding RSD, we concluded that: 1- we should first enhance the RSD prediction ability of model developed in the Labby before applying this model to predict the RSD performance of other machines, 2- there is a need to further investigate what characteristic of Planeta leads to lower RSD values (considering that process control was off).

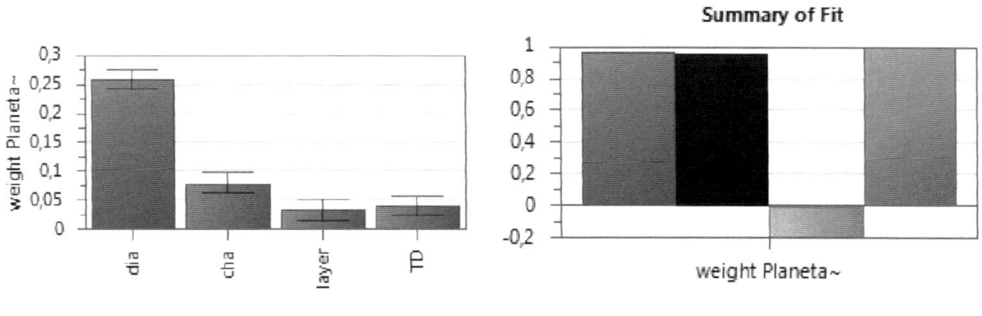

Figure 1: CPPs and CMAs (left figure) and Model fit for fill weight (right figure) for Planeta. The four columns in the Summary of fit indicate R^2, Q^2, Model Validity and Reproducibility.

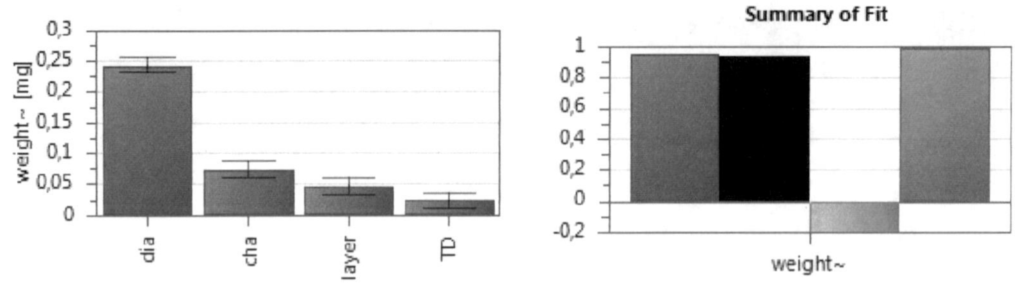

Figure 2: CPPs and CMAs (left figure) and Model fit for fill weight (right figure) for Labby. The four columns in the Summary of fit indicate R^2, Q^2, Model Validity and Reproducibility.

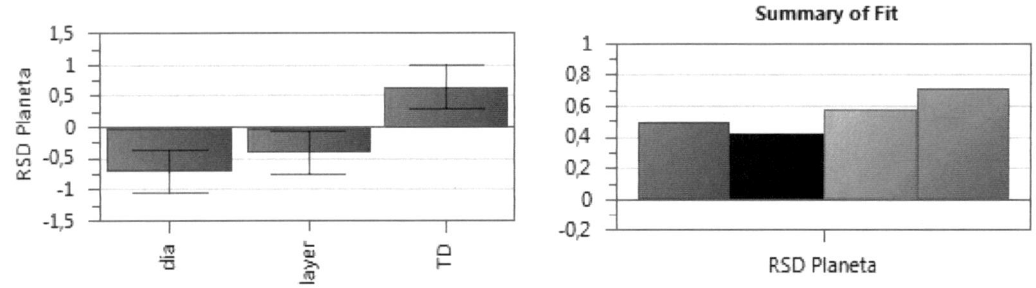

Figure 3: CPPs and CMAs (left figure) and Model fit for weight variability (right figure) for Planeta. The four columns in the Summary of fit indicate R^2, Q^2, Model Validity and Reproducibility.

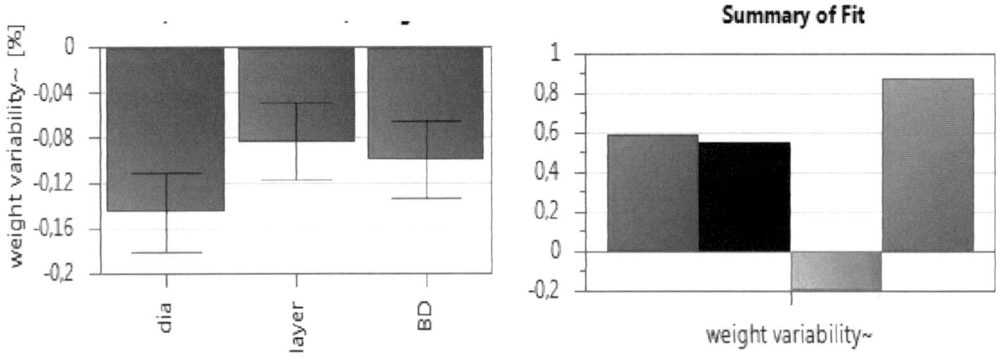

Figure 4: CPPs and CMAs (left figure) and Model fit for weight variability (right figure) for Labby. The four columns in the Summary of fit indicate R^2, Q^2, Model Validity and Reproducibility.

References

1- M. Llusá, E. Faulhammer, M. Fink, G. Scharrer, V. Calzolari, S. Biserni, S. Lawrence, M. Bresciani and J.G. Khinast; 2013. Manufacturing of Low Fill Weight Capsules Using a Nozzle Dosator Capsule Filling Machine. https://aiche.confex.com/aiche/2013/webprogram/Paper306359.html

2- L. Augsburger, 2011. Practical considerations in the scale-up of powder-filled hard shell capsule formulations. Chapter 15, pp. 422-443. Book: Pharmaceutical Process Scale-Up, Third Edition. Informa Healthcare.

362894 A Computational Approach to Predict Lyophilizaiton Performance and Monitor Process Robustness for Biologics

Abstract:

Lyophilization of injectable pharmaceutical products is an approach commonly undertaken to formulate drugs that are chemically or physically unstable to be commercialized as ready to use solutions. Transferring the process parameters with appropriate design space from laboratory scale lyophilizer to the commercial scale without a loss in product quality and process robustness is one of the important aspects of commercializing a lyophilized product. This process is often accomplished by costly engineering runs or through an iterative process at the commercial scale. Here, we are highlighting an experimental process followed by a computational approach to predict commercial process parameters for the primary drying phase of lyophilization.

The objective of this research is to characterize the primary drying process using a combination of laboratory scale lyophilization experiments and finite element method (FEM) based computational modeling approach. Heat and mass transfer coefficients are determined either by the manometric temperature measurement (MTM) based measurement or gravimetric method based sublimation experiments. This data will be used as inputs for the FEM based software called PASSAGE ®, which computes primary drying duration for critical primary drying process parameters such as shelf temperature and chamber pressure. Since, the heat and mass transfer coefficients will vary at different lyophilization scales. These parameters are especially difficult to measure at commercial scale with the available tools and without significant usage of expensive drug substance. We present an alternative approach to use appropriate factors while scaling-up from laboratory scale to commercial scale. As a result, one can predict commercial scale primary drying duration based on these parameters. Additionally, the computational model is validated by comparing predicted product temperature and primary drying duration with experimental studies at lab-scale. Finally, three applications were presented to demonstrate the capability of the model, the first one is a scale-up model to predict primary drying duration at commercial scale, the second one is a computational model to identify a proven acceptable range for the critical primary drying process parameters and the last one is focused on predicting product temperature due to accidental drifts in process parameters during the primary drying stage of the process. The approach presented here provides a process for robust lyophilization scale-up and because of the simple and minimalistic nature of the approach, it will be less capital intensive. This study also facilitates the implementation of Quality by Design (QbD) concept to biologics manufacturing.

AUTHOR INDEX

AUTHOR INDEX

AUTHOR INDEX

AUTHOR INDEX

AUTHOR INDEX

AUTHOR INDEX

AIChE
120 Wall Street, FL 23
New York, NY 10005-4020

ISBN 978-1-5108-1269-7

CURRAN ASSOCIATES INC.
proceedings.com